AGROECOSYSTEMS
in a
Changing Climate

Advances in Agroecology
Series Editor: Clive A. Edwards

Advisory Board

AGROECOSYSTEMS

in a

Changing Climate

Edited by

Paul C. D. Newton
R. Andrew Carran
Grant R. Edwards
Pascal A. Niklaus

Taylor & Francis
Taylor & Francis Group
Boca Raton   London   New York

CRC is an imprint of the Taylor & Francis Group,
an informa business

CRC Press
Taylor & Francis Group
6000 Broken Sound Parkway NW, Suite 300
Boca Raton, FL 33487-2742

International Standard Book Number-10: 0-8493-2088-7 (Hardcover)
International Standard Book Number-13: 978-0-8493-2088-0 (Hardcover)

Library of Congress Cataloging-in-Publication Data

Agroecosystems in a changing climate / editors: Paul C.D. Newman ... [et al.].
    p. cm. -- (Advances in agroecology)
  Includes bibliographical references and index.
  ISBN-13: 978-0-8493-2088-0 (alk. paper)
    1. Agricultural ecology. 2. Climatic changes. 3. Agriculture--Environmental aspects. I. Newman, Paul C. D. II. Series.

S589.7.A479 2006
577.5'522--dc22   2006010178

Visit the Taylor & Francis Web site at
http://www.taylorandfrancis.com

and the CRC Press Web site at
http://www.crcpress.com

Preface

This book is the 12th in the continuing CRC Series on Advances in Agroecology. While other volumes have discussed aspects of climate change, this is the first to deal directly with this topic. In this book we employ a broader definition of climate change to include changes not only in climatic factors per se (temperature and rainfall) but also in the composition of the atmosphere (carbon dioxide in particular but also ozone).

Climate change is an issue that engages many more participants than just the scientific research community. The issue is highly politicised and widely presented and discussed in a range of media and fora. It is therefore not surprising that the opinions people hold about climate change are informed by a range of material of which original scientific research might be only a small component. We are introducing this book into this lively arena because as experimental scientists our experience demonstrates to us that changes in temperature, precipitation, and atmospheric carbon dioxide have the potential to profoundly alter terrestrial ecosystems and the delivery of the services they provide. We do not have to wait for accurate projections of a future climate to make progress here. An important task is to develop our understanding of the effects of the climate change drivers and their interactions on biological systems; from this base of knowledge we will be much better placed to consider the range of future environments that may arise and the range of agroecosystems we will need to cover. Consequently, we have organised the book so that a fundamental understanding of processes is presented; we have then asked applied scientists to consider the consequences of a change in these processes for agroecosystems.

This book has taken a long while to prepare — a period sufficiently long for atmospheric CO_2 to increase by 6 ppm — but this event has allowed our authors to include the most recent findings and views and we would like to thank them for their patience and for sharing their ideas as well as their knowledge of their particular subject areas. We would like to thank the editor of the "Agroecology" series, Clive Edwards, and Taylor & Francis editor, John Sulzycki, for their invitation to prepare this book and for their advice and encouragement. Pat Roberson and Linda Manis have provided the essential publishing expertise and a number of colleagues gave up their time to referee chapters and we thank them all for these essential contributions.

Paul C.D. Newton, R. Andrew Carran,
Grant R. Edwards, and Pascal A. Niklaus

Editors

Paul C.D. Newton, Ph.D., is senior research scientist in the Land and Environmental Management group of AgResearch based in Palmerston North, New Zealand. After completing a diploma in communications studies at the Central London Polytechnic he worked in a number of areas including agriculture and entertainment. He then earned a degree in agricultural botany at the University College of North Wales, Bangor and subsequently earned a Ph.D. in a joint project between Bangor and the Weed Research Organisation. Dr. Newton moved to New Zealand in 1986 for a postdoctoral fellowship and has remained there, studying the effects of global change on grazed pastoral systems. His research has included locating and working at natural CO_2 springs and the construction of a Free Air Carbon Dioxide Enrichment (FACE) experiment in 1997. This experiment is continuing and his work concentrates on how the relationships between plant community structure and ecosystem function is modified by elevated atmospheric CO_2.

R. Andrew Carran, is a senior research scientist at AgResearch located in Palmerston North, New Zealand. After graduating with a M. Agric. Sci (Hons soil science) he worked at research stations of the Grasslands Division of the New Zealand Department of Scientific and Industrial Research until 1991. During this period, he researched soil processes including nitrogen fixation, acidification, and ammonia volatilisation in sheep production systems. Through 1990, he held a Visiting Researcher position at the Leopold Center for Sustainable Agriculture at Iowa State University. Since 1991 his research programme has expanded to include source sink studies of nitrous oxide and methane in soils, and the impacts of elevated CO_2 on soil processes influencing nutrient availability and loss in grazed pastures.

Grant R. Edwards, Ph.D., is a senior lecturer in Pasture Ecology in the Agriculture and Life Sciences Division, Lincoln University, New Zealand. He completed his undergraduate degree in Agriculture Science at Lincoln University, before being awarded a Rhodes Scholarship to study at Oxford University. He was awarded a DPhil from Oxford University in 1994 for his thesis titled: *The Creation and Maintenance of Spatial Heterogeneity in Grasslands: The Role of Plant-Herbivore Interactions.* Thereafter, he had two postdoctoral research positions at Imperial College, London and at AgResearch, Hamilton, New Zealand. From 2000 to 2004, he was a lecturer at Imperial College, London. In 2004 he obtained his present position at Lincoln University. His research interests include the ecology and management of temperate grasslands under a changing climate. Current research conducted by his group addresses how an understanding of livestock diet selection and plant population dynamics can be used to manipulate the botanical composition of pastures, be it for weed control or livestock production reasons.

Pascal A. Niklaus, Ph.D., is currently a senior scientist at the Institute of Plant Sciences of the Swiss Federal Institute of Technology in Zürich, Switzerland. He studied physical, inorganic, and organic chemistry at the University of Basel where he obtained his diploma in chemistry in 1992; investigating model systems for oxygenating enzymes. He obtained his Ph.D. in botany in 1997 at the Institute of Botany of the University of Basel with investigations of global change-effects on grassland ecosystems. During his postdoctoral years, he also was visiting scientist at the Institute of Arable Crops Research in Rothamsted, United Kingdom and at Landcare Research, New Zealand where he investigated soil trace gas relations.

A main focus in Pascal Niklaus' research is on the consequences of environmental change for plant communities and the biogeochemical cycling of nutrients and water in ecosystems; special emphasis is on interactions with soils and effects on the ecosystem's greenhouse gas balance (CO_2, CH_4, N_2O).

Contributors

Vincent Allard
INRA-Agronomie
Fonctionnement et Gestion de
 l'Ecosystème Prairial
France

Daniel J. Archambault
Laurentian University
Sudbury, Ontario, Canada

Andrew Ash
CSIRO Sustainable Ecosystems
St. Lucia, Queensland, Australia

Joseph C. Blankinship
Department of Biological Sciences
Northern Arizona University
Flagstaff, Arizona

R. Andrew Carran
Land and Environmental Management
 Group
AgResearch
Palmerston North, New Zealand

Sukumar Chakraborty
CSIRO Plant Industry
Queensland Bioscience Precinct
St. Lucia, Queensland, Australia

Jann P. Conroy
Centre for Plant and Food Science
University of Western Sydney
Penrith South, Australia

Grant R. Edwards
Agriculture Group
Agriculure and Life Sciences
 Division
Lincoln University
Canterbury, New Zealand

Jürg Fuhrer
Air Pollution/Climate Group
Agroscope FAL Reckenholz
Swiss Federal Research Station for
 Agroecology and Agriculture
Zürich, Switzerland

Oula Ghannoum
Centre for Plant and Food Science
University of Western Sydney
Penrith South, Australia

Stephen L. Goldson
AgResearch
Lincoln, New Zealand

Jo E. Hossell
ADAS Gleadthorpe
Meden Vale, Mansfield,
 Nottingham
United Kingdom

Mark J. Hovenden
School of Plant Science
University of Tasmania
Hobart, Tasmania, Australia

Bruce A. Hungate
Department of Biological Sciences and
  Merriam-Powell Center for
  Environmental Research
Northern Arizona University
Flagstaff, Arizona

Jonathan A. Newman
Department of Environmental Biology
University of Guelph
Guelph, Ontario, Canada

Paul C.D. Newton
Land and Environmental Management
  Group
AgResearch
Palmerston North, New Zealand

Pascal A. Niklaus
Institute of Plant Sciences
ETH Zürich
Zürich, Switzerland

Matthias C. Rillig
Microbial Ecology Program
Division of Biological Sciences
The University of Montana
Missoula, Montana

G. Brett Runion
Soil Dynamics Laboratory
USDA-ARS
Auburn, Alabama

William H. Schlesinger
Nicholas School of the Environment
  and Earth Sciences
Duke University
Durham, North Carolina

Matthew J. Searson
Australian Greenhouse Office
Department of the Environment and
  Heritage
Australia

Chris Stokes
CSIRO Sustainable Ecosystems and
  CRC for Tropical Savannas
  Management
Aitkenvale, Australia

Richard B. Thomas
Department of Biology
West Virginia University
Morgantown, West Virginia

Skip J. Van Bloem
Depto Agronomîa y Suelos
Universidad de Puerto Rico
Mayagüez, Puerto Rico

Lewis H. Ziska
Crop Systems and Global Change
  Laboratory
USDA-ARS
Beltsville, Maryland

Table of Contents

Chapter 1

PART II   Pests, Weeds, and Diseases

PART III   Capacity to Adapt

PART IV Special Examples

In Parts I–III the authors have set out general principles determining agroecosystem responses to global change and the consequences of these have been explored. In this Part we present five Special Examples that bring the focus down to explore the impacts of global change at the agroecosystem, technology, population, and regional level.

1 Introduction

Paul C.D. Newton, R. Andrew Carran,
Grant R. Edwards, and Pascal A. Niklaus

CONTENTS

1.1 THE CONTEXT

Agricultural systems (agroecosystems) are enormously diverse in their biological structure, in the climatic and socioeconomic drivers that shape them, and in the services they provide. Agroecosystems may be critical in sustaining social, economic, and cultural fabrics as well as playing a key role in community health. The challenge we face is to maintain these services and roles during a period of rapid environmental change — perhaps producing combinations of environmental conditions that modern agriculture has not previously experienced. In the first instance, this is an issue for biologists; it is essential to improve our understanding of the potential impacts of environmental change. However, it is also the case that agroecosystems are the site of intense interaction between humans and the natural world. In some situations this offers the prospect of effective adaptation to climatic change — either amelioration of negative effects or exploitation of positive effects; however, at the other extreme there will be many situations in which the effects of climate change will be exacerbated by socioeconomic factors such as income inequality or credit availability (Patt et al. 2005). As biologists we cannot ignore this reality, and if we want our work to be relevant, then we need to consider carefully who and how we are targeting with our knowledge. We consider some of these issues later in this chapter, but first we review the main drivers of environmental change.

1.2 THE ENVIRONMENTAL CHANGES

The changes in the environment we are most concerned with involve temperature, precipitation, and atmospheric CO_2 concentration. Here we consider some of the issues surrounding these *drivers of change*, which we feel are of central importance in determining biological responses. In particular we draw attention to variation in current and predicted temperature and precipitation trends, and to the likelihood of further increases in atmospheric CO_2 concentration.

1.2.1 ATMOSPHERIC CO_2 CONCENTRATION

As well as its indirect effect on ecosystems through its role as a greenhouse gas, atmospheric CO_2 has a direct effect on the biosphere because it is the primary raw material for plant growth. Direct measurements of atmospheric CO_2 have been made since 1958 in Hawaii (Keeling et al. 1982), and new monitoring stations have been established since that time to give a more global coverage. The data from these stations (CDIAC 2005), together with historical records extracted from ice cores (CDIAC 2005), show the atmospheric concentration has increased exponentially since the late 19th century, that it increases each year, and that it is now at a level higher than at any time in the last 650,000 years (Siegenthaler et al. 2005).

Because CO_2 is such an important driver of biological systems, our picture of the future must include the rate of change in atmospheric CO_2 and the level at which the concentration will stabilize; these both depend upon the balance between CO_2 emissions and the rate at which CO_2 can be sequestered into sinks. A recent carbon (C) budget for the 1990s (Schimel et al. 2001) calculates emissions from fossil fuels to be 6.3 Gt C yr^{-1}, with an additional 1.7 Gt C added from land use change. Approximately half of CO_2 emissions are removed by oceanic and terrestrial sinks leaving an annual addition of approximately 3.2 Gt C to the atmosphere. During this period, the CO_2 concentration in the atmosphere increased at a rate of about 1.5 ppm or 0.44% per year. If this balance of sources and sinks were to continue, then by 2050 the concentration in the atmosphere would reach 460 ppm. As neither sink nor source activity is expected to stay constant over this period, prediction becomes a difficult task.

On the source side, prediction of future emissions depends upon a range of assumptions about technology change and population growth. Emissions projected for 2050 range from 11.0 to 23.9 Gt C (Prentice et al. 2001). To put emissions reduction in perspective, we can make a crude calculation: Assuming current emissions of 6.3 Gt yr^{-1} (Schimel et al. 2001) and a world population of 6 billion, then our current emissions rate is 1 t of C per capita yr^{-1}. To maintain a balance with the sinks, which currently absorb only half of these emissions, and to have no net increase in the atmosphere, we need to have emissions of 500 kg per capita yr^{-1}. If we look at current rates of per capita emissions of C from fossil fuel use (Marland et al. 2004), we find that emissions from the United States and Western Europe are, respectively, 10 times and 4 times the 500 kg rate that would balance current sinks. The current per capita emission rate in China also exceeds the "stabilising" output level, and it is only in countries such as Africa and India that emissions are below the 500 kg per capita level. These figures suggest that major changes in our energy creation and use will be necessary to achieve significant reductions in emissions (Hoffert et al. 2002).

The second half of the equation is the activity and size of sinks. Again, we can make a rough calculation to establish the size of the problem by considering how much sink activity would need to change by 2050 to keep net emissions to the atmosphere at the current 3.2 Gt C level; that is, to constrain the annual increase to 1.5 ppm. Using the low prediction of CO_2 emissions for 2050 of 11 Gt C (Prentice et al. 2001) would require sinks to remove 7.8 Gt C to maintain net emissions at 3.2 Gt C. If half of the sink activity is oceanic and half terrestrial, then terrestrial sinks would need to absorb 3.9 Gt C or 2.8 times their current rate (assuming a current terrestrial sink of 1.4 Gt yr^{-1}, Prentice et al. 2001). The stimulation of plant growth due to the rising concentration of CO_2 offers the promise of enhanced sink strength; however, in the absence of significant changes in the ratio of C to nitrogen (N) in terrestrial pools, such an increase in C sequestration would require substantial increases in N availability, perhaps beyond the capacity of ecosystems to provide (Hungate et al. 2003). In fact, elevated CO_2 may exacerbate this constraint, as a common response appears to be a progressive decline in the availability of N to plants (Luo et al. 2004). The potential sink capacity of the terrestrial biosphere remains a critical value if we are to predict future CO_2 concentrations. However, a doubling of sink capacity would be required to absorb even current emissions, let alone those expected in the next decades, placing an unrealistic expectation on the absorbing capacity of this sink.

The average annual CO_2 concentration of the well-mixed atmosphere does not differ greatly among monitoring stations, although there is a slightly lower average in the Southern Hemisphere. However, at different scales there can be considerable variation in concentration, and it is relevant to consider whether these variations are likely to change in the future in response to the changing climate and atmosphere. The net CO_2 exchange of the biosphere results in marked seasonal differences (15 to 20 ppm, Keeling et al. 1996) in atmospheric concentration in the Northern Hemisphere; interestingly, the amplitude of this difference is increasing over time (Keeling et al. 1996), probably because of disturbance and a change in the identity and activity of the vegetation (Zimov et al. 1999). Regional differences in atmospheric CO_2 concentration can arise from urban development where large sources of fossil fuel use can dominate the concentration profile. For example, Ziska et al. (2004) measured average concentrations of 466 at 0.5 km from the city centre of Baltimore, Maryland — 401 ppm 10 km from the centre and 385 ppm at a distance of 50 km. Temperature gradients are also established by urbanization, and clearly both CO_2 and temperature gradients will be determined by future urban development and energy use.

Plants also experience large differences in CO_2 concentration between day and night (often > 100 ppm; e.g., Ziska et al. 2001); we are not aware of data considering trends in this difference, but certainly one aspect of climate change has been a reduction in the diurnal temperature range (Prentice et al. 2001), and it may be that this could influence the biological processes of C fixation and respiration that largely govern the differences in CO_2 concentration near the surface. These biological processes also result in considerable spatial variation in concentrations of CO_2 within plant canopies; plants growing close to the soil surface experiencing concentrations of CO_2 perhaps 100 ppm greater than plants with foliage higher in the canopy

(Bazzaz and Williams 1991). It is not certain whether this spatial variation will alter under climate change, but as soil respiration is sensitive to both temperature and elevated CO_2 (e.g., King et al. 2001) there is a strong likelihood of different canopy profiles in the future. The spatial and temporal variation occurs at scales relevant to plant growth and has been shown to influence plant responses to elevated CO_2 (Ziska et al. 2001).

1.2.2 TEMPERATURE AND PRECIPITATION

The global mean near-surface temperature record shows an increase over the 20th century of about 0.6°C (Folland et al. 2001), which is consistent with satellite data for tropospheric temperatures (Tett et al. 1999; Vinnikov and Grody 2003). Further evidence for a temperature change are the "fingerprints" of increasing temperature that can be seen in a range of biological data such as phenological records (Parmesan and Yohe 2003; Root et al. 2003). There are a number of *forcing agents* that can modify climate, some of which are natural (solar radiation and volcanic aerosols) and some anthropogenic, including greenhouse gases, tropospheric aerosols, cloud changes, and changes in the land surface characteristics altering albedo (Hansen et al. 1998). When climate models are run to simulate long-term temperature trends, the prediction for the latter part of the last century requires greenhouse gas effects be included in order to adequately simulate the observed changes (e.g., Karoly et al. 2003); studies such as these are part of the argument that anthropogenic emissions of greenhouse gases are resulting in a change in our climate (Mitchell et al. 2001). Predictions for the change in mean temperature over the next 100 years range between 1.4 and 5.8°C (Cubasch et al. 2001). Spatial variation in temperature and precipitation trends are widely observed and predicted. Rainfall has increased by 10 to 40% over the past 100 years in northern Europe, but has decreased by 20% in southern Europe. In the United States, soil temperatures (1 m over the period 1967 through 2002) show a positive warming trend at stations in the north and northwestern United States, but a strong cooling trend in the southeastern part of the country. Because part of the spatial variation is driven by land use (Pielke et al. 2002; Stone and Weaver 2003; Feddema et al. 2005), variation within regions is also apparent. For example, Pielke et al. (2002) examined the long-term records from a cluster of stations in eastern Colorado and found "enormous" differences, defying attempts to calculate regional trends.

Any trend in mean annual temperature or precipitation is unlikely to be evenly distributed across seasons. For example, in Australia there has been a trend since 1950 for an increase in minimum temperatures in all seasons in Queensland, but a cooling of maximum summer temperature in northwestern Australia (Anonymous 2005). Differences can be a matter of degree, such as the long-term increase in European summer temperatures over the past 100 years of 0.7°C compared to an increase of 1.1°C in winter temperatures; or can be quite strikingly different even in sign, such as the trends in soil temperatures (40 cm) at Irkutsk, Russia, where there was a marked positive trend in annual average temperature over the past 120 years, but a decline in the average summer temperature of 4°C over this time (Zhang et al. (2001).

Variation is also evident at shorter timescales with a reduction in the diurnal temperature range being frequently and widely observed (Prentice et al. 2001). As one of the major factors implicated in this change is soil moisture content (Stone and Weaver 2003), there is a direct link back to vegetation responses to the changing environment. Diurnal patterns in precipitation have been less thoroughly investigated, but here again trends have been identified (Dai 1999).

1.3 THE STRUCTURE OF THIS BOOK

In summary, we can anticipate a continuing increase in the global mean atmospheric CO_2 concentration, and in global mean temperature with a variety of changes in temperature, precipitation, and CO_2 occurring at different scales relevant to biological activity. How can we deal with this complexity?

We suggest that two aspects are particularly important. First, an improved understanding of the biological consequences achieved through greater integration of basic and applied knowledge; and second, a clearer focus on the audience for this research, as this should enable us to ask and address more targeted and relevant questions.

Our perception is that climate change impact research often occurs in two ways. On the one hand there are studies of the direct effects of a particular set of climate drivers on a specific crop, often concentrating on the agricultural outputs, such as yield and quality (e.g., Reddy and Hodges 2000). On the other hand, there is a more ecological literature that seeks to find some general principles of response (e.g., Körner and Bazzaz 1996, p. 4). In this volume we hope to draw these two approaches together so that ecologists can provide the "theoretical underpinning that informs them (agriculturalists) what might be happening, what to look for, and what to build on" (Lawton 1996, p. 4), and agriculturalists can interpret these ecological insights and general theory in relation to agroecosystem performance. Consequently, each section of this book combines general principles of response leading to applied consequences. We have sections considering (1) the supply of resources necessary to sustain agriculture in the future, which we identify from an understanding of how climate change will modify biogeochemical cycles and changes in plant nutrient demands; (2) the incidence of pests, weeds, and diseases and their control for which we need an understanding of how the population biology of organisms will change; and (3) the adaptations that might be possible, including plant breeding solutions, for which we need an understanding of the capacity for adaptation that exists in plant populations. In addition to the full chapters, we have included Special Example chapters that deal in more detail with specific issues.

Having collected the best information, we are then faced with the issue of communicating it effectively to interested groups and, in particular, to those groups that can act effectively in leading or implementing adaptive measures. On this basis we suggest that while farmers and landowners will likely be interested in projections that consider, for example, changes in yield or the incidence of pests, their behaviour is unlikely to be modified by such predictions, as they tend to be responsive to current conditions. Consequently, we imagine that the issues in this book will resonate most strongly with other researchers and with agribusiness because here are important messages about potential opportunities for the development of new technologies. These are important sectors to reach, as it is through the development of new, adaptive technologies that we can imagine making a difference in agroeco-system performance in a changing environment.

REFERENCES

Anonymous (2005) Annual and seasonal temperature trends since 1950. Australian Govern-ment Bureau of Meteorology, http://www.bom.gov.au/climate/change/seatrends.shtml, accessed August 10, 2005.

Bazzaz, F.A. and Williams, W.E. (1991) Atmospheric CO_2 concentrations within a mixed forest: implications for seedling growth. *Ecology*, 72, 12–16.

CDIAC (Carbon Dioxide Information Analysis Center) (2005) Trends online: a compendium of data on global change. Oak Ridge National Laboratory, U.S. Department of Energy, Oak Ridge, TN. http://cdiac.esd.ornl.gov/trends/co2/contents.htm, accessed August 10, 2005.

Cubasch, U. et al. (2001) Projections of future climate change. In: *Climate Change 2001: The Scientific Basis*. Contribution of Working Group 1 to the Third Assessment Report of the Intergovernmental Panel on Climate Change (Houghton, J.T. et al., Eds.), Cambridge University Press, Cambridge, U.K., 527–582.

Dai, A. (1999) Recent changes in the diurnal cycle of precipitation over the United States. *Geophysical Research Letters*, 26, 341–344.

FAO (Food and Agricultural Organization of the United Nations) (2004) *State of Food Insecurity in the World 2004*. http://www.fao.org/documents/show_cdr.asp?url_file=/docrep/007/y5650e/y5650e00.htm

Feddema, J.J., Oleson, K.W., Bonan, G.B., Mearns, L.O., Buja, L.E., Meehl, G.A. and Washington, W.M. (2005) The importance of land-cover change in simulating future climates. *Science*, 310, 1674–1678.

Folland, C.K. et al. (2001) Observed climate variability and change. In: *Climate Change 2001: The Scientific Basis*. Contribution of Working Group 1 to the Third Assessment Report of the Intergovernmental Panel on Climate Change (Houghton, J.T. et al., Eds.), Cambridge University Press, Cambridge, U.K., 101–181.

Hansen, J.E., Sato, M., Lacis, A., Ruedy, R., Tegen, I. and Matthews, E. (1998) Climate forcings in the industrial era. *Proceedings of the National Academy of Science, USA*, 95, 12753–12758.

Hoffert, M.I. et al. (2002) Advanced technology paths to a global climate stability: energy for a greenhouse planet. *Science*, 298, 981–987.

Hungate, B.A., Dukes, J.S., Shaw, M.R., Luo, Y. and Field, C.B. (2003) Nitrogen and climate change. *Science*, 302, 1512–1513.

Karoly, D.J., Braganza, K., Stoot, P.A., Arblaster, J.M., Mehl, G.A., Broccoli, A.J. and Dixon, K.W. (2003) Detection of a human influence on North American climate. *Science*, 302, 1200–1203.

Keeling, C.D., Bacastow, R.B. and Whorf, T.P. (1982) Measurements of the concentration of carbon dioxide at Mauna Loa Observatory, Hawaii. In: *Carbon Dioxide Review*, Clark, W.C., Ed., Oxford University Press, New York, 377–385.

Keeling, C.D., Chin, J.F.S. and Whorf, T.P. (1996) Increased activity of northern vegetation inferred from atmospheric CO_2 measurements. *Nature*, 382, 146–149.

King, J.S., Pregitzer, K.S., Zak, D.R., Sober, J., Isebrands, J.G., Dickson, R.E., Hendrey, G.R. and Karnosky, D.F. (2001) Fine-root biomass and fluxes of soil carbon in young stands of paper birch and trembling aspen as affected by elevated CO_2 and tropospheric O_3. *Oecologia*, 128, 237–280.

Körner, C. and Bazzaz, F.A. (1996) *Carbon Dioxide, Populations and Communities*. Academic Press, San Diego, CA.

Lawton, J.H. (1996) Corncrake pie and prediction in ecology. *Oikos*, 76, 3–4.

Luo, Y. et al. (2004) Progressive nitrogen limitation of ecosystem responses to rising atmospheric carbon dioxide. *Bioscience*, 54, 731–739.

Marland, G., Boden, T.A. and Andres, R.J. (2004) Global, regional and national fossil fuels CO_2 emissions in Trends: a compendium of data on global change. http://cdiac.esd.ornl.gov/trends/emis/em_cont.htm, accessed August 12, 2005.

Mitchell, J.F.B., Karoly, D.J., Hegerl, G.C., Zwiers, F.W., Allen, M.R. and Marengo, J. (2001) Detection of climate change and attribution of causes. In: *Climate Change 2001: The Scientific Basis*. Contribution of Working Group 1 to the Third Assessment Report of the Intergovernmental Panel on Climate Change, (Houghton, J.T. et al., Eds.), Cambridge University Press, Cambridge, U.K., 697–738.

Parmesan, C. and Yohe, G. (2003) A globally coherent fingerprint of climate change impacts across natural systems. *Nature*, 421, 37–42.

Patt, A., Klein, R.J.T. and de la Vega-Leinert, A. (2005) Taking the uncertainty in climate-change vulnerability assessment seriously. *C.R. Geoscience* 337, 411–424.

Pielke, Sr., R.A. et al. (2002) The influence of land-use change and landscape dynamics on the climate system: relevance to climate-change policy beyond the radiative effect of greenhouse gases. *Philosophical Transactions of the Royal Society of London Series A*, 360, 1705–1719.

Prentice, I.C. et al. (2001) The carbon cycle and atmospheric carbon dioxide. In: *Climate Change 2001: The Scientific Basis*. Contribution of Working Group 1 to the Third Assessment Report of the Intergovernmental Panel on Climate Change (Houghton, J.T. et al., Eds.), Cambridge University Press, Cambridge, U.K., 183–237.

Reddy, K.R. and Hodges, H.F. (2000) *Climate Change and Global Crop Productivity*, CAB International, Wallingford, U.K.

Root, T.L., Price, J.T., Hall, K.R., Schneider, S.H., Rozensweig, C. and Pounds, J.A. (2003) Fingerprints of global warming on wild animals and plants. *Nature*, 421, 57–60.

Schimel, D.S. et al. (2001) Recent patterns and mechanisms of carbon exchange by terrestrial ecosystems. *Nature*, 414, 169–172.

Siegenhaler, U., Stocker, T.F., Monnin, E., et al. (2005) Stable carbon cycle-climate relationship during the late Plesitocene. *Science*, 310, 1313–1317.

Stone, D.A. and Weaver, A.J. (2003) Factors contributing to diurnal temperature range trends in twentieth and twenty-first century simulations of the CCCma coupled model. *Climate Dynamics*, 20, 435–445.

Tett, S.F.B., Stott, P.A., Allen, M.R., Ingram, W.J. and Mitchell, J.F.B. (1999) Causes of twentieth-century temperature change near the Earth's surface. *Nature*, 399, 569–572.

Vinnikov, K.Y. and Grody, N.C. (2003) Global warming trend of mean tropospheric temperature observed by satellite. *Science*, 302, 269–272.

Zhang, T., Barry, R.G., Gilichinsky, D., Bykhovets, S.S., Sorokovikov, V.A. and Jingping, Y.E. (2001) An amplified signal of climatic change in soil temperature during the last century at Irkutsk, Russia. *Climatic Change*, 49, 41–76.

Zimov, S.A., Davidov, S.P., Zimova, G.M., Davidova, A.I., Chapin, III, F.S., Chapin, M.C. and Reynolds, J.F. (1999) Contribution of disturbance to increasing seasonal amplitude of atmospheric CO_2. *Science*, 284, 1973–1976.

Ziska, L.H., Bunce, J.A. and Goins, E.W. (2004) Characterization of an urban-rural CO_2/tem-
    perature gradient and associated changes in initial plant productivity during secondary
    succession. *Oeocologia*, 139, 454–458.
Ziska, L.H., Ghannoum, O., Baker, J.T., Conroy, J., Bunce, J.A., Kobayashi, K. and Okada,
    M. (2001) A global perspective of ground level, "ambient" carbon dioxide for assess-
    ing the response of plants to atmospheric CO_2. *Global Change Biology*, 7, 789–796.

Part I

Resource Supply and Demand

2 Climate Change Effects on Biogeochemical Cycles, Nutrients, and Water Supply

Pascal A. Niklaus

CONTENTS

2.1   INTRODUCTION

The atmospheric CO_2 concentration has increased by $\approx 30\%$ relative to the preindustrial concentration of 280 µL L^{-1} and is projected to reach 540 to 970 µL L^{-1} by the end of this century, depending on emission scenarios and climate feedback (IPCC [Intergovernmental Panel on Climate Change], 2001). CO_2 and other atmospheric gases of anthropogenic origin are radiatively active, and increases in global temperatures in the range of 1.4 to 5.8° C are predicted, depending on emission scenarios and climate sensitivity (IPCC, 2001). As a consequence of warming, changes in the global distribution of precipitation are anticipated, with projected increases at medium to high latitudes, but decreases in other areas (e.g., the European Mediterranean).

In this chapter, I analyse how these global changes might affect the biogeochemical cycling of nutrients and hydrology, and how this ultimately may impact on agricultural ecosystems.

An important distinction to be made is between agroecosystems with high fertiliser input and relatively open nutrient cycles on one hand, and low input and seminatural systems in which nutrient cycles are relatively closed on the other hand. In intensified agriculture, relatively large amounts of nutrients are removed from the ecosystem with the crop and need to be resupplied in the form of mineral or organic fertiliser. Effects of global change on soil nutrient cycling are less likely to be of importance in these systems. However, soil processes such as trace gas emissions may change, which can strongly feed back on the climate system. Also, the nutritional composition of crops may change, altering their nutritional quality and possibly requiring changes in fertiliser composition. In natural ecosystems as well as in extensively managed systems, such as low-input crop cultivation, pastures, rangelands, and low-intensity forestry, a significant fraction of plant nutrient demand is met by the internal mineral nutrient cycles of soils, and effects of climate change on soil processes may directly feed back on plant growth.

Despite several decades of global change research, available data on the effects on soil nutrient cycling are surprisingly limited, especially for agroecosystems and nutrients other than N. In many areas, we are still in the stage of pattern searching without having a very detailed understanding of the mechanisms underlying responses. In the attempt not to unnecessarily narrow the scope of this chapter to what is already well known, included is data from all available sources, including studies in natural ecosystems. Whether and to what extent these findings can be extrapolated to typical agricultural situations remains to be explored.

2.1.1   Essential Elements

Plant tissue is primarily composed of carbon (C), hydrogen (H), and oxygen (O). These elements are derived from the fixation of atmospheric CO_2 and from the uptake of soil H_2O, and are generally available in ample quantities. However, virtually all naturally occurring elements are also found in plants, and more than 10 are essential for growth (Welch, 1995). Mineral nutrients are generally classified into macronutrients, required by plants at relatively large concentrations (nitrogen [N], phosphorus [P], potassium [K], sulfur [S], calcium [Ca] and magnesium [Mg]; Epstein, 1965); and micronutrients, which are required in much lower quantities (chlorine [Cl], iron [Fe], boron [B], manganese [Mn], zinc [Zn], copper [Cu], molybdenum [Mo] and nickel [Ni]). Still other elements are beneficial to plants but probably not essential for growth (sodium [Na], silicon [Si], cobalt [Co] and selenium [Se]). Micronutrients are predominantly bound in enzymes, where they often have important functional roles at the active sites, whereas macronutrients are constituents of organic macromolecules (e.g., N, P, and S in proteins and nucleic acids) or act as osmotica (e.g., K).

Most studies of global change effects on nutrient cycling have so far focused on nitrogen. One reason may be that N is the nutrient required in the largest quantity; another reason may be the relative ease with which N can be measured. The nitrogen cycle is also clearly the most complex of all cycles of essential elements because N occurs at a wide range of oxidation states; is involved in a vast array of microbial transformations; and also occurs in gaseous, solid, and dissolved forms, endowing it with exceptional mobility. There is, therefore, a large potential for climate change to interfere with N nutrition. However, the N cycle is also special in that a biological pathway exists with N_2 fixation by which ecosystems can adjust to altered N demand. This is not the case for the other elements. Indeed, while N clearly is often limiting (Vitousek and Howarth, 1991), the level at which N becomes limiting in the long term is frequently determined by the availability of other mineral nutrients (e.g., P: Cole and Heil, 1981; McGill and Cole, 1981; Niklaus and Körner, 2004, Mo: Hungate et al., 2004).

Many animals, both wild and domestic, forage on plants and accommodate their mineral nutrient needs from plant sources. Animals require many of the same elements as plants, but additionally require various complex organic molecules. Plant chemical composition, therefore, can determine animal growth, but the limiting component is often not easy to determine. For example, herbivores are often more limited by N than by carbohydrates. An important consideration is that nutrient concentrations that are sufficient for plants may be too low for the animals that feed on them. For example, many New Zealand and Australian soils are very low in cobalt; cobalt is not essential to plants* and, thus, does not limit their growth. However, sheep foraging on these plants exhibit severe cobalt-deficiency symptoms (Lee et al., 1999). The accumulation of nonplant-essential elements in plant tissue (e.g., iodine or cobalt) and the accumulation of plant-essential elements beyond limiting concentrations, therefore, can be ecologically and economically important.

Human nutrition ultimately also depends on plant chemical composition, whether plants are consumed directl - indirectly as animals that previously fed on plants (Underwood and Mertz, 1987). Besides insufficient total energy and protein input, the World Health Organization (WHO) identified micronutrient deficiency as

---

* Except for N_2 fixation.

a major cause of malnutrition (WHO, 2000). Over 2 billion people are currently affected by deficiencies in iodine, zinc, iron, selenium, and calcium, but also of more complex phytochemicals such as vitamins A and E, niacin, and folate (Grusak and DellaPenna, 1999; WHO, 2000). Clearly, human malnutrition at the global level is a complex phenomenon involving political, sociological, and economic aspects beyond the scope of this chapter. However, these data emphasize that micronutrient effects of global change may have important implications beyond the functioning of plants (Allaway, 1987).

2.1.2 MECHANISMS AND KEY PROCESSES

In the following analysis, I focus on a number of key processes in nutrient cycles that are experimentally accessible. I review experimental evidence on how these respond to simulated climate change, and analyse, from a theoretical point of view, how they may be affected, directly or indirectly, by elevated CO_2, elevated temperatures and altered precipitation. While these drivers are quite different and the details of the mechanisms involved clearly are complex, two principal groups of mechanisms can be identified (Figure 2.1).

A first group of effects is related to changes in carbon cycling. A primary effect of elevated CO_2 is to increase photosynthesis and plant growth; warming also increases primary production of most ecosystems, but it also affects respiration and decomposition. Elevated CO_2 and warming, therefore, result in alterations of the carbon balance of plants, soils, and soil organisms.

The second group of effects is related to the hydrological cycle. Elevated CO_2 reduces stomatal conductance in virtually all vascular plant species, and this can reduce evapotranspiration and water use, at least per unit of plant biomass. Warming, on the other hand, increases evapotranspiration and generally results in a drier environment. Precipitation, finally, has a direct effect on the water balance of the

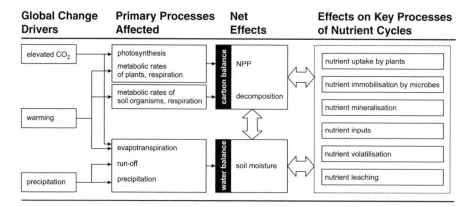

FIGURE 2.1 Schematic of effects of global change drivers via alterations of carbon and water balance. See text for a detailed discussion.

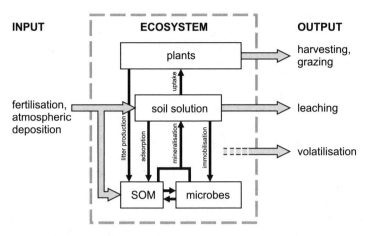

FIGURE 2.2 Key processes in nutrient cycles that might be affected by global environmental change (SOM = soil organic matter).

ecosystem. Alterations of the ecosystem water balance affect many processes including the diffusion of gases and nutrients, sorption processes, and leaching rates. Redox potential is also strongly affected by the soil water balance, primarily by controlling gas-filled pore space and, therefore, the diffusion of O_2; this can induce shifts in the activity of different microbial groups involved in nutrient transformations, but operating under different redox conditions, for example, between nitrifying and denitrifying bacteria.

Effects via carbon and water balance interact because in many ecosystems productivity and decomposition are limited by low soil water content (but by high soil moisture in water-logged soils). There are also more complex feedback mechanisms involved that will be discussed. Some key quantities and processes that are focused on in this chapter are schematically shown in Figure 2.2.

2.1.2.1 Nutrient Balance of the Whole Ecosystem

Summing all nutrient fluxes across an ecosystem's boundaries theoretically allows calculation of the nutrient balance of the whole ecosystem. If nutrient outputs are larger than inputs, nutrient stocks within the ecosystem will become depleted and productivity may decrease in the long run. Data allowing for a complete balance are scarce; however, there is a substantial body of literature reporting the main fluxes, and these may give an indication of the changes to expect in a future climate.

Climate change may affect the nutrient balance of ecosystems in several ways:

1. Nutrient deposition from the atmosphere could change with altered weather patterns (precipitation, range of transport, anthropogenic emissions of pollutants [N, S], etc.).
2. Biological fixation of atmospheric N_2 may change (addressed by Thomas et al., Chapter 4, this volume).
3. Weathering rates of minerals may change.

4. The amount of nutrients exported may change due to changes in amount and elemental composition of plant material.

5. Losses by leaching and erosion could change due to effects on nutrient cycling and alterations in climate, especially in precipitation.

In the long run, changes in the nutrient balance may necessitate a corresponding change in the elemental ratio of nutrient inputs to maintain the status quo.

2.1.2.2 Mineralisation of Soil Organic Matter

Plants take up nutrients from soil solution. One process delivering nutrients to soil solution is the mineralisation of organic matter. The mineralisation of different elements does not occur in concert because mineral nutrients differ in the way they are bound in organic matter (McGill and Cole, 1981). Nitrogen is predominantly covalently bound to carbon. Sulphur is covalently bound both directly to C and via ester linkages. Phosphorus is essentially ester-bound, while potassium does not bind covalently at all, but forms ionic bonds. The mineralisation of N is, therefore, largely coupled to that of C, while ester-linked P and S can be mineralised independently of C by the hydrolytic action of soil exoenzymes.

Cycling rates of mineral nutrients are controlled by complex interactions between plants, soil microbes, and abiotic factors (Schlesinger, 1996). These controls differ between mineral nutrients: The cycling rate of limiting nutrients will control plant productivity (e.g., N and P: Güsewell, 2004; Niklaus and Köorner, 2004; Vitousek and Howarth, 1991; Vitousek et al., 1993), whereas plant productivity will generally control the cycling of nutrients that are available in excess (e.g., sulphur). The cycling of still other elements is predominantly geochemically controlled and relatively independent of plant growth (e.g., chlorine).

Global change can alter organic matter mineralisation rates in several ways. For example, the quality and quantity of organic matter produced by plants can be affected, resulting in altered decomposition. Microbial activity and decomposition can be affected by soil fluxes of plant-derived C (*priming effect*), and abiotic conditions such as temperature and moisture, which are important controls of decomposition rates, may also be changed.

2.1.2.3 Immobilisation in Soil Microbial Biomass

The soil microbial biomass is a highly dynamic organic matter pool and its nutrient content often exceeds that of plants. Changes in amounts of mineral nutrients bound in microbial cells can substantially alter the within-ecosystem nutrient distribution and affect plant growth. The soil microbial community is predominantly saprophytic and, therefore, depends on inputs of plant-derived organic material. As a consequence, soil microbial community biomass may respond to alterations in soil C fluxes under global change. Apart from C supply, soil microbial biomass is also controlled top–down by grazing by protozoa and microfauna (Blankinship and Hungate, Chapter 5, this volume), by the availability of nutrients such as P or N, which can (co-)limit microbial growth, and by soil conditions that can have strong effects on microbial populations (e.g., soil moisture, temperature, freeze–thaw and drying–rewetting cycles, soil mineral, and aggregate structure).

2.1.2.4 Leaching and Volatilisation of Nutrients

Significant amounts of nutrients can be lost from ecosystems by leaching and, in the case of N, by volatilisation of NH_3, NO, N_2O, and N_2 (Barnard et al., 2005). Leaching losses will generally be more important for compounds such as NO_3^- and K^+, which are highly soluble and show low sorption to the solid phase. Leaching rates are related to many factors that are likely to change in a future climate, for example to soil solution concentrations of nutrients and the amount of water drained. However, water flow often follows preferential flowpaths in the soil, and leaching events are erratic, so that predictions are difficult.

2.2 EFFECTS OF ELEVATED CO_2 ON NUTRIENT CYCLING

Elevated CO_2 concentrations often stimulate plant growth. One mechanism for this response is that photosynthesis is stimulated by elevated CO_2, at least in C_3 plants. A second mechanism is that in almost all species, stomatal closure is induced in response to the increased availability of CO_2. Lower stomatal conductance can result in reduced evapotranspiration, which in turn can result in comparably higher soil moisture at any given plant biomass, or to the maintenance of higher plant biomass at any given level of soil H_2O (Jackson et al., 1998; Niklaus and Körner, 2004; Owensby et al., 1999). A recent analysis has demonstrated that this latter indirect effect (increased water use efficiency) accounts for much of the biomass stimulation observed in arid and semiarid but also in some mesic grasslands exposed to elevated CO_2 (Morgan et al., 2004b), and both C_3 and C_4 plants, therefore, generally exhibit growth responses to elevated CO_2 (Wand et al., 1999). Nutrient dynamics may be altered by both increased C fixation and increased soil moisture.

2.2.1 Evidence for Changes in Soil Carbon Fluxes

Higher plant productivity under elevated CO_2 will ultimately increase organic matter inputs to soils unless all the extra production is removed as harvested plant material (hay, crop).* The pathways by which extra C enters the soil under elevated CO_2, however, still remain elusive. Only limited field data is available (see, for example, Hungate et al., 1997b; Niklaus et al., 2001a; Ross et al., 1995), in part due to methodological difficulties to assess plant–soil C fluxes *in situ* (Darrah, 1996; Hilbert et al., 1987; Hungate et al., 1996; Lund et al., 1999; Niklaus et al., 2000). Exudation is a major component of soil C inputs, but there are virtually no data on flux rates and chemical composition of exudates under field conditions. There have been early indications from laboratory studies, mostly conducted under hydroponic conditions, that rhizodeposition per unit root length would increase under elevated CO_2 (review

---

* Even then, soil organic matter inputs will increase due to biomass turnover between harvests.

by Darrah, 1996), and this effect has been interpreted as plants being saturated with carbohydrates and passively leaking or actively "disposing" belowground the extra C not needed. However, this notion could not be corroborated in a number of field systems, probably because soil matrix effects and resorption of organic compounds reduce or eliminate the effects observed in hydroponics (Darrah, 1996; Jones and Darrah, 1996). Field data are mostly based on observations of fine root turnover (Arnone et al., 2000; Fitter et al., 1996; Pregitzer et al., 2000) and carbon tracer studies (e.g., Hungate et al., 1997b; Newton et al., 1995; Niklaus et al., 2001a). Nevertheless, even in the absence of a carbon overflow-type effect on rhizodeposition under elevated CO_2, and in the absence of increased root-to-shoot ratios, increased plant productivity should result in increased soil C inputs. However, the ultimate magnitude of this effect is still unknown, and so are the pathways by which extra C enters the soil.

2.2.2 NUTRIENT IMMOBILISATION AND MINERALISATION RESPONSES TO ELEVATED CO_2

Increased soil C fluxes under elevated CO_2 could lead to higher soil microbial biomass and immobilisation of mineral nutrients. This *negative* feedback on plant growth has been demonstrated in a pot CO_2 experiment where microbial biomass N increased and plant responses to elevated CO_2 were negative (Diaz et al., 1993), presumably due to increased input of high C:N compounds to soils. Extra C inputs to soils can, however, increase microbial activity and, thus, prime the mineralisation of organic matter. This *positive* feedback on plant growth has been proposed by Zak et al. (1993). Priming effects on mineralisation and increased nutrient immobilisation in microbial biomass are not mutually exclusive and microbial biomass and the availability of nutrients to plants may increase concurrently (higher net immobilisation plus increased net mineralisation rates).

Experimental evidence of elevated CO_2 effects on microbial biomass is equivocal. While some greenhouse studies reported increases in microbial biomass, responses in field studies of grassland appear to be smaller and often even absent (Table 2.1). Various experimental protocols were used; nitrogen mineralisation, for example, was measured in the field or using isolated soil samples, with methods that ranged from buried bags, aerobic or anaerobic laboratory incubations, to short- and long-term ^{15}N isotopic pool dilution methods. Despite all these differences, some broad response patterns emerged.

Microbial C often does not respond to elevated CO_2, even after several years of increased plant productivity. However, microbial N and N mineralisation increase in many studies, suggesting that microbial N responds more than C, and that this extra microbial N originates from increased mineralisation of soil organic matter rather than from immobilisation of soil mineral N. Two different mechanisms may explain these observations: First, microbial N may have been primed by extra C inputs under elevated CO_2 (mechanism proposed by Zak et al., 1993); second, increased soil moisture at elevated CO_2 may have led to increased N mineralisation, at least in (temporarily) H_2O-limited ecosystems. CO_2 effects may effectively be indirect, that is, soil moisture effects in disguise. It is very difficult to disentangle these two components of elevated CO_2 unless an experimental design involves a factorial irrigation treatment.

I have previously argued that cases in which microbial sequestration of extra N under elevated CO_2 occurs may be restricted to systems where N supply is abundant and nutrient cycles are not in equilibrium with plant demand, that is, the cycling of C and N are not strongly coupled (Niklaus and Körner, 1996); others arrived at similar conclusions (Hu et al., 1999). The data that have become available in the meanwhile, however, indicate that increased net microbial N immobilisation occurs in many (though not all) systems exposed to elevated CO_2. High N supply seems to favour this response, probably via increased plant growth under elevated CO_2 and associated increased soil inputs of C and N from plants; however, this increase in N is generally paralleled by an increase in N mineralisation rates, so that reductions in plant N availability (the negative feedback mechanism identified by Diaz et al., 1993) appear not to happen. One reason why this is not the case may be that overall the C:N in organic matter inputs to soils does not change much under elevated CO_2.

Dynamics of P and S might respond to elevated CO_2 in a fashion similar to that of N; after all, increases in soil microbial biomass will be accompanied by the immobilisation of these nutrients as well, and the decomposition of soil organic matter will release the mineral nutrients that it contained. However, an important difference is that a significant portion of soil organic P and S is ester-bound and, therefore, can be mineralised independently of C as has been discussed above.

Microbial biomass P and S can be measured with techniques analogous to the ones used for N (e.g., chloroform fumigation-extraction; Brookes et al., 1982; Wu et al., 1994); however, to my knowledge, absolutely no data is currently available from elevated CO_2 studies. At first approximation, microbial P and S pools could be assumed to change in concert with microbial biomass, but this assumption can be poor, as data for N demonstrate.

An indicator of P status of plants and soil microbes is the activity of soil extracellular phosphatases; these enzymes are released in response to growth limitations by plants, and their activity is especially high in P-depleted zones in the immediate vicinity of plant roots (Barrett et al., 1998; Spiers and McGill, 1979). Experimental data under elevated CO_2 is quite limited. In a grassland, Ebersberger et al. (2003) found a 32% increase in alkaline phosphatase activity in a N–P-colimited calcareous meadow (Niklaus and Körner, 2004) exposed to elevated CO_2 for 6 years (but no increase in enzymes related to the N cycle). Moorhead and Linkins (1997) reported increased phosphatase activities in tussock tundra exposed to elevated CO_2 for 3 years, and Dhillion et al. (1996) found increased acid phosphatase activity in soil turves with the annual Mediterranean grass *Bromus madritensis* (but no change in extractable soil inorganic P and N). For trees and forest ecosystems, increases in acid phosphatase activity have been found in soil below *Quercus ilex* exposed to twice-ambient CO_2 for 5 years (Moscatelli et al., 2001), but no effects were found in a plantation of 16-year-old *Pinus taeda* exposed to elevated CO_2 for 2 to 3 years. Matamala and Schlesinger (2000) and de Lucia et al. (1997) reported signs of decreased phosphatase activity under ponderosa pine seedlings exposed to elevated CO_2 (but increased concentrations of P-chelating oxalate).

TABLE 2.1
Effects of Elevated CO_2 on Microbial Biomass C and N, N-Mineralisation and Nitrification

System/Species	CO_2 Treatment Conc. (ppmv)	Years for Which Effects Are Listed	Fertiliser Inputs	Effects of CO_2 Enrichment				Reference
				Microbial Biomass C	Biomass N	N-Mineralisation	Nitrification	
Grassland								
Lolium perenne	600	2	14 and 56 g N m^{-2}a^{-1}	n.s.	n.s.		n.s.[1]	Schortemeyer et al. 1996
Trifolium repens	600	2	14 and 56 g N m^{-2}a^{-1}	n.s.	n.s.		n.s.[1]	Schortemeyer et al. 1996
Lolium perenne	600	1 to 10	14 g N m^{-2}a^{-1}			n.s.[2]		Schneider et al. 2004
	600	1 to 10	56 g N m^{-2}a^{-1}			↑[2]		Daepp et al. 2000
Annual grassland (sandstone soil)	A+350	2 to 5	unfertilised	↑/=		↑[3]		Hungate 1997a,b, 2000
Annual grassland (serpentine soil)	A+350	2 to 5	unfertilised	↑/=		↑[3]		
Perennial pasture	475	1 to 5	0 kg N m^{-2}a^{-1}; P,K,Mg,S supplied	n.s.	↑/=	(↑)	n.s.	Ross et al. 2004; pers. comm.
Perennial meadow	600	6	unfertilised	n.s.	↑/=	↑/=	n.s.[5]	Niklaus 1998; Niklaus et al. 2001d, 2003
Planted perennial communities	600	5	disturbed,unfertilised	n.s.	n.s.	=	↓[5]	Niklaus et al. 2001b,c
Alpine meadow	680	4	unfertilised	n.s.	n.s.			Niklaus and Körner 1996; Körner et al. 1997
Alpine meadow	680	4	full fertiliser (equiv. 4.5 g N m^{-2}a^{-1})	n.s.	n.s.			

	CO₂	Years	N status					Reference
Tallgrass prairie	2 × A	2 to 8	unfertilised	↑/=[4]	↑/=[4]	↑/=[4]		Williams et al. 2000; Rice et al. 1994
Arid desert	550	3 to 5	unfertilised		n.s.[6]	n.s.		Billings et al. 2004
Shortgrass steppe	720		unfertilised	n.s.	n.s.			Stahl, pers. comm.; Morgan et al. 2004a
Lolium perenne/ Holcus lanatus	600	3	full fertiliser (equiv. 24 g N m^{-2}a^{-1})		n.s.		n.s.[5]	Barnard et al. 2004
Woody								
Pine forest	A+200	1 to 5	unfertilised	n.s.	n.s.	n.s.		Finzi and Schlesinger 2003
Populus tremuloides Betula papyrifera Acer saccharum	550	3	unfertilised	n.s.	≠	n.s.[3]	n.s.[3]	Holmes et al. 2003
Liquidambar styraciflua	565	2 to 3	unfertilised	n.s.	n.s.	n.s.[3]	n.s.[3]	Sinsabaugh et al. 2003
Crop Production								
Oryza sativa	A+200	1 to 2	8 to 9.4 g N m^{-2}a^{-1}	≠	↑	↑		Inubushi et al. 2001

Note: Data was only included from field studies for which effects on microbial N were reported. Effects listed in parentheses are marginally significant.

[1] Number of autotrophic ammonia oxidising bacteria.
[2] Measured by pool dilution using ^{15}N-labelled fertiliser; effect increases with time.
[3] Gross rates measured by ^{15}N pool dilution.
[4] No effect on average across all years; there was a significant increase of C in 1 and a significant increase in N in 2 out of 6 years.
[5] Nitrification potential measured in incubations.
[6] Data in wetted soils as required for chloroform-fumigation-extraction method (Sparling et al. 1990).

Soil P availability has been measured more directly in a number of studies, mainly by bicarbonate extraction (Olsen-P). For example, Johnson et al. (2003) reported reduced soil P availability in scrub oak forest exposed to elevated CO_2 for several years on some but not all sampling dates. However, no differences in P collected on anion exchange resins were found in the same study.

The interpretation of soil phosphatase data is not straightforward for several reasons. First, the production of extracellular phosphatases is supposedly induced by a lack of available P. Increased phosphatase activities could increase mineralisation rates and ameliorate this condition; however, if the production of phosphatases has been induced by a particularly low availability of P, mineralisation rates may effectively still be reduced. Elevated phosphatase activities, therefore, can be interpreted either as signs of increased P mineralisation (because this is the purpose for which they are produced) or of reduced supply to plants (because this is what induces their production). Second, P can also be rendered plant-available by other avenues; organic acids such as oxalate and citrate can effectively chelate phosphate and improve its solubility, mobility, and plant availability. Third, measurement of soil P pools by sequential extraction procedures yield indicators of plant-available forms of P. However, low concentration can be a result of increased uptake by plants, or the reason for low uptake, and active uptake mechanisms via organic acid exudation and mycorrhizal networks are important and not accounted for by these extractions for P (Rouhier and Read, 1998).

Enzyme activity and soil concentration measurements provide valuable information, but in order to achieve a more conclusive understanding of CO_2 effects on P cycling, several methods would ideally be combined. These include a complete assessment of ecosystem P pools (plants, soil microbial biomass, soil pool measurements by sequential extraction procedures) and isotopic measurements of exchange kinetics and microbial immobilisation ($^{32}P/^{33}P$).

Effects of elevated CO_2 on S-cycling have not received much attention so far. Mineralisation of S has been found to correlate with soil microbial activity in some studies (e.g., Eriksen et al., 1995) and effects of elevated CO_2 therefore appear possible. Ebersberger et al. (2003) measured arylsulphatase activity in a calcareous grassland and did not find any change under elevated CO_2, but enzyme activities may be poor indicators of S mineralisation rates. Supply of S has generally not been a concern for agricultural production because atmospheric inputs from air pollution are sufficient to cover plant demand, at least in industrialised areas. However, SO_2 emission control has now led to a negative S balance in many agricultural soils (e.g., Knudsen and Pedersen, 1993). Plant S supply from mineralisation and potential effects of elevated CO_2 may become more important in the future.

2.2.3 PLANT TISSUE QUALITY

A significant part of plant biomass is exported as yield in many ecosystems (hay, crop, animal biomass). The elemental composition of plant tissue is therefore a critical determinant of the nutrient balance of the ecosystem. A shift in elemental ratios under global change thus may shift the coupling of nutrient cycles and affect nutrient limitations, also higher up the food chain. This may necessitate adaptations in fertiliser use, and also alter the nutritional quality of plants.

Elevated CO_2 affects plant tissue quality by several mechanisms:

1. Carbohydrate levels of green plant tissue increase, primarily in the form of starch (Penuelas and Estiarte, 1998; Wong, 1990).
2. Leaf nitrogen concentrations decrease in many studies, mainly due to increased carbohydrate contents and reduced amounts of photosynthetic enzymes in C_3 plants (e.g., Ainsworth et al., 2002; Rogers and Ellsworth, 2002; Sage et al., 1997; Sage, 2002; Seneweera et al., 2002; Stitt and Krapp, 1999).
3. Leaf P decreases often less than leaf N, presumably because more P is required in phosphorylated intermediates when photosynthetic capacity increases under elevated CO_2 (Ghannoum et al., Chapter 3, this volume; and Gifford et al., 2000).
4. Allocation to secondary compounds may increase under elevated CO_2 due to reduced C and N limitation (Herms and Mattson, 1992; Penuelas and Estiarte, 1998).

Not much is known about effects on elements other than N and P. What is the experimental evidence that shifts in elemental composition occur under elevated CO_2, especially for elements other than N? In a search for patterns, I have compiled data on pools and concentrations of nutrients in plants exposed to elevated CO_2 (Table 2.2). I have mostly focused on multiyear field studies and excluded experiments for which only N concentration was reported because the aim here is to assess differential responses of mineral elements.

A remarkable finding is that foliar K concentrations decreased under elevated CO_2 in many studies conducted under relatively infertile conditions: Newbery et al. (1995) reported decreases in shoot K concentration in *Agrostis capillaris* grown under low K supply; specific root absorption, measured as [85]Rb absorption of excised roots, increased by over 100%, indicating increased demand for K. Decreased shoot K concentrations were also reported by Schenk et al. (1997) for *Lolium perenne/Trifolium repens* swards. In trees, reductions in foliar K concentration were reported for *Quercus alba* and *Picea abies* (Norby et al.,1986) and in mixed stands of *Quercus germinata/myrtifolia* (Johnson et al., 2003). It is noteworthy that the decreases in K were often larger than the decreases in N. These reductions in K translated into reduced litter K concentrations where this was measured, and reductions in K again were larger than reductions in N (mixed stands of *Quercus germinata/myrtifolia*: Johnson et al., 2003; ombrotrophic *Sphagnum* bogs: Hoosbeek et al., 2002). This decrease in foliar K occurs less frequently in well-fertilised systems, though there are exceptions (e.g., cotton: Prior et al., 1998; *Picea sitchensis*: Murray et al., 1996).

It remains to be explored whether this decrease in K concentrations is ecologically significant. Soils differ greatly with respect to K availability. Fine-textured soils have generally larger exchange capacities, which prevent K leaching, and K is constantly resupplied from mineral weathering. In these soils, K is generally not limiting. However, sandy soils with low exchange capacity can result in significant K leaching,

TABLE 2.2
Nutrient Pools and Concentrations in pLants Grown under Elevated CO_2

Ecosystem	Type of Study	Soil	CO₂ Treatment Conc. (ppm)	Duration	Nutrient Status	Fraction	Parameter	N	P	K	Mg	Ca	S	Zn	Mn	Fe	Cu	B	Si	Reference
Grasslands																				
Calcareous grassland	field	natural soil	600	6 yrs	unfertilised N/P colimited	shoot	conc.	decr.[2]	decr.[2]	–	–	–	–	–	–	–	–	–	–	Niklaus et al. 1998; Niklaus & Körner 2004
Tallgrass prairie	field	natural soil	2 x A	3 yrs	unfertilised	shoots	conc.	decr.[4]	decr.[4]	–	–	–	–	–	–	–	–	–	–	Owensby et al. 1993; Kemp et al. 1994
				2 yrs		litter (Andropogon)	conc.	n.s.	–47%*	–	–	–	–	–	–	–	–	–	–	
				2 yrs		litter (Sorghastrum)	conc.	n.s.	n.s.	–	–	–	–	–	–	–	–	–	–	
				2 yrs		litter (Poa)	conc.	–14%*	n.s.	–	–	–	–	–	–	–	–	–	–	
Lolium perenne/ Trifolium repens swards	pots	natural soil	670	2 yrs	full fertiliser at 200 kgN /ha	shoot	conc.	n.s.	n.s.	–18%**	n.s.	10%***	n.s.	–	–	–	–	–	–	Schenk et al. 1997
						shoot	pool	22%**	20%*	3%	22%**	37%**	20%**	–	–	–	–	–	–	
Agrostis capillaris	pots		680	43 d	fertilised	shoot	pool	+122%*	+193%*	+135%*	+82%*	+175%*	–	–	–	–	–	–	–	Baxter et al. 1994
Poa alpina	in			105 d			pool	+13%	+23%*	+15%	+25%*	–48%*	–	–	–	–	–	–	–	
Festuca vivipara	OTC			189 d			pool	–73%*	–48%*	–38%*	–37%*	–45%*	–	–	–	–	–	–	–	
Agrostis capillaris				43 d			conc.	n.s.	n.s.	n.s.	n.s.	n.s.	–	–	–	–	–	–	–	
Poa alpina				105 d			conc.	n.s.	n.s.	n.s.	n.s.	n.s.	–	–	–	–	–	–	–	
Festuca vivipara				189 d			conc.	n.s.	n.s.	n.s.	n.s.	n.s.	–	–	–	–	–	–	–	
Agrostis capillaris	pots	sand	A+250	23 wks	Modified hoagland solution with variable concentrations of NPK	shoot	conc.	–9%	–13%	–38%**	–	–	–	–	–	–	–	–	–	Newbery et al. 1995
							pool	+8%	–3%	–23%	–	–	–	–	–	–	–	–	–	
						roots (excised)	absorption	n.s.	n.s.	≥+100% ***	–	–	–	–	–	–	–	–	–	

Agricultural Crops

		Soil	CO_2	Duration	Nutrition	Tissue	Basis											Reference
Oryza sativa	field	rice paddy	A+200	123 d	NPK–fertiliser	grain	conc.	−10%*	−4%	−9%**	−6%	−1%	–	–	–	–	+4%	Yamakawa et al. 2003
Triticum aestivum	pots	sieved arable	700	116d	NPKCa–fertiliser	whole plant[5]	conc.	−4%**	3%	−11%***	–	–	–	–	–	–	–	van Vuuren et al. 1997
							pool	5%	+12%*	−3%#	–	–	–	–	–	–	–	
Gossypium hirsutum	field	arable soil	550	150 d	full fertiliser	leaves	conc.	−7%#	+2%	+4%*	+5%#	–	+15%#	+3%	+12%#	+10%#	–	Prior et al. 1989
						whole plant incl. root	conc.	−11%*	−2%	−7%#	−6%#	–	−3%#	−9%#	−8%#	−9%#	–	
						leaves	pool	+7%#	+17%#	+19%#	+22%	+12%	+33%#	+20%	+33%	+28%#	–	
						whole plant incl. root	pool	+21%	+33%*	+26%#	+27%#	+17%	+31%*	+24%	+26%	+24%#	–	

Trees

		Soil	CO_2	Duration	Nutrition	Tissue	Basis											Reference
Quercus rubra (3 yr old)	OTC	natural soil	700	2 yrs	natural	leaves	conc.	−4%*	+8%	−15%*	+13%	−3%*	+6%	–	−25%*	–	–	Le Thiec et al. 1995
Picea abies (5 yr old)	OTC	natural soil			natural	needles	conc.	−4%*	+21%	−6%*	−36%*	−28%*	+8%	–	−30%*	–	–	
Picea sitchensis	pots in	artificial	A+350	3 yrs	full fertiliser	current year branch needles	conc.	−15%*	n.s.	n.s.	n.s.							Murray 1996
	OTC	mixture				previous year branch needles	conc.	−7%	n.s.	n.s.								
Citrus aurantium	OTC	natural soil	A+300	4 yrs	full fertiliser	leaves	conc.	decr.	n.s.	decr.	decr.	decr.	n.s.	n.s.	decr.	n.s.	–	Gries et al. 1993
Quercus germinata/ Quercus myrtifolia stands	field	natural soil	A+350	5 yrs	unfertilised	shoot (*Q. myrtifolia* + *Q. germinata*)	pool	13%	18%	3%	44%	33%	10%	53%	58%	60%	15%	Johnson et al. 2003
						shoot (*Q. myrtifolia* + *Q. germinata*)	conc.	−17%***	−14%**	−23%**	+0%	−4%	−23%**	+6%	+13%	+14%	−18%	
						litterfall	conc.	+1%	+11%	−9%**	+8%	+8%	−20%**	−3%	+53%***	−63%	−22%***	
						standing litter	conc.	+4%	−20%	−10%**	+0%	+10%	−16%***	+30%*	+37%**	+67%	−18%*	

TABLE 2.2 (CONTINUED)
Nutrient Pools and Concentrations in pLants Grown under Elevated CO_2

Ecosystem	Type of Study	Soil	CO_2 Treatment Conc. (ppm)	Duration	Nutrient Status	Fraction	Parameter	N	P	K	Mg	Ca	S	Zn	Mn	Fe	Cu	B	Si	Reference
Fagus sylvatica/ Picea abies	OTC	acidic loam soil				shoots (Fagus + Picea)	pool	-2%	-1%	+4%	-3%	+5%	+2%	+8%	+6%	+12%	–	–	–	Hagedorn et al. 2002 .
Fagus sylvatica/ Picea abies		calcareous sand				shoots (Fagus + Picea)	pool	+13% **	+63% ***	+38% ***	+30% ***	+29% ***	+17% ***	+56% ***	+32% **	+60% **	–	–	–	
Fagus sylvatica		acidic loam soil				foliage	conc.	-7%	-6%	3%	-9%	4%	-10%	0%	15%	4%	–	–	–	Hagedorn, pers. comm
		calcareous sand				foliage	conc.	-10%	18%	1%	-14%	1%	-7%	29%	-24%	-11%	–	–	–	
Picea abies		acidic loam soil				foliage	conc.	-19%	-18%	-17%	-12%	15%	-13%	15%	6%	-10%	–	–	–	
		calcareous sand				foliage	conc.	-12%	5%	-6%	-6%	9%	-7%	28%	-3%	1%	–	–	–	
Pinus ponderosa	OTC	natural soil	700	2–3 yrs[6]	unfertiliser, low and high $(NH_4)_2SO_4$ application	needles	conc.	-8%#	-4%#	-12%	-17%	-1%	-9%#	-13%	5%	+22%#	-7%	-11%	–	Walker et al. 2000
Other Ecosystems																				
Ombrotrophic bogs (4 sites across Europe)	field	natural soil	560	3 yrs	unfertilised	Sphagnum litter	conc.	-4%	-10%	-14%*	–	–	–	–	–	–	–	–	–	Hoosbeek et al. 2003
						Eriophorum leaf litter	conc.	-13%*	-3%	-8%***	–	–	–	–	–	–	–	–	–	
						Eriophorum root litter	conc.	-9%	+3%	+23%	–	–	–	–	–	–	–	–	–	

Note: Only studies also reporting nutrients other than N were included, and preference was given to multiyear field experiments. When studies contained factorial treatments, responses were calculated for each treatment combination and average responses are listed; the same was done for multiyear data sets. Significant effects are indicated by * P < 0.05, ** P < 0.01 and, *** P < 0.001, is used to indicate P < 0.1 and is also used when significances are not obvious, for example, for the average response of multiyear data sets where significant effects were found in some but not all years. Refer to the original publications for details.

1 There were transient responses that vanished at end of experiment.
2 Decrease in P was smaller than decrease in N.
3 Measured with 15N, 32P, and 86Rb.
4 Effect on [N] not present in all species at all dates but larger than on [P], which decreased in 1 species and 1 out of 3 years.
5 Average of wet and dry treatment shown.
6 Total study duration was 5 years; average effects across 2nd and 3rd year of treatment and all CO_2 and N fertiliser levels given here; # indicates significant effect in at least one year.

especially in high rainfall areas. For example, rangelands of western Australia often exhibit severe K limitation (Bolland et al., 2002; Cox, 1973) and effects of elevated CO_2 on K contents of grass, therefore, may be important. Moreover, elevated CO_2 may affect leaching rates and associated K losses in these ecosystems.

Data for all other nutrients is far too limited to attempt any generalisations; remarkable effects are the decrease in leaf S and B concentrations in the mixed *Quercus* stands investigated by Johnson et al. (2003) and decreased Mn concentrations in *Quercus alba* and *Picea abies* (Norby et al., 1986).

2.2.4 LEACHING AND VOLATILISATION

Higher soil moisture has been reported in many ecosystems exposed to elevated CO_2 (Morgan et al., 2004b) and this will favour the leaching of nutrients because more water will drain when saturation is exceeded. However, other factors that may affect leaching have also been found to change under elevated CO_2: Soil aggregation was found to increase (Rillig et al., 1999) or decrease (Niklaus et al., 2003, 2001a) under elevated CO_2, and Newton et al. (2004) reported reductions in soil hydrophobicity. Dry soils also tend to shrink and form cracks, a phenomenon that may be less pronounced under elevated CO_2 because soils might be moister. Increased root production under elevated CO_2 might be another factor reducing crack formation. Finally, detritus is an important determinant of soil hydrophobic properties. What the combined effect of all these changes on preferential flowpaths and water retention remains elusive.

Direct evidence for nutrient leaching in elevated CO_2 studies is both scarce and controversial. Körner and Arnone (1992), for example, have reported increased NO_3^- leaching from tropical communities, while Torbert et al. (1996) found decreased NO_3^- concentrations below the rooting zone. Niklaus et al. (2001b) observed strong and persistent reductions in soil NO_3^- concentrations in calcareous grassland communities exposed to elevated CO_2 for several years, but it is not clear how much NO_3^- was leached from this system. Soil NO_3^- concentrations are regulated by many interacting processes, including nitrification and denitrification, immobilisation of NH_4^+ and NO_3^- by soil microbes, and rooting patterns and root uptake kinetics for NH_4^+ and NO_3^-. There is also evidence that plant secondary compounds (terpenes and tannins) can inhibit nitrification (Northup et al., 1998, 1995; Olson and Reiners, 1983; Reiners, 1981; Rice and Pancholy, 1973, 1974; White, 1986, 1988; White and Gosz, 1987) and this process may be more important at elevated CO_2. Therefore, predictions are difficult, but a principal control is certainly plant uptake of mineral N, which will reduce NH_4^+ available for nitrification or remove the NO_3^- produced.

Andrews and Schlesinger (2001) studied a forest site exposed to elevated CO_2 and reported increased soil CO_2 concentrations, which were caused by higher plant and microbial respiration. These increased soil CO_2 concentrations accelerated soil acidification and mineral weathering, and there was evidence of increased leaching of bicarbonate base cations. This is another mechanism by which extra nutrient leaching losses may occur under elevated CO_2.

2.3 EFFECTS OF ELEVATED TEMPERATURE AND PRECIPITATION ON NUTRIENT CYCLING

Elevated CO_2, together with other radiatively active atmospheric gases, will lead to higher global mean temperatures and altered precipitation patterns (IPCC, 2001). These climatic drivers differ in an important aspect from the increase in atmospheric CO_2. For any practical purpose considered here, atmospheric CO_2 concentrations are globally and temporally uniform. Therefore, effects found in different studies are relatively easy to compare. In contrast, even in the absence of global change, temperature and humidity change constantly in any ecosystem, even at the time scale of hours, and the ecosystem is constantly responding to these alterations. In the long term, temperature and precipitation define large-scale vegetation distribution and biomes; in the short term, however, ecosystems are very persistent to even extreme episodes such as droughts or exceptionally cold spells.

Many ecosystem processes exhibit a pronounced temperature optimum, and ecosystems often operate above or below this optimum — this may even change in the course of the day or season. Consequently, a superimposed experimental or "natural" future warming may have positive or negative effects, depending on the ecosystem studied and the actual climatic conditions prevailing. This greatly exacerbates difficulties in the attempt to identify common response patterns across studies.

Several avenues have been pursued to study warming effects on ecosystems. A number of studies have capitalised on existing geographical gradients in temperature (e.g., Jenny, 1980; Raich and Schlesinger, 1992; Saleska et al., 1999; Trumbore et al., 1996) and on natural inter-annual variability (e.g., Chapin et al., 1995; Oechel et al., 2000). While this approach is largely correlative, it has the advantage that it addresses changes over longer time scales and largely avoids experimental artifacts. Alternatively, ecosystem temperature has been manipulated by heating soils with buried electric cables (e.g., McHale et al., 1998; Peterjohn et al., 1993; Rustad and Fernandez, 1998), with infrared lamps suspended over vegetation (e.g., Saleska et al., 1999), and by using passive open-top-like structures that increase temperatures by acting as a small greenhouse (e.g., Arft et al., 1999; Robinson et al., 1995). While these manipulative studies allow us to causally attribute detected effects to the warming treatment, they are often associated with experimental problems. For example, infrared heating, as used in many studies, is characterised by largely lacking a convective heating component, and buried cables may lead to heterogenous temperature distribution, especially in dry soils.

Effects of soil moisture and temperature on microbial activities often are interactive because moisture optima become more narrow at elevated temperatures (Figure 2.3). A good example is the data by Goncalves and Carlyle (1994) who studied N mineralisation *in vitro* at different temperatures and soil moisture contents. The reduction in mineralisation rate when reducing soil moisture from 60% to 10% of field capacity was much larger relatively at 25°C ($\approx$70%) than at 5°C ($\approx$50%). Warming, both experimentally and naturally, is often accompanied by increased evapotranspiration and, therefore, decreased soil moisture and plant water availability (e.g., Luo et al., 2001; Saleska et al., 1999), which results in a further reduction

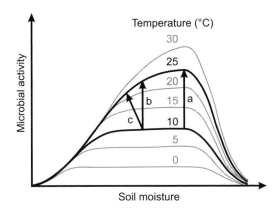

FIGURE 2.3 Schematic representation of interactive effects of temperature and moisture on microbial activity. Arrows show hypothetical increases in microbial activity as caused by a temperature rise. Three different scenarios, a through c, with respect to soil moisture are shown. At the low temperature, soil moisture is optimal in all three scenarios and microbial activity equal. Arrows a and b show effects of temperature increases without concomitant change in soil moisture; note that the effect size in scenario b is smaller than in scenario a because the optimal soil moisture range becomes narrower at elevated temperature. Scenario c shows the situation that is typical of many manipulative warming studies in which temperature increases are accompanied by a concomitant drying of soils or litter layer. In this case, activities are further reduced by the actual drying of soils. (Drawing based on data by Bunnell and Tate [1974] as shown in Paul, E.A. and Clark, F.E. [1996]; scenarios added by author.)

of microbial activity (arrow c in Figure 2.3). As discussed, elevated CO_2 also affects water relations (though in the opposite direction) and this is clearly a very important mechanism of responses to elevated CO_2. In warming studies, however, this indirect effect via alterations of water relations is even more important. It is often very difficult to separate these two factors, and confounding interpretations may be the result.

A further point to consider is that soil temperatures are determined by a number of processes, including radiative heating, convective energy exchange with air layers above the soil, and by heat conductance within soils (Figure 2.4). As a consequence, soil temperatures do not necessarily track ambient air temperatures. On the contrary, several whole-ecosystem studies have shown that *increased* ambient air temperatures can actually result in *decreased* soil temperatures (Coulson et al., 1993; Robinson et al., 1995; Wookey et al., 1993). While this appears to be counterintuitive at first glance, the phenomenon becomes understandable when considering that increased biomass of plants at elevated ambient temperatures can effectively insulate soils from solar radiation; furthermore, taller-stature species will absorb solar radiation farther off the ground and convective heating of soils will effectively be reduced (T_1 and T_2 in Figure 2.4). During the cold season, a related temperature reversal can happen as a consequence of snowpack (Groffman et al., 2001a). A warmer climate is likely to be associated with less frequent and shorter snow cover. Because snow very effectively insulates soil from convective and radiative heat loss, more frequent

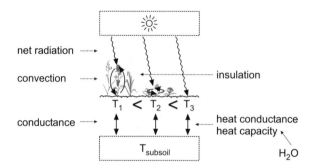

FIGURE 2.4 Schematic representation of vegetation effects on soil temperature. Incident radiation is converted to heat where it intercepts plant canopies. In taller plant canopies, this will happen farther off the ground and convective heating of soils will be smaller than in low-stature canopies. Soils under tall canopies, therefore, will be cooler (T_1) than under small canopies (T_2), which will be cooler than bare ground (T_3). Conductive cooling of top soils by cooler subsurface soils will depend on heat conductivity and heat capacity of soils, which are strongly influenced by moisture contents. (Latent heat fluxes due to evapotranspiration are not shown.)

soil freezing will occur. This is all the more important because root and microbial processes are still active at temperatures near or even below 0°C, and because freeze–thawing cycles can dramatically affect these processes (Groffman et al., 2001b).

2.3.1 NET PRIMARY PRODUCTION, DECOMPOSITION, AND THE CARBON AND NITROGEN BALANCE OF ECOSYSTEMS

Soil organic matter levels are determined by the balance between primary production and decomposition. An important question is whether productivity and decomposition respond similarly to climate change; if they do not, soil organic matter pools will change, resulting in concomitant long-term release or immobilisation of nutrients. It has been hypothesised that soil organic matter decomposition will increase more strongly at elevated temperatures than net primary productivity will, resulting in an imbalance between inputs and outputs to the ecosystem and, therefore, in soil organic matter losses. But what is the available evidence to support this claim?

Plant growth is limited by temperature in many ecosystems (Churkina and Running, 1998). Several growth-determining processes (e.g., leaf expansion rates or photosynthesis) are directly dependent on temperature, and the length of the growing season may also increase in a warmer climate. Soil microbial respiration, at least when measured on isolated samples and, in the short term, generally increases with temperature, often exponentially, at least when soil moisture is optimal and temperatures are not too high (cf. Figure 2.3). When both production and decomposition are considered separately, one might conclude that the temperature dependence of respiration is stronger (i.e., exponential), and that a net decomposition of new organic matter will happen at elevated temperatures.

However, under steady-state conditions, long-term plant productivity and decomposition are approximately balanced. If there is a disequilibrium, substrate available for decomposition will increase or decrease until equilibrium is reached again (decomposition generally follows first-order kinetics). From that, one might conclude that respiration is determined by net primary productivity and this notion is often found in the literature (e.g., Kirschbaum, 1995). However, it ignores feedback mechanisms of decomposition on primary productivity, for example via the nitrogen cycle. If primary production is N-limited and N availability controlled by decomposition, one might as well contend that decomposition controls primary productivity; this shows that the involved feedback mechanisms clearly are more complex and multidirectional.

Natural temperature gradients offer an opportunity to avoid circularity and to study production and decomposition in an integrated context. Jenny (1980), in his now classical work, compared tall grass prairie hay yields across a north–south transect through the midwestern United States. Yields remained very constant, although the gradient covered a mean annual temperature range from below 5°C to 20°C. Kirschbaum (2000) argued that there probably was increased moisture limitation at the warmer sites, and that Jenny might have found a positive effect of warming on productivity if water had not been limiting. Jenny also investigated a similar gradient (5 to 14°C; Canada to Texas) in the more western short grass belt, but again no temperature dependency of productivity was found. Post et al. (1982) analysed the dependence of soil carbon content on climate. Climate and corresponding life zones were classified according to temperature, precipitation and potential evapotranspiration (PET) (Holdridge classification; Holdridge, 1947). Soil carbon stores were negatively correlated with temperature, but positively correlated with precipitation and precipitation:PET (which is an index of water availability). However, Post et al. (1982) found a large variation of soil C within individual life zones, indicating that local conditions and other factors were also very important determinants. Townsend et al. (1995) studied soil organic matter dynamics and respiration rates along an elevation gradient covering ≈10°C on the island of Hawaii; soil organic matter contents increased with altitude (i.e., decreased with temperature), and mean annual respiration rates showed the opposite effect. Other gradient studies with similar results exist, and generally suggest that soil organic matter might decrease in a warmer climate. However, there is also evidence for the opposite. For example, Liski and Westman (1997) studied soil carbon contents along a climosequence across Finland spanning 800 to 1300°C·days*; topography, parent material, and soil age were standardised carefully. Two site types differing in the type of coniferous forest were sampled: *Pinus sylvestris/Calluna* sites had low productivity, whereas *Picea abies/Myrtillus* sites had higher productivity. Soils were sampled to 1 meter depth and organic and mineral soil C analysed separately; at the *warmer* end of the sequence, mineral soil C was 57% (low productivity sites) and 28% (high productivity sites) *higher* than at the cool end, and temperature accounted for 33 to 41% of the observed variation in soil C. Organic soil C pools did not vary with temperature.

---

* A threshold of +5°C was used.

Matters are more complicated with respect to soil nitrogen because soil N does not change in concert with soil organic C. Post et al. (1985), in their global survey of soil profiles, found that soil nitrogen pools followed similar trends as they had previously found for C (i.e., N was negatively correlated with temperature and positively correlated with precipitation and precipitation:PET). However, soil C:N ranged from <10 in tropical deserts, over intermediate values in the temperate zone (10 to 20), to >20 in wet forests. Interestingly, the reason why wet forests had high C:N depended on temperature was that in the wet tropics, C:N was high because SOM (soil organic matter) was stored as relatively recalcitrant humic substances in advanced stages of decay; in wet tundra C:N was high because decomposition was slow. Jenny (1980) analysed N pools in the top 20 centimetres of soils collected in the North American Great Plains; N pools decreased with temperature. However, Jenny found a strong dependence on soil moisture, with decreasing nitrogen contents at elevated mean annual NSQ.* Jenny also reported decreasing mineral soil C:N along an altitude transect at Kaiser Pass (near Fresno, CA) along which temperature decreased from 16 to 6°C and precipitation doubled from 450 to 900 mm a^{-1}.

Manipulative soil warming experiments have shown increased soil respiration at elevated temperature (e.g., Peterjohn et al., 1994); an important insight, however, was that respiration often acclimates relatively quickly to warmer conditions, that is, respiration responses decrease with time (Luo et al., 2001; McHale et al., 1998; Melillo et al., 2002). Soil respiration–temperature dependencies as found in short-term warming studies, therefore, cannot be extrapolated to longer time frames, or an overestimation of soil organic matter losses will result (Eliasson et al., 2005). Simulation studies also indicate that what is experimentally observed as acclimation may, in fact, not be an acclimation of microbial communities but a depletion in young organic matter fractions (Eliasson et al., 2005). New inputs of plant organic matter decompose quicker than old soil organic fractions; this is evidenced in radiocarbon age of soil respiration, which is much smaller than the age of bulk soil C (Trumbore, 2000). At low temperatures, fresh (active) soil organic matter turns over more slowly than under warmer conditions (Trumbore et al., 1996); as a consequence, acclimation responses to warming should be smaller and slower in cold than in warmer climates, and this indeed was found in several studies (e.g., Luo et al., 2001; Oechel et al., 2000).

At the ecosystem level, a number of important feedback mechanisms can alter soil organic matter decomposition, sometimes in rather unexpected ways. As already noted, several studies have shown that *increased* ambient air temperatures can result in *decreased* soil temperatures (Figure 2.4 and Coulson et al., 1993; Robinson et al., 1995; Wookey et al., 1993). Another important ecosystem-level feedback mechanism is that plant community composition may change in response to the warming treatment; a good example is the study by Saleska et al., (1999), which showed that warming of a montane meadow induced soil drying, which in turn led to a shift from productive to less productive but more drought-tolerant species. The warmed plots had lower soil respiration rates than control plots because soils were drier, but also because the input of decomposable material to soils was lowered due to the reduced productivity of the

---

* NSQ equals precipitation divided by the absolute H_2O saturation deficit of air, and correlates with precipitation:potential evaporation.

drought-tolerant species. Such species responses can be slow: Chapin et al. (1995) studied tussock tundra under increased temperatures and found that responses after 3 years were poor predictors of longer-term (9 years) changes in community composition. Responses after nine years showed closer resemblance to patterns of vegetation distribution along natural environmental gradients.

Finally, increased temperatures may lead to increased decomposition rates and thus increased nutrient availability; plant productivity, therefore, may increase if nutrients are limiting. Because C:N of plant material is, on average, larger than C:N of soils, soil organic matter decomposition may effectively lead to ecosystem-level C sequestration. However, if a substantial fraction of the nutrients mineralised is lost, for example, by leaching, the ecosystem will become a net source of CO_2 (e.g., Melillo et al., 2002; Shaver et al., 1992).

To conclude, predicting effects of warming on soil organic matter levels is very difficult. Correlative studies have the advantage that they assess responses near steady state, and thus also account for slow soil and vegetation adjustments; they may, however, miss transient effects that are *real*, that is, those that occur over decades to a century, which is the time frame over which temperature increases in the range of several degrees are expected. The other extreme involves warming studies conducted with isolated soil samples; it appears very dangerous to extrapolate from such data because plant production and microbial decomposition are intimately intertwined at many levels (cf. previous discussion of ecosystem-level feedback), and these ecological mechanisms are effectively eliminated by separating soils from plants and natural climate variability. Laboratory incubations, however, are very helpful when combined with ecosystem-level field studies because they allow for a better process-level understanding of effects observed in the field. Long-term whole-ecosystem warming experiments (Chapin et al., 1995, e.g.,) or reciprocal transplant studies along natural environmental gradients (e.g., Ineson et al., 1998) are probably a good middle ground in that they allow for important ecological feedback mechanisms (that may currently be unknown), can be combined with additional treatments (e.g., manipulations of soil moisture), and allow us to follow the trajectory along which the ecosystem adjusts to the new climate.

2.3.2 NUTRIENT IMMOBILISATION AND MINERALISATION

As discussed, increased temperatures might change organic matter contents and associated nutrient pools in the long term; but what is the available direct evidence that plant-available nutrients (resulting from net mineralisation) change in the short term, and what is known about the underlying processes?

The strong temperature dependency of organic matter decomposition may accelerate N mineralisation in a warmer climate. At the same time, a number of processes may affect mineralisation in the opposite direction:

1. Warming often results in decreased soil moisture, at least in manipulative experiments, which could decrease mineralisation rates (e.g., Robinson et al., 1995).

2. Some plant species respond to increased temperature with increased lignification, which could reduce decomposability of litter and nutrient
   release rates.
3. The composition of plant communities may shift in response to warming,
   with more lignified species prevailing at elevated temperatures.
4. Increased biomass production may cool the litter layer and top soil (Figure
   2.4), which may reduce decomposition rates.

Net mineralisation is the outcome of gross mineralisation and immobilisation,
each of which is controlled by a complex of factors. A meta-analysis by Rustad et
al., (2001) revealed, on average, an increase in N mineralisation rates in response
to elevated temperatures, although there were exceptions. Nadelhoffer et al. (1991)
incubated different soils from Alaskan tundra at 3, 9, and 15°C, and periodically
leached soils to measure NH_4^+, NO_3^-, and PO_4^{3-} in the leachates collected. An
important finding was that the increase in N mineralisation was much greater from
9 to 15°C than from 3 to 9°C, despite the marked increase in temperature dependency,
which is often found at low temperatures (Kirschbaum, 1995). Soil respiration
behaved similarly, and the authors attributed this unexpected temperature dependence to the different temperature sensitivities of the enzymes involved in cellulose
degradation. Interestingly, the release of PO_4^{3-} was often *reduced* at *elevated* temperatures. In some soils, PO_4^{3-} release was 5 to 10 times greater at 3°C than at 9°C
and 15°C, and Nadelhoffer et al. (1991) concluded that P immobilisation increases
more rapidly than gross P mineralisation, and that this might result in increased P
limitation under warmer conditions.

Tscherko et al. (2001) measured soil enzyme activities in microcosms subjected
to experimental warming. Microbial biomass N did not change in this study, and
the activities of soil enzymes related to N mineralisation remained mostly unaltered
(with the exception of arginine deaminase, which was increased).

Rustad and Fernandez (1998) studied the decomposition of red spruce and red
maple litter in temperate forest floor electrically warmed by 5°C. All litter had grown
under ambient conditions, so that the initial quality did not differ between treatments.
Interestingly, both litter types decomposed quicker in the heated treatments, but
decomposition (mass loss) was accelerated during the first 6 months in maple, and
during the 18- to 30-month period in spruce. K and B were released quicker in
maple, while Ca, Mg, K, and Zn were released more quickly in spruce. No differences were found for the other elements (Cu, Mn, N, Al, Fe). This study nicely
demonstrates that warming effects on decomposition are time and element specific.

Chapin et al. (1995) increased ambient temperatures by 3.5°C over tundra vegetation plots for 9 years with the help of plastic tents. Soil NH_4^+ and PO_4^{3-} concentrations were increased in the heated treatment and increased amounts of NO_3^- were
collected on soil-buried ion exchange resins. N and P availability actually were
increased during winter after the greenhouses had been removed from heated plots,
and Chapin et al. therefore concluded that changes in mineralisation were not just
due to direct warming effects on soil processes, but rather a consequence of changes
in species composition and litter quality. Total N and P pools in vegetation did not
change at elevated temperature when plots were not fertilised. However, mosses and

lichens, which initially accounted for $\approx \frac{1}{3}$ of vegetation N and P, had virtually disappeared from the heated plots, so that N and P in vascular (rooted) plants had increased. Chapin et al. argued that N and P uptake from soil had in fact increased, which would correspond to the soil solution and ion exchange resin data. Interestingly, fertilising the same plots with 10 g N and 5 g P m^{-2} a^{-1} reversed the temperature effects; N and P were reduced by $\approx \frac{1}{3}$ by heating, which compares with a reduction of $\approx \frac{1}{4}$ in biomass.* This reduction in plant growth was not due to a lack of soil nutrients because soil NH_4^+, NO_3^- and PO_4^{3-} actually increased severalfold when fertilised plots were warmed. K, Ca, and Mg pools were not altered by warming when no nutrients were added, but closely followed the plant biomass reductions when fertiliser was added.

2.3.3 PLANT TISSUE AND LITTER QUALITY

Plant tissue quality may be affected directly by effects on physiological processes within plants, and indirectly by effects on ecosystem-level processes such as nutrient mineralisation, soil moisture, and so forth. An important effect of warming is that lignification of many plants increases; lignin content is important because it limits digestion of fibre by animals. Also, the amount of dry matter that can be consumed by an herbivore is reduced because undigested fibres pass slowly through its digestive system and, thus, contribute to filling (Moore and Jung, 2001). Lignin content of forage also depends on moisture availability, but effects can be both positive and negative, depending on conditions. Moore and Jung discuss that lignification also depends on mineral nutrition, and that the ratio of available N:S may be a critical determinant of fibre content and lignification in some species. The mechanisms involved are complex, however, because (1) nutrient deficiencies delay plant growth, which results in lower lignification, and (2) nutrient status may have more direct effects on physiological processes (see Moore and Jung, 2001, and references therein).

Many studies have addressed climate change effects on *yield* of forage and grain crops (e.g., Amthor, 2001; Fuhrer, 2003) but the effects on *quality* have been considered less frequently, despite their importance. Xu et al. (2002) studied herbage quality along a natural temperature gradient in a Chinese high-altitude ecosystem and found significant reductions in crude protein content, ether extract, and nitrogen-free extract. Acid detergent fibre and acid detergent lignin indicated an increase in indigestible herbage components. Buse et al. (1998) exposed *Quercus robur* trees to elevated temperatures and analysed leaf tissue quality. Warming decreased N content, accelerated the accumulation of condensed tannins, and leaves became tougher, indicating reduced nutritional quality for insects feeding on these leaves. Lignin and fibre concentrations also are a critical determinant of decomposition rates of litter, which often correlate with lignin:N concentrations of the substrate, and changes in concentration and structure of these polymers may also affect nutrient recycling rates within the ecosystem.

Not much is known about warming effects on micronutrient tissue concentrations. These concentrations in seed and grain are controlled by many processes,

---

* It should be noted that soil moisture was not affected by any treatment.

including root membrane transport, transport via xylem, storage in plant tissues, and phloem loading and unloading (Grusak and DellaPenna, 1999; Welch, 1995). Elevated temperatures and changes in soil moisture affect development rate and phenology of species, and there is also the possibility that the physiological processes involved will respond more directly to temperature changes. Changes in micronutrient concentrations therefore appear possible, even if the availability of these elements does not change in soil solution.

2.3.4 LEACHING AND EROSION

Leaching of nutrients and gaseous losses are important both from a nutrient conservation perspective and because of their impact on stream and groundwater quality. Not much experimental data is available, however. For example, Lukewille and Wright (1997) warmed boreal forest floor by 3 to 5°C using electric cables and found increased NO_3^- and NH_4^+ in runoff. Joslin and Wolfe (1993) took advantage of natural topographical differences, which created differences in soil temperatures in *Picea rubens* stands, and found higher soil NO_3^-, Mg, and Al concentrations at the warmer spots. On the other hand, Ineson et al. (1998) reported decreased NO_3^- concentrations in leachates from sheep pasture subjected to warming by several °C (transplant experiment). In this latter study, increased N uptake by the vegetation in response to the warming was probably responsible for the reduction in N leaching rates.

This latter example underlines that leaching primarily occurs when the temporal synchrony between mineralisation and plant uptake is disturbed. N leaching from agricultural ecosystems, therefore, is most important when plant demand is low or plants are absent. Increased wintertime soil temperatures may lead to increased net N mineralisation rates and the buildup of inorganic nitrogen. Concomitantly increasing precipitation, as predicted for many temperate areas, will exacerbate this phenomenon. A snow manipulation study at the Hubbard Brook Experimental Forest (New Hampshire) has further shown that the loss of snow cover and resulting mild frosts can boost nitrogen mineralisation, again at a time when plant nitrogen uptake is minimal, and that this can result in greatly increased nitrification and N leaching losses (Groffman et al., 2001b).

Elevated temperatures during summer can increase evapotranspiration; in areas with relatively dry climate, drier soils may be subject to increased erosion by wind and rain; also, the frequency of droughts may increase, further supporting erosive losses (Schär et al., 2004). Plant canopies reduce erosive power of rain by interception, subsurface roots hold the soil in place, and crop residues and surface litter dramatically reduce rill detachment rates and sediment transport capacities (Nearing et al., 2004). Increased drought may lead to decreased plant biomass, and this might therefore further exacerbate erosion. Moreover, Nearing et al. point out that in a warmer climate, more precipitation may fall as rain and less as snow, which would further enhance erosion.

Increased rainfall may also increase erosion, not just because of the increased total amount of precipitation, but especially because of an increased likelihood of extreme rainfall events as is predicted for many areas (IPCC, 2001). Rainfall erosivity is strongly determined by rainfall intensity; as discussed by Nearing et al. (2004), a 1.5 to 2.0% increase in erosion rates can be expected per percent increase in precipitation.

2.4 THE HYDROLOGICAL CYCLE AND SCALING ISSUES

Manipulations of precipitation, atmospheric CO_2 concentration, and of air and soil temperature alter the water balance of ecosystems, either directly or indirectly; in the case of many elevated CO_2 and temperature field studies, indirect effects on plant communities via alterations of water availability have proven to be as important as, and sometimes even more important than, the direct effects of the treatment applied (as discussed in this chapter; see also Jackson et al., 1998; Morgan et al., 2004b; Niklaus and Körner, 2004; Owensby et al., 1999; Robinson et al., 1995).

The complexity of natural ecosystems and of the interactions among ecosystem components and global change drivers is enormous by any standards. Thus, studying effects of simulated global change in field experiments, under conditions as natural as possible and allowing for potentially important interactions and feedback, plays an important part in the attempt to predict the performance of ecosystems under future climatic conditions. However, a number of questions arise: (1) How are experimental treatments applied so as not to create experimental artifacts, and (2) how can plot-scale effects be extrapolated to the landscape level?

The first question relates to the technical installations and the way in which experimental treatments are applied. For example, CO_2 can be applied within chamber structures; elevated temperature treatments often consist of heating air and canopies with infrared heaters hung over the plots, and soils are sometimes heated with buried electric heating cables. Chamber structures and blowers may, however, alter the coupling of vegetation to the atmosphere, and radiative heating differs from the convective increase in temperature expected in a future climate. A further question is whether a step increase in atmospheric CO_2 or in temperature will evoke the same ecosystem responses as a gradual increase would. While these issues are important, they will not be discussed further here (see Klironomos et al., 2005; Luo and Reynolds, 1999; Thornley and Cannell, 2000, for examples).

The second question concerns the process of scaling up plot-level responses. Experimental treatments are in practice only applied to relatively small experimental units (usually plots one to a maximum of several 10 meters in diameter; any subplot treatments applied, for example, within a FACE ring, will lead to an effective further reduction in unit size). A critical question is: Would experimental effects detected at the plot scale be different if the simulated environmental changes were applied to much larger areas? Such a discrepancy is to be expected when feedback mechanisms operating at scales above the plot level are in effect. An area where such feedback appears particularly likely are plot-level changes in evapotranspiration; plant transpiration and atmospheric dynamics are coupled at time scales of hours or shorter, and the mutual dependence of both processes suggests that their interaction cannot be ignored.

At the plot level, a number of factors drive changes in evapotranspiration (ET) (Field et al., 1995):

1. Air, leaf, and soil surface temperatures affect the water pressure deficit of ambient air, the water vapour pressure in leaf intercellular space and above soil and litter surface.
2. Stomatal aperture and density, combined with leaf area, affects the total area over which saturated intercellular air is in contact with the atmosphere.
3. Canopy architecture affects the aerodynamic coupling of leaves and soils to the plot's atmosphere.

The extent to which these effects then translate into alterations in ET depends on gradients in water vapour mole fraction driving diffusion and the diffusive resistance of the atmospheric layers through which water vapour is transported. At the plot level, diffusive resistance includes stomatal and cuticular conductance, leaf boundary layer and canopy conductance; equivalent terms apply for soil evaporation. However, only a negligible amount of water vapour is stored in the canopy air volume itself; on a diurnal basis, evapotranspirational water diffuses into the air volume above the ecosystem — the well-mixed planetary boundary layer (PBL). The dynamics of the PBL is driven by radiative heat-up of the Earth's surface, which convectively warms air layers close to the surface, which rise due to their lowered density, leading to turbulent mixing; mixing of the PBL is further driven by shearing forces due to surface drag and wind. The PBL is topped by an inversion layer that separates it from the warmer and drier air above. The PBL is only a few 10 to 100 meters thick in the morning, but driven by radiative heat-up during the day, the inversion layer rises up to several kilometers above ground. After sunset, a nonturbulent nocturnal layer develops above the land's surface. The following day, the turbulent PBL builds up again when ground temperatures increase after sunrise.

Elevated CO_2, temperature, and precipitation change the surface energy balance of ecosystems, either directly (effect of temperature) or indirectly via alterations of sensible and latent heat flux (for example, as a result of altered stomatal conductance, leaf area, or canopy architecture). As a consequence, these drivers interact with PBL dynamics. Jarvis and McNaughton (1986), Jacobs and de Bruin (1992), and Amthor (1999) provide good reviews on effects of elevated CO_2, but the same reasoning (with opposite signs) applies to warming and irrigation treatments.

Some of these interactions will reduce apparent effects of global change treatments, while other interactions will work in the opposite direction. The mechanisms of these interactions revolve around effects on the surface energy balance of ecosystems and on the dynamics of diurnal PBL growth. I will only give a brief summary of the mechanisms proposed and refer the reader to Amthor (1999) and Jacobs and de Bruin (1997) for details. In essence, a reduced evapotranspiration will result in a drier PBL. This will reduce latent and increase sensible heat flux from canopies, which will result in a warmer and drier PBL. A drier PBL will then increase evapotranspiration, thus counteracting the original effect. On the other hand, a warmer PBL will grow more quickly and this may affect the mixing dynamics with dry air layers above, which will affect PBL water content and, therefore, also feed back on evapotranspiration. Changes in evapotranspiration can feed back on stomatal regulation; for example, increased evapotranspiration might induce stomatal closure, which would further increase PBL temperature and water pressure deficit, thus further accelerating PBL growth. Increased ecosystem temperature may also result in higher outgoing long-wave radiation and increase heat transport into the soil; under these conditions, less energy would have to be transferred into the PBL, which would reduce its temperature and water pressure deficit, thus reducing evapotranspiration.

The difficulty now is that leaf-level physiology affects PBL dynamics at regional scales, but that PBL dynamic also feeds back on leaf-level responses via altered canopy temperature and water pressure deficit; these interactions take place at time scales of hours. In any practical experimental setting, the PBL part of this interaction is effectively eliminated because only the plot-level part of the system is manipulated. PBL dynamics will effectively be determined by ambient conditions, because PBL air layers are too distant from the plot. From a PBL perspective, the treated plot is effectively a small invisible island in an ambient ocean.

How can these experimental limitations be overcome?

1. To effectively affect PBL dynamics, land areas several hundreds kilometers across would have to be manipulated, even under relatively calm conditions. Clearly, this is neither feasible for CO_2 enrichment nor for elevated temperature field studies. It may, however, be feasible for irrigation treatments. To assess potential PBL feedback, elevated CO_2 or elevated temperature-induced responses of ET might experimentally be replaced by irrigation or natural rainfall events (increasing ET) or by clipping (decreasing ET) of large areas. PBL dynamics can be followed with radar techniques and related tools. However, these surrogate treatments would be relatively crude approximations because canopy conductance and surface energy balance would also be modified, and biological feedback would not be operating. They could, however, serve to test ET effects on the PBL.

2. As a first approximation, PBL responses to the ET changes found experimentally could be modelled. If PBL dynamics are not altered, the experimental findings should be valid without correction. If they are altered, however, an assessment of whether and how this might feed back on ET and other ecosystem responses is required. It is important to know at least qualitatively to what degree treatment effects on ET may be alleviated or exacerbated by regional-scale atmospheric feedback.

3. As a second step, biological feedback to altered PBL dynamics and ET could be included, that is, accounting for the bidirectionality of the PBL–ecosystem interaction. This might be achieved with an integrated ecosystem-PBL model, or by testing effects on the ecosystem under climatic conditions modified as expected based on PBL modelling. In the simplest case, this feedback might be slow. For example, in an open-top chamber study of Mediterranean annual grassland responses to elevated

CO_2, Jackson et al. (1994) have found a slowed-down exhaustion of winter precipitation under elevated CO_2, which allowed for an extension of the growing season. Ignoring the possibility that the chamber/blower combination altered the coupling of vegetation to the atmosphere, a consequence of PBL feedback might just be that this extension in the growing season is smaller than expected. More complicated cases of PBL–ecosystem interaction would occur if this interaction changes on a diurnal basis (e.g., as in Jacobs and de Bruin,1997).

4. In cases in which indirect treatment effects via alterations of ET and thus water availability are expected, it might be wise to conduct experiments under a range of water availabilities (e.g., $CO_2 \times$ irrigation treatments; Volk et al., 2000). This would allow the separation of direct and indirect treatment effects, and to assess the sensitivity of the results to water availability, and to potential modifications of water availability via PBL feedback.

5. As another possibility, the system could be simulated in a kind of virtual reality setup (C. Körner, pers. comm.). Growth conditions could be simulated in a growth chamber. The atmospheric conditions (temperature and humidity) would then be modified in real time based on a model of PBL dynamics; this model would be fed with leaf conductivity or transpiration data measured online. This appears tempting because it does not require the simulation of the relatively complex biological part of the system, but the technical difficulties in simulating the climatic conditions and the atmospheric coupling appropriately may well be overwhelming.

2.5 CONCLUSIONS

It is clear that elevated CO_2 and temperature, the principal climatic drivers considered here, will have important effects on the biogeochemical cycling of nutrients in agroecosystems. Available data suggests that direct effects of elevated CO_2 and warming will occur, but many observations are equivocal so that generalisations are difficult. An important conclusion, however, is that indirect effects of elevated CO_2 and warming via alterations of the water balance of ecosystems may be more important than the direct impacts of these changes.

Several decades of experimental global change research have substantially advanced our understanding of plant responses, but it also is evident that our knowledge of responses of nutrient cycles lags behind by a great deal. In my view, the following key areas deserve particular attention.

2.5.1 MULTIPLE ELEMENT INTERACTIONS

The majority of elevated CO_2 and warming studies have so far focused on the cycling of carbon. Data on many processes of the nitrogen cycle (leaching, mineralisation and immobilisation rates, gaseous losses) is relatively limited, but data on the cycling of other elements is virtually nonexistent. This is all the more unfortunate because elements other than N frequently limit or colimit plant production, for example,

P and K (Bolland et al., 2002; Cox, 1973; Vitousek and Howarth, 1991). On intensively weathered soils, Ca, Mg, and S can also become limiting; cation exchange capacity is reduced and increased sorption of phosphorus and molybdenum can dramatically reduce the availability of these elements.

It appears possible that some inconsistencies among reported responses to simulated global change could be due to interactions with elements that simply have been ignored so far. A good example of such an interaction is the study by Hungate et al. (2004), which found that molybdenum availability ultimately limited N_2 fixation responses to elevated CO_2 in scrubland. Another example is an N-limited calcareous meadow in which N_2 fixation was colimited by P and CO_2 (Niklaus and Körner, 2004; Niklaus et al., 1998).

With scientists increasingly having access to rapid automatic analysers and modern analytical tools such as inductively coupled plasma emission spectroscopy (ICP), the simultaneous analysis of a wide range of elements in plants and soil fractions appears increasingly feasible.

2.5.2 Soil Types

Most global change experiments have been conducted on relatively young and fertile soils of the temperate zone, and not much data are available for the often highly weathered soils of the tropics. Because nutrient dynamics and limitations are quite different in these soils, and because agricultural practices often rely to a lesser extent on fertilisers (e.g., low intensity traditional farming, rangelands), global change effects on nutrient cycling in these systems may be different and deserve more attention.

2.5.3 Controls of Crops and Forage Nutrient Contents

Nutrient contents of crops are crucial both because they are important for human and animal nutrition and because they affect ecosystem nutrient cycling. However, the mechanisms controlling plant tissue nutrient concentrations are not well understood. Concentrations of some nutrients depend on supply rates; this has been shown for Zn, Ni, and Se. However, other micronutrient concentrations often do not respond to increased supply (e.g., Fe). The accumulation of these micronutrients in seed and grain is controlled by many processes, including root membrane transport, transport via xylem, possibly temporary storage in plant tissues, and phloem loading and unloading (Grusak and DellaPenna, 1999; Welch, 1995). Whether and how these processes are affected by global change is largely unclear, but the data on CO_2 effects on K concentrations (Table 2.2), for example, suggests that this happens.

2.5.4 Integrated Studies along Food Chains

Nutrients are transferred along food chains, and global change effects on the composition of primary production or on the metabolism of consumers may propagate to higher trophic levels. These effects are difficult or impossible to predict by simple elemental analysis. With respect to proteins, for example, the composition of amino acids, their digestibility, absorption in the digestive system of animals, and physiological utilisation after digestion may change (Friedman, 1996). Similar reasonings apply also to other nutrients. It, therefore, appears imperative that effects on consumers be investigated experimentally in studies covering whole food chains.

2.5.5 MULTIYEAR ECOSYSTEM-LEVEL STUDIES

Feedback mechanisms at the whole ecosystem level can be very complex, and it appears that for many processes, this feedback is more important than the actual direct effect of elevated CO_2 and warming. Three compelling examples discussed are (1) indirect effects of elevated CO_2 and warming on soil moisture (Sections 2.2 and 2.3), (2) feedback via vegetation change that led to other soil responses than expected due to warming alone (Section 2.3.1), and (3) cooling of soils due to increased plant growth at warmer temperatures (Section 2.3.1, Figure 2.4).

These examples are still relatively simple, and it appears definite that yet unknown more complex feedback, for example, via soils and soil organisms, also are operative. This underlines the importance of ecosystem-level field studies, and these should preferentially be run for at least several years because many feedback mechanisms, such as those operating via vegetation change, are long-term consequences.

2.5.6 HYDROLOGICAL FEEDBACK

Effects of elevated CO_2 and warming on hydrology can be more important than direct effects on plants and soils. However, this often does not receive sufficient attention. Many studies do not report soil moisture data, and when effects of soil moisture are found, it often is not possible to separate direct and indirect effects of the treatment. This latter problem can be solved by using factorial irrigation treatments or by studying interannual variability in soil moisture. An open question is that of scaling (Section 2.4): Will plot-level effects on hydrology persist when treatments are scaled to regional levels? Addressing this scaling issue seems crucial and will require collaborative efforts of ecophysiologists, meteorologists, and climate modellers.

ACKNOWLEDGMENTS

The author is indebted to the Swiss National Science Foundation for a Fellowship for Advanced Researchers during which this chapter was written. Matthias Rillig, Romain Barnard, and Tanya Handa are acknowledged for their valuable input on this manuscript.

REFERENCES

Ainsworth, E.A., Davey, P.A., Bernacchi, C.J., et al. (2002) A meta-analysis of elevated CO_2 effects on soybean (Glycine max) physiology, growth and yield. *Global Change Biology*, 8, 695–709.

Allaway, W.H. (1987) Soil-plant-animal and human interrelationships in trace element nutrition. In *Trace Elements in Human and Animal Nutrition*, 5th ed., Vol. 2, Mertz, W., Ed., Academic Press, Orlando, FL, 465–485.

Amthor, J.S. (1999) Increasing atmospheric CO_2, water use, and water stress: scaling up from the plant to the landscape. In *Carbon Dioxide and Environmental Stress*, Luo, Y. and Mooney, H.A., Eds., Academic Press, San Diego, CA, 33–59.

Amthor, J.S. (2001) Effects of atmospheric CO_2 concentration on wheat yield: review of results from experiments using various approaches to control CO2 concentration. *Field Crops Research,* 73, 1.

Andrews, J.A. and Schlesinger, W.H. (2001) Soil CO_2 dynamics, acidification, and chemical weathering in a temperate forest with experimental CO_2 enrichment. *Global Biogeochemical Cycles,* 15, 149–162.

Arft, A.M., Walker, M.D., Gurevitch, J., et al. (1999) Responses of tundra plants to experimental warming: meta-analysis of the international tundra experiment. *Ecological Monographs,* 69, 491–511.

Arnone, J.A., Zaller, J.G., Spehn, E., Niklaus, P.A., Wells, C.A., and Körner, C. (2000) Dynamics of roots systems in intact native grasslands: effects of elevated atmospheric CO_2. *New Phytologist,* 147, 73–85.

Barnard, R., Barthes, L., Le Roux, X., et al. (2004) Atmospheric CO_2 elevation has little effect on nitrifying and denitrifying enzyme activity in four European grasslands. *Global Change Biology,* 10, 488–497.

Barnard, R., Leadley, P.W., and Hungate, B.A. (2005) Global change, nitrification, and denitrification: a review. *Global Biogeochemical Cycles,* 19.

Barrett, D.J., Richardson, A.E., and Gifford, R.M. (1998) Elevated atmospheric CO_2 concentrations increase wheat root phosphatase activity when growth is limited by phosphorus. *Australian Journal of Plant Physiology,* 25, 87–93.

Baxter, R., Gantley, M., Ashenden, T.W., and Farrar, J.F. (1994) Effects of elevated carbon-dioxide on 3 grass species from montane pasture. 2. Nutrient-uptake, allocation and efficiency of use. *Journal of Experimental Botany,* 45, 1267–1278.

Billings, S.A., Schaeffer, S.M., and Evans, R.D. (2004) Soil microbial activity and N availability with elevated CO_2 in Mojave desert soils. *Global Biogeochemical Cycles,* 18.

Bolland, M.D.A., Cox, W.J., and Codling, B.J. (2002) Soil and tissue tests to predict pasture yield responses to applications of potassium fertiliser in high-rainfall areas of southwestern Australia. *Australian Journal of Experimental Agriculture,* 42, 149–164.

Brookes, P.C., Powlson, D.S., and Jenkinson, D.S. (1982) Measurement of microbial biomass phosphorus in soil. *Soil Biology and Biochemistry,* 14, 319–329.

Buse, A., Good, J.E.G., Dury, S., and Perrins, C.M. (1998) Effects of elevated temperature and carbon dioxide on the nutritional quality of leaves of oak (*Quercus robur L.*) as food for the winter moth (*Operophtera brumata L.*). *Functional Ecology,* 12, 742.

Chapin, F.S., Shaver, G.R., Giblin, A.E., Nadelhoffer, K.J., and Laundre, J.A. (1995) Responses of arctic tundra to experimental and observed changes in climate. *Ecology,* 76, 694–711.

Churkina, G. and Running, S.W. (1998) Contrasting climatic controls on the estimated productivity of global terrestrial biomes. *Ecosystems,* 1, 206–215.

Cole, C.V. and Heil, R.D. (1981) Phosphorus effects on terrestrial nitrogen cycling. In *Terrestrial Nitrogen Cycles*, Vol. 33, Clark, F.E. and Rosswall, T., Eds., *Ecological Bulletin*, Stockholm, Sweden, 363–374.

Coulson, S., Hodkinson, I.D., Strathdee, A., Bale, J.S., Block, W., Worland, M.R., and Webb, N.R. (1993) Simulated climate change — the interaction between vegetation type and microhabitat temperatures at Ny-Alesund, Svalbard. *Polar Biology,* 13, 67–70.

Cox, W.J. (1973) Potassium for pastures. *Journal of Agriculture Western Australia*, 14, 215–219.

Daepp, M., Suter, D., Almeida, J.P.F., et al. (2000) Yield response of *Lolium perenne* swards to free air CO_2 enrichment increased over six years in a high N input system on fertile soil. *Global Change Biology*, 6, 805–816.

Darrah, P.R. (1996) Rhizodeposition under ambient and elevated CO_2 levels. *Plant and Soil*, 187, 265–275.

de Lucia, E.H., Callaway, R.M., Thomas, E.M., and Schlesinger, W.H. (1997) Mechanisms of phosphorus acquisition for ponderosa pine seedlings under high CO_2 and temperature. *Annals of Botany*, 79, 111–120.

Dhillion, S.S., Roy, J., and Abrams, M. (1996) Assessing the impact of elevated CO_2 on soil microbial activity in mediterranean model ecosystem. *Plant and Soil*, 187, 333–342.

Diaz, S., Grime, J.P., Harris, J., and McPherson, E. (1993) Evidence of a feedback mechanism limiting plant response to elevated carbon dioxide. *Nature*, 364, 616–617.

Ebersberger, D., Niklaus, P.A., and Kandeler, E. (2003) Long-term CO_2 enrichment stimulates N-mineralisation and enzyme activities in calcareous grassland. *Journal Soil Biology & Biochemistry*, 35, 965–972.

Eliasson, P.E., McMurtrie, R.E., Pepper, D.A., Stromgren, M., Linder, S., and Agren, G.I. (2005) The response of heterotrophic CO_2 flux to soil warming. *Global Change Biology*, 11, 167–181.

Epstein, E. (1965) Mineral metabolism. In *Plant Biochemistry*, Bonner, J. and Warner, J.E., Eds., Academic Press, London, 438–466.

Eriksen, J., Mortensen, J.V., Nielsen, J.D., and Nielsen, N.E. (1995) Sulphur mineralisation in five Danish soils as measured by plant uptake in a pot experiment. *Agriculture Ecosystems & Environment*, 56, 43–51.

Field, C.B., Jackson, R.B., and Mooney, H.A. (1995) Stomatal responses to increased CO_2: implications from the plant to the global scale. *Plant, Cell and Environment*, 18, 1214–1225.

Finzi, A.C. and Schlesinger, W.H. (2003) Soil-nitrogen cycling in a pine forest exposed to 5 years of elevated carbon dioxide. *Ecosystems*, 6, 444–456.

Fitter, A.H., Self, G.K., Wolfenden, J., et al. (1996) Root production and mortality under elevated atmospheric carbon dioxide. *Plant and Soil*, 187, 299–306.

Friedman, M. (1996) Nutritional value of proteins from different food sources. A review. *Journal of Agricultural and Food Chemistry*, 44, 6–29.

Fuhrer, J. (2003) Agroecosysterm responses to combinations of elevated CO_2, ozone, and global climate change. *Agriculture Ecosystems & Environment*, 97, 1–20.

Gifford, R.M., Barrett, D.J., and Lutze, J.L. (2000) The effects of elevated CO_2 on the C:N and C:P mass ratios of plant tissues. *Plant and Soil*, 224, 1–14.

Goncalves, J.L.M. and Carlyle, J.C. (1994) Modeling the influence of moisture and temperature on net nitrogen mineralization in a forested sandy soil. *Journal Soil Biology & Biochemistry*, 26, 1557–1564.

Gries, C., Kimball, B.A., and Idso, S.B. (1993) Nutrient-uptake during the course of a year by sour orange trees growing in ambient and elevated atmospheric carbon-dioxide concentrations. *Journal of Plant Nutrition*, 16, 129–147.

Groffman, P.M., Driscoll, C.T., Fahey, T.J., Hardy, J.P., Fitzhugh, R.D., and Tierney, G.L. (2001a) Colder soils in a warmer world: a snow manipulation study in a northern hardwood forest ecosystem. *Biogeochemistry*, 56, 135.

Groffman, P.M., Driscoll, C.T., Fahey, T.J., Hardy, J.P., Fitzhugh, R.D., and Tierney, G.L. (2001b) Effects of mild winter freezing on soil nitrogen and carbon dynamics in a northern hardwood forest. *Biogeochemistry*, 56, 191.

Grusak, M.A. and DellaPenna, D. (1999) Improving the nutrient composition of plants to enhance human nutrition and health. *Annual Review of Plant Physiology and Plant Molecular Biology*, 50, 133–161.

Güsewell, S. (2004) N:P ratios in terrestrial plants: variation and functional significance. 164, 243–266.

Hagedorn, F., Landolt, W., Tarjan, D., Egli, P., and Bucher, J.B. (2002) Elevated CO_2 influences nutrient availability in young beech-spruce communities on two soil types. *Oecologia*, 132, 109–117.

Herms, D.A. and Mattson, W.J. (1992) The dilemma of plants — to grow or defend. *Quarterly Review of Biology*, 67, 283–335.

Hilbert, D.W., Prudhomme, T.I., and Oechel, W.C. (1987) Response of tussock tundra to elevated carbon dioxide regimes: analysis of ecosystem CO_2 flux through nonlinear modelling. *Oecologia*, 72, 466–472.

Holdridge, L.R. (1947) Determination of world plant formations from simple climatic data. *Science*, 105, 367–368.

Holmes, W.E., Zak, D.R., Pregitzer, K.S., and King, J.S. (2003) Soil nitrogen transformations under *Populus tremuloides*, *Betula papyrifera* and *Acer saccharum* following 3 years exposure to elevated CO_2 and O_3. *Global Change Biology*, 9, 1743–1750.

Hoosbeek, M.R., Van Breemen, N., Vasander, H., Buttler, A., and Berendse, F. (2002) Potassium limits potential growth of bog vegetation under elevated atmospheric CO_2 and N deposition. *Global Change Biology*, 8, 1130–1138.

Hu, S., Firestone, M.K., and Chapin, F.S. (1999) Soil microbial feedbacks to atmospheric CO_2 enrichment. *Trends in Ecology and Evolution*, 14, 433–437.

Hungate, B.A., Chapin, F.S., Zhong, H., Holland, E.A., and Field, C.B. (1997a) Stimulation of grassland nitrogen cycling under carbon dioxide enrichment. *Oecologia*, 109, 149–153.

Hungate, B.A., Holland, E.A., Jackson, R.B., Chapin, F.S., Mooney, H.A., and Field, C.B. (1997b) The fate of carbon in grasslands under carbon dioxide enrichment. *Nature*, 388, 576–579.

Hungate, B.A., Jackson, R.B., Field, C.B., and Chapin, F.S. (1996) Detecting changes in soil carbon in CO_2 enrichment experiments. *Plant and Soil*, 187, 135–145.

Hungate, B.A., Jaeger, C.H., Gamara, G., Chapin, F.S., and Field, C.B. (2000) Soil microbiota in two annual grasslands: responses to elevated atmospheric CO_2. *Oecologia*, 124, 589–598.

Hungate, B.A., Stiling, P.D., Dijkstra, P., et al. (2004) CO_2 elicits long-term decline in nitrogen fixation. *Science*, 304, 1291–1291.

Ineson, P., Taylor, K., Harrison, A.F., Poskitt, J., Benham, D.G., Tipping, E., and Woof, C. (1998) Effects of climate change on nitrogen dynamics in upland soils. 1. A transplant approach. *Global Change Biology*, 4, 143–152.

Inubushi, K., Hoque, M., Miura, S., Kobayashi, K., Kim, H.Y., Okada, M., and Yabashi, S. (2001) Effect of free-air CO_2 enrichment (FACE) on microbial biomass in paddy field soil. *Soil Science and Plant Nutrition*, 47, 737–745.

IPCC (Intergovernmental Panel on Climate Change) (2001) Climate change 2001: third assessment report of the Intergovernmental Panel on Climate Change, Cambridge University Press, Cambridge, U.K.

Jackson, R.B., Sala, O.E., Field, C.B., and Mooney, H.A. (1994) CO_2 alters water use, carbon gain, and yield for the dominant species in a natural grassland. *Oecologia*, 98, 257–262.

Jackson, R.B., Sala, O.E., Paruelo, J.M., and Mooney, H.A. (1998) Ecosystem water fluxes for two grasslands in elevated CO_2: a modeling analysis. *Oecologia*, 113, 537–546.

Jacobs, C.M.J. and de Bruin, H.A.R. (1997) Predicting regional transpiration at elevated atmospheric CO_2: influence of the PBL-vegetation interaction. *Journal of Applied Meteorology,* 36, 1663–1675.

Jacobs, C.M.J. and de Bruin, H.A.R. (1992) The sensitivity of regional transpiration to land-surface characteristics — significance of feedback. *Journal of Climate,* 5, 683–698.

Jarvis, P.G. and McNaughton, S.J. (1986) Stomatal control of transpiration: scaling up from leaf to region. *Advances in Ecological Research,* 15, 1–49.

Jenny, H. (1980) *The Soil Resource: Origin and Behaviour,* Vol. 37, Ecological Studies: Analysis and Synthesis, Springer, New York.

Johnson, D.W., Hungate, B.A., Dijkstra, P., Hymus, G., Hinkle, C.R., Stiling, P., and Drake, B.G. (2003) The effects of elevated CO_2 on nutrient distribution in a fire-adapted scrub oak forest. *Ecological Applications,* 13, 1388–1399.

Jones, D.L. and Darrah, P.R. (1996) Re-sorption of organic compounds by roots of *Zea mays* L. and its consequences in the rhizosphere. *Plant and Soil,* 178, 153–160.

Joslin, J.D. and Wolfe, M.H. (1993) Temperature increase accelerates nitrate release from high-elevation red spruce soils. *Canadian Journal of Forest Research (Revue Cana-dienne De Recherche Forestiere),* 23, 756–759.

Kemp, P.R., Waldecker, D.G., Owensby, C.E., Reynolds, J.F., and Virginia, R.A. (1994) Effects of elevated CO_2 and nitrogen-fertilization pretreatments on decomposition on tallgrass prairie leaf-litter. *Plant and Soil,* 165, 115–127.

Kirschbaum, M.U.F. (1995) The temperature-dependence of soil organic-matter decomposition, and the effect of global warming on soil organic-C storage. *Journal Soil Biology & Biochemistry,* 27, 753–760.

Kirschbaum, M.U.F. (2000) Will changes in soil organic carbon act as a positive or negative feedback on global warming? *Biogeochemistry,* 48, 21–51.

Klironomos, J.N., Allen, M.F., Rillig, M.C., Piotrowski, J., Makvandi-Nejad, S., Wolfe, B.E., and Powell, J.R. (2005) Abrupt rise in atmospheric CO_2 overestimates community response in a model plant-soil system. *Nature,* 433, 621–624.

Knudsen, L. and Pedersen, C.A. (1993) Sulphur fertilization in Danish agriculture. *Sulphur in Agriculture,* 17, 29–31.

Körner, C. and Arnone, J.A. (1992) Responses to elevated carbon dioxide in artificial tropical ecosystems. *Science,* 257, 1672–1675.

Körner, C., Diemer, M., Schäppi, B., Niklaus, P.A., and Arnone, J. (1997) The responses of alpine grassland to four seasons of CO_2 enrichment: a synthesis. *Acta Oecologica,* 18, 165–175.

Le Thiec, D., Dixon, M., Loosveldt, P., and Garrec, J.P. (1995) Seasonal and annual variations of phosphorus, calcium, potassium and manganese contents in different cross-sections of *Picea abies* L. Karst needles and *Querens rubra* L. leaves exposed to elevated CO_2. *Trees-Structure and Function,* 10, 55–62.

Lee, J., Masters, D.G., White, C.L., Grace, N.D., and Judson, G.J. (1999) Current issues in trace element nutrition of grazing livestock in Australia and New Zealand. *Australian Journal of Agricultural Research,* 50, 1341–1364.

Liski, J. and Westman, C.J. (1997) Carbon storage in forest soil of Finland. 1. Effect of thermoclimate. *Biogeochemistry,* 36, 239–260.

Lukewille, A. and Wright, R.F. (1997) Experimentally increased soil temperature causes release of nitrogen at a boreal forest catchment in southern Norway. *Global Change Biology,* 3, 13–21.

Lund, C.P., Riley, W.J., Pierce, L.L., and Field, C.B. (1999) The effects of chamber pressurization on soil-surface CO_2 efflux and the implications for NEE measurements under elevated CO_2. *Global Change Biology,* 5, 269–281.

Luo, Y.Q. and Reynolds, J.F. (1999) Validity of extrapolating field CO_2 experiments to predict carbon sequestration in natural ecosystems. *Ecology,* 80, 1568–1583.

Luo, Y.Q., Wan, S.Q., Hui, D.F., and Wallace, L.L. (2001) Acclimatization of soil respiration to warming in a tall grass prairie. *Nature,* 413, 622–625.

Matamala, R. and Schlesinger, W.H. (2000) Effects of elevated atmospheric CO_2 on fine root production and activity in an intact temperate forest ecosystem. *Global Change Biology,* 6, 967–979.

McGill, W.B. and Cole, C.V. (1981) Comparative aspects of cycling of organic C, N, S and P through soil organic-matter. *Geoderma,* 26, 267–286.

McHale, P.J., Mitchell, M.J., and Bowles, F.P. (1998) Soil warming in a northern hardwood forest: Trace gas fluxes and leaf litter decomposition. *Canadian Journal of Forest Research (Revue Canadienne De Recherche Forestiere),* 28, 1365–1372.

Melillo, J.M., Steudler, P.A., Aber, J.D., et al. (2002) Soil warming and carbon-cycle feedbacks to the climate system. *Science,* 298, 2173–2176.

Moore, K.J. and Jung, H.J.G. (2001) Lignin and fiber digestion. *Journal of Range Management,* 54, 420.

Moorhead, D.L. and Linkins, A.E. (1997) Elevated CO_2 alters belowground exoenzyme activities in tussock tundra. *Plant and Soil,* 189, 321–329.

Morgan, J.A., Mosier, A.R., Milchunas, D.G., LeCain, D.R., Nelson, J.A., and Parton, W.J. (2004a) CO_2 enhances productivity, alters species composition, and reduces digestibility of shortgrass steppe vegetation. *Ecological Applications,* 14, 208–219.

Morgan, J.A., Pataki, D.E., Körner, C., et al. (2004b) Water relations in grassland and desert ecosystems exposed to elevated atmospheric CO_2. *Oecologia,* 140, 11–25.

Moscatelli, M.C., Fonck, M., De Angelis, P., Larbi, H., Macuz, A., Rambelli, A., and Grego, S. (2001) Mediterranean natural forest living at elevated carbon dioxide: soil biological properties and plant biomass growth. *Soil Use and Management,* 17, 195–202.

Murray, M.B., Leith, I.D., and Jarvis, P.G. (1996) The effect of long term CO_2 enrichment on the growth, biomass partitioning and mineral nutrition of Sitka spruce (*Picea sitchensis* (Bong) Carr). *Trees-Structure and Function,* 10, 393–402.

Nadelhoffer, K.J., Giblin, A.E., Shaver, G.R., and Laundre, J.A. (1991) Effects of temperature and substrate quality on element mineralization in 6 arctic soils. *Ecology,* 72, 242–253.

Nearing, M.A., Pruski, F.F., and O'Neal, M.R. (2004) Expected climate change impacts on soil erosion rates: a review. *Journal of Soil and Water Conservation,* 59, 43–50.

Newbery, R.M., Wolfenden, J., Mansfield, T.A., and Harrison, A.F. (1995) Nitrogen, phosphorus and potassium uptake and demand in Agrostis capillaris: the influence of elevated CO_2 and nutrient supply. *New Phytology,* 130, 565–574.

Newton, P.C.D., Carran, R.A., and Lawrence, E.J. (2004) Reduced water repellency of a grassland soil under elevated atmospheric CO_2. *Global Change Biology,* 10, 1–4.

Newton, P.C.D., Clark, H., Bell, C.C., et al. (1995) Plant growth and soil processes in temperate grassland communities at elevated CO_2. *Journal of Biogeography,* 22, 235–240.

Niklaus, P.A. (1998) Effects of elevated atmospheric CO_2 on soil microbiota in calcareous grassland. *Global Change Biology,* 4, 451–458.

Niklaus, P.A., Alphei, J., Ebersberger, D., Kampichler, C., Kandeler, E., and Tscherko, D. (2003) Six years of *in situ* CO_2 enrichment evoke changes in soil structure and biota of nutrient-poor grassland. *Global Change Biology,* 9, 585–600.

Niklaus, P.A., Glöckler, E., Siegwolf, R., and Körner, C. (2001a) Carbon allocation in calcareous grassland under elevated CO_2: a combined 13_C pulse-labelling/soil physical fractionation study. *Functional Ecology,* 15, 43–50.

Niklaus, P.A., Kandeler, E., Leadley, P.W., Schmid, B., Tscherko, D., and Körner, C. (2001b) A link between elevated CO_2, plant diversity and soil nitrate. *Oecologia,* 127, 540–548.

Niklaus, P.A. and Körner, C. (1996) Responses of soil microbiota of a late successional alpine grasslands to long term CO_2 enrichment. *Plant and Soil,* 184, 219–229.

Niklaus, P.A. and Körner, C. (2004) Synthesis of a six-year study of calcareous grassland responses to *in situ* CO_2 enrichment. *Ecological Monographs,* 74, 491–511.

Niklaus, P.A., Leadley, P.W., Schmid, B., and Körner, C. (2001c) A long-term field study on biodiversity × elevated CO_2 interactions in grassland. *Ecological Monographs,* 71, 341–356.

Niklaus, P.A., Leadley, P.W., Stöcklin, J., and Körner C (1998) Nutrient relations in calcareous grassland under elevated CO_2. *Oecologia,* 116, 67–75.

Niklaus, P.A., Stocker, R., Körner, C., and Leadley, P.W. (2000) CO_2 flux estimates tend to overestimate ecosystem carbon sequestration at elevated CO_2. *Functional Ecology,* 14, 546–559.

Niklaus, P.A., Wohlfender, M., Siegwolf, R., and Körner, C. (2001d) Effects of six years atmospheric CO_2 enrichment on plant, soil, and soil microbial C of a calcareous grassland. *Plant and Soil,* 233, 189–202.

Norby, R.J., Oneill, E.G., and Luxmoore, R.J. (1986) Effects of atmospheric CO_2 enrichment on the growth and mineral nutrition of *Quercus alba* seedlings in nutrient-poor soil. *Plant Physiology,* 82, 83–89.

Northup, R.R., Dahlgren, R.A., and McColl, J.G. (1998) Polyphenols as regulators of plant-litter-soil interactions in northern California's pygmy forest: a positive feedback? *Biogeochemistry,* 42, 189–220.

Northup, R.R., Yu, Z.S., Dahlgren, R.A., and Vogt, K.A. (1995) Polyphenol control of nitrogen release from pine litter. *Nature,* 377, 227–229.

Oechel, W.C., Vourlitis, G.L., Hastings, S.J, Zulueta, R.C., Hinzman, L., and Kane, D. (2000) Acclimation of ecosystem CO_2 exchange in the Alaskan Arctic in response to decadal climate warming. *Nature,* 406, 978–981.

Olson, R.K. and Reiners, W.A. (1983) Nitrification in subalpine balsam fir soils — tests for inhibitory factors. *Journal Soil Biology & Biochemistry,* 15, 413–418.

Owensby, C.E. (1993) Potential impacts of elevated CO_2 and above- and below-ground litter quality of a tallgrass prairie. *Water, Air, and Soil Pollution,* 70, 413–424.

Owensby, C.E., Ham, J.M., Knapp, A.K., and Auen, L.M. (1999) Biomass production and species composition change in a tallgrass prairie ecosystem after long-term exposure to elevated atmospheric CO_2. *Global Change Biology,* 5, 497–506.

Paul, E.A. and Clark, F.E. (1996) *Soil microbiology and biochemistry,* 2nd ed., Academic Press, San Diego, CA.

Penuelas, J. and Estiarte, M. (1998) Can elevated CO_2 affect secondary metabolism and ecosystem function? *Trends in Ecology and Evolution,* 13, 20–24.

Peterjohn, W.T., Melillo, J.M., Bowles, F.P., and Steudler, P.A. (1993) Soil warming and trace gas fluxes — experimental-design and preliminary flux results. *Oecologia,* 93, 18–24.

Peterjohn, W.T., Melillo, J.M., Steudler, P.A., Newkirk, K.M., Bowles, F.P., and Aber, J.D. (1994) Responses of trace gas fluxes and N availability to experimentally elevated soil temperatures. *Ecological Applications,* 4, 617–625.

Post, W.M., Emanuel, W.R., Zinke, P.J., and Stangenberger, A.G. (1982) Soil carbon pools and world life zones. *Nature,* 298, 156–159.

Post, W.M., Pastor, J., Zinke, P.J., and Stangenberger, A.G. (1985) Global patterns of soil-nitrogen storage. *Nature,* 317, 613–616.

Pregitzer, K.S., Zak, D.R., Maziasz, J., DeForest, J., Curtis, P.S., and Lussenhop, J. (2000) Interactive effects of atmospheric CO_2 and soil-N availability on fine roots of *Populus tremuloides*. *Ecological Applications,* 10, 18–33.

Prior, S.A., Torbert, H.A., Runion, G.B., Mullins, G.L., Rogers, H.H., and Mauney, J.R. (1998) Effects of carbon dioxide enrichment on cotton nutrient dynamics. *Journal of Plant Nutrition,* 21, 1407–1426.

Raich, J.W. and Schlesinger, W.H. (1992) The global carbon-dioxide flux in soil respiration and its relationship to vegetation and climate. *Tellus Series B-Chemical and Physical Meteorology,* 44, 81–99.

Reiners, W.A. (1981) Nitrogen cycling in relation to ecosystem succession. In *Terrestrial Nitrogen Cycles*, Vol. 33, Clark, F.E. and Rosswall, T., Eds., *Ecological Bulletin*, Stockholm, Sweden, 507–528.

Rice, C.W., Garcia, F.O., Hampton, C.O., and Owensby, C.E. (1994) Soil microbial response in tallgrass prairie to elevated CO_2. *Plant and Soil,* 165, 67–74.

Rice, E.L. and Pancholy, S.K. (1973) Inhibition of nitrification by climax vegetation. II. Additional evidence and possible role of tannins. *American Journal of Botany,* 60, 691–698.

Rice, E.L. and Pancholy, S.K. (1974) Inhibition of nitrification by climax vegetation. III. Inhibitors other than tannins. *American Journal of Botany,* 61, 1095–1103.

Rillig, M.C., Wright, S.F., Allen, M.F., and Field, C.B. (1999) Rise in carbon dioxide changes soil structure. *Nature,* 400, 628.

Robinson, C.H., Wookey, P.A., Parsons, A.N., et al. (1995) Responses of plant litter decomposition and nitrogen mineralisation to simulated environmental change in a high arctic polar semi-desert and a subarctic dwarf shrub heath. *Oikos,* 74, 503–512.

Rogers, A., and Ellsworth, D.S. (2002) Photosynthetic acclimation of *Pinus taeda* (loblolly pine) to long-term growth in elevated CO_2 (FACE). *Plant Cell and Environment,* 25, 851–858.

Ross, D.J., Newton, P.C.D., and Tate, K.R. (2004) Elevated CO_2 effects on herbage production and soil carbon and nitrogen pools and mineralization in a species-rich, grazed pasture on a seasonally dry sand. *Plant and Soil,* 260, 183–196.

Ross, D.J., Tate, K.R., and Newton, P.C.D. (1995) Elevated CO_2 and temperature effects on soil carbon and nitrogen cycling in ryegrass/white clover turves of an Endoaquept soil. *Plant and Soil,* 176, 37–49.

Rouhier, H. and Read, D.J. (1998) The role of mycorrhiza in determining the response of *Plantago lanceolata* to CO_2 enrichment. *New Phytologist,* 139, 367–373.

Rustad, L.E., Campbell, J.L., Marion, G.M., et al. (2001) A meta-analysis of the response of soil respiration, net nitrogen mineralization, and aboveground plant growth to experimental ecosystem warming. *Oecologia,* 126, 543–562.

Rustad, L.E. and Fernandez, I.J. (1998) Soil warming: consequences for foliar litter decay in a spruce-fir forest in Maine, USA. *Soil Science Society of America Journal,* 62, 1072–1080.

Sage, R.F. (2002) How terrestrial organisms sense, signal, and respond to carbon dioxide. *Integrative and Comparative Biology,* 42, 469–480.

Sage, R., Schäppi, B., and Körner, C. (1997) Effect of atmospheric CO_2 enrichment on Rubisco content in herbaceous species from high and low altitude. *Acta Oecologica,* 18, 183–192.

Saleska, S.R., Harte, J., and Torn, M.S. (1999) The effect of experimental ecosystem warming on CO_2 fluxes in a montane meadow. *Global Change Biology,* 5, 125–141.

Schär, C., Vidale, P.L., Lüthi, D., Frei, C., Haberli, C., Liniger, M.A., and Appenzeller, C. (2004) The role of increasing temperature variability in European summer heatwaves. *Nature,* 427, 332–336.

Schenk, U., Jager, H.J., and Weigel, H.J. (1997) The response of perennial ryegrass/white clover swards to elevated atmospheric CO_2 concentrations. 1. Effects on competition and species composition and interaction with N supply. *New Phytologist,* 135, 67–79.

Schlesinger, W.H. (1996) *Biogeochemistry. An Analysis of Global Change,* 2nd ed., Academic Press, San Diego, CA.

Schneider, M.K., Luscher, A., Richter, M., et al. (2004) Ten years of free-air CO_2 enrichment altered the mobilization of N from soil in *Lolium perenne L.* swards. *Global Change Biology,* 10, 1377–1388.

Schortemeyer, M., Hartwig, U.A., Hendrey, G.R., and Sadowsky, M.J. (1996) Microbial community changes in the rhizospheres of white clover and perennial ryegrass exposed to free air carbon dioxide enrichment (FACE). *Soil Biology and Biochemistry,* 28, 1717–1724.

Seneweera, S.P., Conroy, J.P., Ishimaru, K., et al. (2002) Changes in source-sink relations during development influence photosynthetic acclimation of rice to free air CO_2 enrichment (FACE). *Functional Plant Biology,* 29, 945–953.

Shaver, G.R., Billings, W.D., Chapin, F.S., Giblin, A.E., Nadelhoffer, K.J., Oechel, W.C., and Rastetter, E.B. (1992) Global change and the carbon balance of Arctic ecosystems. *Bioscience,* 42, 433–441.

Sinsabaugh, R.L., Saiya-Corka, K., Long, T., Osgood, M.P., Neher, D.A., Zak, D.R., and Norby, R.J. (2003) Soil microbial activity in a *Liquidambar* plantation unresponsive to CO_2-driven increases in primary production. *Applied Soil Ecology,* 24, 263–271.

Sparling, G.P., Feltham, C.W., Reynolds, J., West, A.W., and Singleton, P. (1990) Estimation of soil microbial C by a fumigation-extraction method: use on soils of high organic matter content, and a reassessment of the k_{EC}-factor. *Journal Soil Biology and Biochemistry,* 22, 301–307.

Spiers, G.A. and McGill, W.B. (1979) Effects of phosphorus addition and energy supply on acid-phosphatase production and activity in soils. *Journal Soil Biology & Biochemistry,* 11, 3–8.

Stitt, M. and Krapp, A. (1999) The interaction between elevated carbon dioxide and nitrogen nutrition: the physiological and molecular background. *Plant Cell and Environment,* 22, 583–621.

Thornley, J.H.M. and Cannell, M.G.R. (2000) Dynamics of mineral N availability in grassland ecosystems under increased [CO_2]: hypotheses evaluated using the Hurley Pasture Model. *Plant and Soil,* 224, 153–170.

Torbert, H.A., Prior, S.A., Rogers, H.H., Schlesinger, W.H., Mullins, G.L., and Runion, G.B. (1996) Elevated atmospheric carbon dioxide in agroecosystems affects ground-water quality. *Journal of Environmental Quality,* 25, 720–726.

Townsend, A.R., Vitousek, P.M., and Trumbore, S.E. (1995) Soil organic-matter dynamics along gradients in temperature and land-use on the island of Hawaii. *Ecology,* 76, 721–733.

Trumbore, S. (2000) Age of soil organic matter and soil respiration: radiocarbon constraints on belowground C dynamics. *Ecological Applications,* 10, 399–411.

Trumbore, S.E., Chadwick, O.A., and Amundson, R. (1996) Rapid exchange between soil carbon and atmospheric carbon dioxide driven by temperature change. *Science,* 272, 393–396.

Tscherko, D., Kandeler, E., and Jones, T.H. (2001) Effect of temperature on below-ground N-dynamics in a weedy model ecosystem at ambient and elevated atmospheric CO_2 levels. *Journal Soil Biology & Biochemistry,* 33, 491–501.

Underwood, E.J. and Mertz, W. (1987) Introduction. In *Trace Elements in Human and Animal Nutrition,* Vol. 1, 5th ed., Mertz, W., Ed., Academic Press, Orlando, FL, 1–19.

VanVuuren, M.M.I., Robinson, D., Fitter, A.H., Chasalow, S.D., Williamson, L., and Raven, J.A. (1997) Effects of elevated atmospheric CO_2 and soil water availability on root biomass, root length, and N, P and K uptake by wheat. *New Phytologist,* 135, 455–465.

Vitousek, P.M. and Howarth, R.W. (1991) Nitrogen limitation on land and in the sea: how can it occur? *Biogeochemistry,* 13, 87–115.

Vitousek, P.M., Walker, L.R., Whiteaker, L.D., and Matson, P.A. (1993) Nutrient limitations to plant growth during primary succession in Hawaii Volcanoes National Park. *Biogeochemistry,* 23, 197–215.

Volk, M., Niklaus, P.A., and Körner C (2000) Soil moisture effects determine CO_2-responses of grassland species. *Oecologia,* 125, 380–388.

Walker, R.F., Johnson, D.W., Geisinger, D.R., and Ball, J.T. (2000) Growth, nutrition, and water relations of ponderosa pine in a field soil as influenced by long-term exposure to elevated atmospheric CO_2. *Forest Ecology and Management,* 137, 1–11.

Wand, S.J.E., Midgley, G.F., Jones, M.H., and Curtis, P.S. (1999) Responses of wild C_4 and C_3 grass (Poaceae) species to elevated atmospheric CO_2 concentration: a meta-analytic test of current theories and perceptions. *Global Change Biology,* 5, 723–741.

Welch, R.M. (1995) Micronutrient nutrition of plants. *Critical Reviews in Plant Sciences,* 14, 49–82.

White, C.S. (1986) Volatile and water-soluble inhibitors of nitrogen mineralization and nitrification in a Ponderosa pine ecosystem. *Biology and Fertility of Soils,* 2, 97–104.

White, C.S. (1988) Nitrification inhibition by monoterpenoids — theoretical-mode of action based on molecular-structures. *Ecology,* 69, 1631–1633.

White, C.S. and Gosz, J.R. (1987) Factors controlling nitrogen mineralization and nitrification in forest ecosystems in New Mexico. *Biology and Fertility of Soils,* 5, 195–202.

WHO (World Health Organization) (2000) Nutrition for health and development: a global agenda for combating malnutrition. Progress report, World Health Organization, Geneva, Switzerland.

Williams, M.A., Rice, C.W., and Owensby, C.E. (2000) Carbon dynamics and microbial activity in tallgrass prairie exposed to elevated CO_2 for 8 years. *Plant and Soil,* 227, 127–137.

Wong, S.C. (1990) Elevated atmospheric partial pressure of CO_2 and plant growth. *Photosynthesis Research,* 23, 171–180.

Wookey, P.A., Parsons, A.N., Welker, J.M., Potter, J.A., Callaghan, T.V., Lee, J.A., and Press, M.C. (1993) Comparative responses of phenology and reproductive development to simulated environmental-change in sub-arctic and high arctic plants. *Oikos,* 67, 490–502.

Wu, J., Odonnell, A.G., He, Z.L., and Syers, J.K. (1994) Fumigation-extraction method for the measurement of soil microbial biomass-S. *Journal Soil Biology & Biochemistry,* 26, 117–125.

Xu, S.X., Zhao, X.Q., Sun, P., Zhao, T.B., Zhao, W., and Xue, B. (2002) A simulative study on effects of climate warming on nutrient contents and *in vitro* digestibility of herbage grown in Qinghai-Xizang Plateau. *Acta Botanica Sinica,* 44, 1357.

Yamakawa, Y., Saigusa, M., Okada, M., and Kobayashi, K. (2004) Nutrient uptake by rice and soil solution composition under atmospheric CO_2 enrichment. *Plant and Soil,* 259, 367–372.

Zak, D.R., Pregitzer, K.S., Curtis, P.S., Teeri, J.A., Fogel, R., and Randlett, D.L. (1993) Elevated atmospheric CO_2 and feedback between carbon and nitrogen cycles. *Plant and Soil,* 151, 105–117.

3 Nutrient and Water Demands of Plants under Global Climate Change

Oula Ghannoum, Matthew J. Searson, and Jann P. Conroy

CONTENTS

ABBREVIATIONS

A: CO_2 assimilation rate (μmol m^{-2} s^{-1})
C_a: ambient CO_2 concentration
C_i: intercellular CO_2 concentration, [CO_2]: CO_2 concentration (ppm)
E: leaf transpiration rate (mmol m^{-2} s^{-1})
ET: evapotranspiration (from soil and plants)
g_b: leaf boundary layer conductance (mol m^{-2} s^{-1})
g_s: stomatal conductance (mol m^{-2} s^{-1})
N: nitrogen
Ω: decoupling coefficient
P: phosphorus
Pi: inorganic phosphate
Rubisco: ribulose-1,5-bisphosphate carboxylase/oxygenase
RuBP: ribulose bisphosphate
VPD$_l$: leaf-to-air vapor pressure difference (kPa)
WUE: water use efficiency (dry mass gain per unit of water transpired)

3.1   INTRODUCTION

By the year 2100, atmospheric CO_2 concentration ([CO_2]) is projected to reach 540 to 970 ppm. The ensuing global warming is estimated to be 1.4 to 5.8°C, with increases in minimum temperature more pronounced than the maximum. The temperature rise is also predicted to be greater in temperate than tropical latitudes. Anticipated changes for precipitation are more variable. It is likely that precipitation will increase between 30° N and 30° S, while tropical and subtropical areas will receive less rainfall. The frequency of low and high temperature extremes, and drought and flood events is likely to increase (IPCC 2001). These projected changes in atmospheric [CO_2], temperature, and rainfall patterns are expected to have profound and complex impacts on the world's agriculture. This industry represents the main form of human land use, with crops accounting for 11% of the Earth's land area. Agriculture also consumes about 70% of total human water usage, and is an important source (CO_2, H_2O, CH_4, NO_3) and sink (CO_2, H_2O) of greenhouse gases (FAO 2002). Therefore, agroecosystems are major targets and effectors of global climate change. Thus, it is crucial to begin the 21st century with a clear understanding of how global climate change will affect agricultural crops. In the past 30 years, a significant body of literature has emerged on the responses of individual C_3 crops to elevated $[CO_2]$ and their interactions with nutrient and water supply. With the advent of open-top chambers (OTC), free air CO_2 enrichment (FACE), and screen-aided CO_2-control (SACC) technologies and their spread in the last decade, more information is becoming available on the long-term (several years) responses of the main crops in field-like situations. Less data are available on the long-term interaction of increased temperature with elevated $[CO_2]$, and on C_4, relative to C_3, crops. In addition, the mechanistic basis of the physiological responses to many components of climate change remain poorly understood and documented.

In this chapter, we will focus on the mechanisms governing the changes in nutrients and water demands of C_3 and C_4 crops to elevated $[CO_2]$. We will then discuss how projected changes in air temperature and precipitation are likely to interact with the response to elevated $[CO_2]$, while identifying the key gaps in our understanding in this area. Our discussion will largely focus on leaf and plant processes because they represent the levels where we have the best understanding and data. We will scale our discussion to the canopy level whenever data permit.

3.2 OVERVIEW OF THE GROWTH RESPONSE OF PLANTS TO GLOBAL CLIMATE CHANGE

3.2.1 THE GROWTH RESPONSE OF C_3 AND C_4 PLANTS TO ELEVATED $[CO_2]$

Most terrestrial plants fix atmospheric CO_2 via the C_3 photosynthetic pathway (Figure 3.1). Some of the main C_3 crops are wheat (*Triticum aestivum*), rice (*Oryza sativa*), potato (*Solanum tuberosum*), grape (*Vitis vinifera*), cotton (*Gossypium hirsutum*), soybean (*Glycine max*), sunflower (*Helianthus annuus*), and rapeseed (*Brassica napus*). Under the current atmospheric $[CO_2]$ (370 ppm), ribulose-1,5-bisphosphate carboxylase/oxygenase (Rubisco), the primary CO_2 fixing enzyme in the mesophyll of C_3 leaves, operates near its K_m for CO_2 (Jordan and Ogren 1984). In addition to its carboxylation function, Rubisco reacts with molecular O_2 (oxygenation), resulting in the loss of CO_2 through photorespiration. Consequently, rising atmospheric $[CO_2]$ enhances C_3 photosynthetic rates by stimulating the carboxylation reaction of Rubisco while suppressing its oxygenation. Long-term exposure to elevated $[CO_2]$ can reduce the stimulation of C_3 photosynthesis in a process termed *photosynthetic acclimation* (Moore et al. 1999). However, the stimulation of photosynthesis by elevated $[CO_2]$ is often sufficient to enhance initial growth rates, resulting in greater vegetative and reproductive yields in controlled environments as well as in natural and managed field situations (Kimball 1983; Poorter 1993; Idso and Idso 1994; Drake et al. 1997; Wand et al. 1999; Kimball et al. 2002; Poorter and Navas 2003).

A small proportion (~4%) of the world's plant species fix CO_2 via the C_4 photosynthetic pathway (Figure 3.1), yet these species contribute about 20% of terrestrial primary productivity (Lloyd and Farquhar 1994). The main C_4 crops are maize (*Zea mays*), sugarcane (*Saccharum officinarum*), and sorghum (*Sorghum bicolor*). The high efficiency of C_4 photosynthesis is due to the operation of a

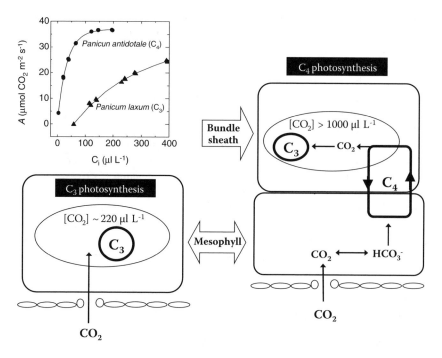

FIGURE 3.1 A schematic representation of C_3 (lower left) and C_4 (right) photosynthesis. CO_2 assimilation response curves of a C_3 (*Panicum laxum*, triangles) and C_4 (*Panicum antidotale*, circles) grass are shown (upper left). CO_2 assimilation rates (*A*) are expressed as a function of intercellular CO_2 concentration (C_i). Gas exchange measurements were made at 30°C and 1200 µmol quanta $m^{-2} s^{-1}$.

CO_2-concentrating mechanism, which serves to concentrate $[CO_2]$ at the site of its fixation to levels high enough to nearly saturate photosynthesis and suppress photorespiration in air (Figure 3.1; Hatch 1987). For this reason, elevated $[CO_2]$ has a small or no effect on C_4 photosynthesis (Ghannoum et al. 2000). Vegetative growth stimulation by a doubling of ambient $[CO_2]$ averages 45% and 12 to 33% for C_3 and C_4 plants, respectively (Wand et al. 1999; Poorter and Navas 2003).

Under well-watered conditions, elevated $[CO_2]$ enhances the growth of C_3 and C_4 plants via two routes. First, by enhancing leaf CO_2 assimilation rates (*A*). This effect is seen in C_4 plants only when C_4 photosynthesis is not CO_2-saturated, such as under high light intensity and high soil nitrogen (N) supply (Ghannoum et al. 1997; Ghannoum and Conroy 1998). Secondly, elevated $[CO_2]$ reduces stomatal conductance (g_s) to water vapor, which in turn reduces the leaf transpiration rate (E). Lower leaf E can improve shoot water relations (Knapp et al. 1993; Seneweera et al. 1998, 2001; Niklaus and Körner 2004) and increase leaf temperature (T_l) (Idso et al. 1987; Wall et al. 2001; Siebke et al. 2002), and both factors can lead to enhanced growth. The latter is particularly important for C_4 plants because C_4 photosynthesis responds positively to increased T_l over a wide range (Berry and Björkman 1980; Ghannoum et al. 2001).

3.2.2 Interaction of Water Availability with the Growth Response to Elevated [CO_2]

Under water stress, the growth response to elevated [CO_2] is maintained or increased in some plants relative to their response under well-watered conditions (Morison and Gifford 1984b; Owensby et al. 1997; Samarakoon and Gifford 1995; Seneweera et al. 1998, 2001; Centritto et al. 1999a, 2002; Serraj et al. 1999; Niklaus and Körner 2004). There are two main explanations for this observation. First, elevated-[CO_2]-induced reductions in plant E lead to soil water conservation. This slows down the development of water stress, thus providing more water and time for photosynthesis and growth. This explanation has been corroborated by many studies (Samarakoon and Gifford 1995; Serraj et al. 1999; Seneweera et al. 2001: Niklaus and Körner 2004). The second explanation is that elevated [CO_2] can directly alleviate the adverse effects of water stress on leaf water relations (Seneweera et al. 1998; Wall 2001) and A (De Luis et al. 1999; Centritto et al. 2002). In the latter case, elevated [CO_2] is thought to alleviate the CO_2 limitation caused by reduced g_s under drought (Lawlor and Cornic 2002). The enhanced growth response to elevated [CO_2] under water stress does not translate necessarily into relatively greater crop yield responses (Amthor 2001).

3.2.3 Interaction of Temperature with the Growth Response to [CO_2]

Higher air temperatures interact with the growth response to elevated [CO_2] by affecting a multitude of processes, such as photosynthesis, photorespiration, respiration, transpiration, vegetative growth, and yield (Morison and Lawlor 1999). In general, the growth response to elevated [CO_2] increases with temperature for most C_3 plants (Kimball et al. 1993; Bunce 1998a; Reddy et al. 1998; Ziska 1998). This response is anchored in the properties of Rubisco and its substrates. Higher temperatures reduce Rubisco specificity for CO_2 and CO_2 solubility relative to O_2 (Long 1991). Both factors increase photorespiration in C_3 plants, leading to a relatively greater impact of high [CO_2] on A at higher temperatures (Long 1991). In addition, by suppressing photorespiration, elevated [CO_2] increases the temperature optimum of C_3 photosynthesis (Long 1991), thus extending the temporal and spatial temperature range for C_3 photosynthesis. High temperatures also speed the development of determinate crops, thus modulating the effect of high [CO_2] on crop yield (Eamus 1991; Polley 2002). In wheat, a moderate increase in temperature can offset the benefits bestowed by high [CO_2] on yield (Amthor 2001). Due to the lack of apparent photorespiration in C_4 photosynthesis, the short-term response of A to increased temperature is unaffected by high [CO_2] (Ghannoum et al. 2001). The long-term interaction of high temperature and high [CO_2] on growth and yield of C_4 crops is not well studied.

3.3 NUTRIENT DEMAND OF PLANTS UNDER GLOBAL CLIMATE CHANGE

3.3.1 OVERVIEW OF MINERAL NUTRIENT DEMANDS UNDER CLIMATE CHANGE

This section will focus on differences between C_3 and C_4 plants in their demand for mineral nutrients and how this demand is moderated by growth at high atmospheric $[CO_2]$. The moderating effect of climate change — particularly increased temperature — will also be discussed, although there is a big gap in the literature on the combined effects of temperature and $[CO_2]$ on nutrient demand. Of the macronutrients, the vast majority of investigations on the inter-action between nutrient supply and high $[CO_2]$ have focused on nitrogen (N), with fewer studies on phosphorus (P). The concentrations of these essential macronutrients are relatively high in leaves because they are major constituents of organic molecules such as proteins and nucleic acids. The average concentrations for all plants of N and P in shoot dry matter that are sufficient for adequate growth are 15 and 2 g kg¹, respectively (Marschner 1995).

In contrast to macronutrients, CO_2 enrichment studies on micronutrients are extremely rare, possibly because their concentration in the leaf is relatively low. For example, the average zinc (Zn) concentration that is sufficient for adequate growth is 0.02 g kg¹ in shoot dry mass (Marschner 1995). Nevertheless, micronutrients are essential for plant growth because of their crucial role in metabolism, such as maintenance of enzyme conformation. For example, Zn is a key component of the enzyme superoxide dismutase, which is thought to be involved in regulating the synthesis of indole acetic acid, thereby influencing apical development (Marschner 1995). Importantly, the currently low concentrations of micronutrients such as Zn and iron (Fe) in cereals are already causing health problems, and it is feared that rising atmospheric $[CO_2]$ may exacerbate this problem (Loladze 2002).

Nitrogen and P are of particular interest when considering the impact of rising $[CO_2]$ and climate change because these nutrients commonly inhibit productivity of crops. N and P also provide an interesting contrast due to the large differences in the leaf concerning their concentrations, distribution between metabolic pools, and function. Furthermore, there is a vast difference between N and P mobility and concentrations in the soil, which affects nutrient uptake by plants. Therefore, this chapter will focus on these macronutrients. Zn is the only micronutrient, as far as we know, to be investigated in a CO_2 enrichment study and will be used to illustrate the interaction between micronutrients and high $[CO_2]$.

In the long run, the nutrient demand for optimum growth of a crop growing under different $[CO_2]$ depends on the leaf nutrient concentration required to support the maximum genetic growth or reproductive potential of that crop — the *critical concentration*. A lower critical concentration reflects a lower nutrient demand by the crop if productivity remains the same and vice versa. A lower leaf critical concentration of a particular nutrient may also mean that the leaf concentration of that nutrient will be lower. This could influence the quality of cereals because leaves are a major source of macro- and micronutrients for grain during filling. While the critical concentration is important for fertilized crops where the aim is to maximize productivity, in situations where nutrient supplies are low, such as in pastures, three key questions require answers. What is the leaf nutrient concentration required to (1) realize a growth response to high $[CO_2]$, (2) allow aboveground growth to continue, and (3) ensure plant survival?

3.3.2 DEMAND FOR NUTRIENTS — THE CRITICAL LEAF NUTRIENT CONCENTRATION

3.3.2.1 Determination of Critical Concentration

Empirical studies have been used to determine the critical leaf concentration for a wide range of crops (Reuter and Robinson, 1997). In these studies, plants are grown in the field or in pots with a range of nutrient supplies, and a specific plant part (usually the youngest fully expanded leaf) is sampled at different times during development for measurement of mineral nutrient concentration (dry mass basis). The biological yield (dry matter production) or the economic yield (mass of grain, fruit, tubers) is also measured. The critical concentration is determined by plotting leaf nutrient concentration against yield. It is defined as the concentration at which 90% of the maximum yield is reached. The 90% level is chosen because it is most cost effective in terms of fertilizer use. Concentrations below the critical level are deficient, and above it are adequate to support maximum productivity. Importantly, the critical concentration is the same for each species under standard environments, regardless of the soil type. Leaf analysis and published standards for critical concentrations are used successfully as a guide to fertilizer management of crops and pastures at sites with a variety of soil types (Reuter and Robinson 1997). The question is whether elevated $[CO_2]$ will alter the critical concentration and whether the response of C_3 and C_4 plants will differ.

3.3.2.2 Influence of High $[CO_2]$ on Critical Nutrient Concentration

CO_2 enrichment has a marked effect on the critical concentrations of the macronutrients N and P, but this differs between the nutrients and between C_3 and C_4 plants (Table 3.1, Figure 3.2). For a range of C_3 crops, the critical N concentrations are reduced while the P concentration remains the same or is increased. In contrast, for C_4 crops and pasture species, the critical N concentrations are unaffected by high $[CO_2]$ (Table 3.1); however, there have been no studies investigating whether elevated $[CO_2]$ affects critical P concentration of C_4 leaves. In the only experiment using a range of micronutrient supplies, there was a dramatic reduction due to high $[CO_2]$ in the critical Zn concentration in rice (Table 3.1). The physiological basis for the changes in critical nutrient concentrations caused by elevated $[CO_2]$ are explored in the next section and the implications are discussed in Section 3.3.5.

TABLE 3.1

Critical Concentration of a Range of Crop Plants Grown at Different Atmospheric [CO$_2$]

C_3/C_4	Species	[CO$_2$] ppm	Critical Nutrient Concentration N g kg^1	P g kg^1	Zn mg kg^1	Reference
C_3	Cotton	350	42	4.1		Rogers et al. 1996
		550	38	5.3		
		900	36	7.8		
		330	46			Wong 1979
		640	23			
	Wheat	340	43			Hocking and Meyer 1991
		1500	33			
	Rice	350	40			Aben et al. 1999
		700	27			
		350		1.8		Seneweera et al. 1994
		700		1.8		
		360			40	Defiani 1999
		700			15	
	Soybean	350	40			Cure et al. 1988b
		700	32			
				4.2		Cure et al. 1988a
				4.1		
C_4	Maize	330	33			Wong 1979
		640	29			
		340	32			Hocking and Meyer 1991
		1500	32			
	Makarikari	380				Rudmann 2000
	panic	860				
		380		4.1[1]		O. Ghannoum, unpublished
	Buffel	380	19			Rudmann 2000
	grass	860	21			
		380		2.6[1]		O. Ghannoum, unpublished

Notes: These critical concentrations in the youngest fully expanded leaves or whole shoot are required to support 90% of the maximum vegetative yield. Nutrients are deficient below the critical concentration and adequate above it. A lower value reflects greater nutrient use efficiency.

[1] Not determined for high [CO$_2$].

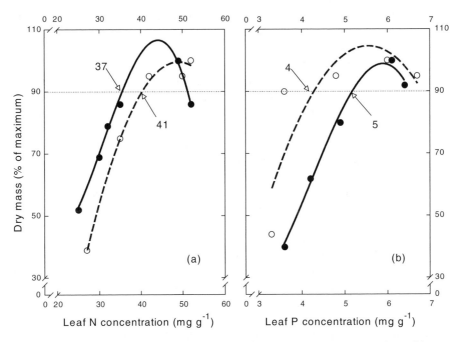

FIGURE 3.2 Influence of elevated [CO_2] on critical N (a) and P (b) concentrations of leaves of wheat. Plants were grown for 36 days in soil with a range of N (a) or P (b) supplies in controlled environments at either a [CO_2] of 350 or 550 ppm. Nutrients other than N and P were supplied at rates that were adequate to sustain maximum growth. Soil water was restored to full capacity daily. Elevated [CO_2] stimulated growth by approximately 35% when leaf N and P concentrations were adequate. (Redrawn from Rogers, G.S. (1996) Influence of N and P nutrition on responses of wheat and cotton to elevated CO_2, Ph.D. thesis, University of Western Sydney, Australia. With permission,).

3.3.2.3 Physiological Basis for Changes in Critical Concentrations at High [CO_2]

Leaves are used to measure critical concentrations because they supply most of the photosynthate required for plant growth and reproductive development. The critical concentration reflects the optimum nutrient concentration needed to maintain maximum metabolic activity of the leaf under standard environmental conditions. This concentration is a function of (1) the nutrient requirements of the biochemical reactions, particularly photosynthesis, in the soluble pools and membranes — the nutrient-rich fractions; and, (2) the dry mass accumulated as structural material (cell wall, vascular tissue) and insoluble and soluble carbohydrate storage compounds (starch, fructans, and so forth) — the nutrient-poor fractions.

At the biochemical level, changes in limitations to photosynthesis of C_3 plants at high [CO_2] may provide an explanation for the reduction in critical N and the increase in critical P concentrations in this photosynthetic type (see Table 3.1, Figure 3.2). Two of these limitations were identified by the model of Farquhar and von Caemmerer (1982): The rate of photosynthesis is limited first by RuBP (ribulose biphosphate) carboxylation capacity at lower $[CO_2]$, and secondly by the capacity of electron transport to regenerate RuBP at higher $[CO_2]$. The biochemical reactions, and accumulation of structural and storage carbohydrates in the leaf (the source of photosynthates) cannot be considered in isolation from the rest of the plant. Photosynthesis is strongly regulated by the demand for carbohydrates, which is driven by the activity of shoot apices, roots, seed, and fruit (the sinks for photosynthate). Consequently, the Farquhar and von Caemmerer (1982) model was modified by Sharkey (1985) to include the limitation imposed on photosynthesis by the sink activity, which can restrict the availability, in the chloroplast, of inorganic phosphate (Pi), which is required for ATP synthesis.

The reason for the RuBP carboxylation limitation at low $[CO_2]$ is that the first enzyme in the photosynthetic CO_2 fixation pathway, Rubisco, has an affinity for O_2 as well as CO_2. As a consequence, there are two cycles competing for energy in the chloroplast: The photorespiratory cycle (PCO) and the photoreductive cycle (PCR). CO_2 is the final product of the PCO cycle, while the PCR cycle results in the production of carbohydrates, which are used for growth. In the chloroplast, the product of the PCO cycle is phosphoglycolate. Importantly, this is dephosphorylated prior to its export from the chloroplast and the Pi is recycled. In contrast, the product of the PCR cycle, triose phosphate, is transported out of the chloroplast in exchange for Pi from the cytoplasm. When sink demand does not match source activity, phosphorylated intermediates build up, the cytoplasmic Pi concentration drops, and there is insufficient Pi for recycling back into the chloroplast. This leads to feedback inhibition of photosynthesis (Paul and Pellny 2003).

Why might the demand for P in C_3 plants increase as the atmospheric $[CO_2]$ rises? At high CO_2, the ratio of CO_2 to O_2 in the chloroplast increases, the PCR is favored over the PCO cycle, and the rate of photosynthetic CO_2 fixation is increased. As a result, the flux of Pi into the chloroplast must be greater to support these higher rates. This could account for the higher demand for P at elevated $[CO_2]$ and explain the difference between the critical P concentrations of C_3 plants due to high $[CO_2]$. However, there may be feedback inhibition of photosynthesis if sink demand does not match the enhanced source activity. This may partly account for acclimation of photosynthesis to high $[CO_2]$, which is discussed in Part III of this book.

What is the explanation for the reduced demand for N in leaves of C_3 plants at high $[CO_2]$? The most important consideration is that up to 25% of the leaf N is allocated to the enzyme, Rubisco (Evans 1989). At elevated $[CO_2]$, the efficiency of photosynthesis is increased because the PCR cycle is favored and photorespiration is reduced. Consequently, the photosynthetic rate can be maintained with a lower Rubisco content and, therefore, lower total leaf N concentration. This would explain the reduction in critical N concentration at high $[CO_2]$ (see Table 3.1, Figure 3.2) and the observation that leaf N concentrations are generally reduced in leaves of C_3 plants at elevated $[CO_2]$ (Cotrufo et al. 1998). The lower Rubisco content in leaves of high-$[CO_2]$-grown plants is also likely to contribute to the acclimation of photosynthesis to high $[CO_2]$ (Part III, this book).

The limitations to C_4 photosynthesis are more complex. C_4 plants generally have a lower critical N concentration than C_3 plants (see Table 3.1), reflecting the higher N use efficiency of C_4 photosynthesis (Long 1999; Makino et al. 2003). The greater efficiency of C_4 photosynthesis is primarily due to the operation of a CO_2 concentrating mechanism, which serves to elevate $[CO_2]$ in the bundle sheath cells, where Rubisco and the PCR cycle are located, to levels high enough to suppress photorespiration in air. In addition, the specific activity of Rubisco is about twice as high in C_4 than C_3 leaves (Seemann et al. 1984). Therefore, raising atmospheric $[CO_2]$ generally has a small effect on C_4 photosynthesis (Ghannoum et al. 2000), and the critical N and P concentrations are little affected (see Table 3.1, Figure 3.2).

The large reduction in the critical concentration of the micronutrient Zn in the C_3 plant, rice, at elevated $[CO_2]$ is puzzling (Defiani 1999). It was expected that there might be a greater demand for Zn in rice grown at high $[CO_2]$ because of the faster development of the shoot apex (Jitla et al. 1997; Defiani 1999). Zn is essential for cell division, and unlike other mineral nutrients, its concentration in the shoot apex is an order of magnitude higher than in the expanded leaves (Kitagishi and Obata 1986). One explanation may be that there are two pools of Zn in leaves, a soluble pool and a bound pool, and at high $[CO_2]$ there may be less bound Zn (Cakmak and Marschner 1987).

Is there any evidence that the changes in structural dry mass or accumulation of insoluble storage compounds at elevated $[CO_2]$ contribute to the change in critical concentrations? The specific leaf area (leaf area per unit mass) of C_3 plants is generally reduced at high $[CO_2]$ with greater changes occurring in C_3 dicots than monocots (Poorter et al. 1996; Poorter and Navas 2003). A large accumulation of starch often occurs at elevated $[CO_2]$ particularly in C_3 dicots, whereas monocots accumulate soluble sugars (Rogers et al. 1996a,b; Wong 1990; Poorter and Navas 2003). In some species, there is an increased layer of mesophyll cells at elevated $[CO_2]$ (Thomas and Harvey 1983). Both higher accumulation of storage carbohydrates and increased layers of cells would reduce the specific leaf area, the latter because the ratio of cell wall material to other tissue is increased. However, the N concentrations in leaves is lower at high $[CO_2]$, irrespective of whether they are expressed on a structural dry mass basis by correcting for concentrations of storage carbohydrates or on an area basis (Rogers et al. 1996a,b; Wong 1990). This supports our suggestion that the lower demand for N is due to biochemical changes at elevated $[CO_2]$. In contrast to C_3 plants, the specific leaf area of C_4 plants is either unaffected or tends to be greater at elevated $[CO_2]$ (Poorter et al. 1996; Wand et al. 1999). The reason for this is not known.

3.3.3 THE LOWER END OF THE SCALE — NUTRIENT DEFICIENCY AND ELEVATED $[CO_2]$

While critical concentrations provide valuable information in fertilized systems about the leaf nutrient requirements for maximum productivity, they may not consistently predict the minimum concentration required to (1) realize a response to elevated $[CO_2]$, (2) sustain growth of above ground biomass, and (3) ensure the survival of the plant species.

The minimum concentration required to obtain a growth enhancement at high $[CO_2]$ depends on the activity of both the source and sink, although there is feedback regulation between the two. Deficiency of a number of nutrients reduces sink generation, and this is thought to be a key factor determining the response of plants to high [CO_2] (Rogers et al. 1996b; Poorter and Navas 2003). For example, tillers are strong carbohydrate sinks during vegetative development of wheat and rice. When nutrients were adequate, high [CO_2] stimulated sink generation in the shoot apex of 16-day-old rice: There were five tiller buds in the shoot apex at a [CO_2] of 700 ppm compared to only three at a [CO_2] of 350 ppm (Jitla et al. 1997). These changes were the forerunners of faster development of the vegetative shoot at elevated [CO_2]. This was responsible for the 42% increase in tillering at the maximum tillering stage, and the 57% increase in final grain yield. The link between high [CO_2] and development of sinks during early growth was demonstrated by delaying exposure to high [CO_2] by 15 days. This totally inhibited the tillering response to high [CO_2] and the 20% increase in grain yield was due entirely to a greater number of grains per panicle at elevated [CO_2] (Jitla et al. 1997).

While N deficiency reduces tiller numbers, it did not preclude a tillering response to high [CO_2] (700 ppm) in rice (Aben et al. 1999). There were three times as many tillers at a [CO_2] of 700 ppm than at 360 ppm when the leaf N concentration was 2.5 g kg^1. Plant dry mass was stimulated by a similar magnitude. The close agreement in values of leaf N concentrations needed to support 90% of maximum growth and photosynthesis for rice grown at ambient and high [CO_2] suggests that source and sink activity were well coordinated in this species, at least during the tillering stage (Aben et al. 1999). In some crops the balance between source and sink activity is coordinated at all stages throughout development, but not in others. Failure to coordinate source and sink at particular stages of development may account for photosynthetic acclimation and the reduction in the photosynthetic response to elevated [CO_2]. For example, acclimation occurred at the grain-filling stage in rice growing at high [CO_2] in a FACE experiment, but not at the maximum tillering stage (Seneweera et al. 2002). In contrast, a FACE study with the C_3 pasture species, ryegrass, spanning a decade showed that there was no significant change in the photosynthetic response to high [CO_2] over time (Ainsworth et al. 2003). Changes in the balance between source and sink activity could influence the N concentration needed to achieve a photosynthetic response to elevated [CO_2], although this has not been tested in experiments using a range of N supplies.

P deficiency often precludes a photosynthetic response to elevated [CO_2], possibly because there is insufficient Pi for recycling into the chloroplast (Paul and Pellny 2003). This may be a direct effect of P deficiency or arise due to feedback caused by limited sink growth (Paul and Pellny 2003). These processes may explain why there was no growth stimulation by elevated [CO_2] in cotton at a leaf P concentration of 3.5 g kg^1 (Rogers et al. 1993). In contrast, growth of wheat was enhanced by 50% by elevated [CO_2] at a leaf P of 3.5 g kg^1 (Rogers et al. 1993). A possible explanation is that source and sink activity are more closely coordinated in wheat than cotton allowing Pi to be recycled more effectively. In the same experiment, the growth of cotton and wheat were stimulated by high [CO_2] even when leaf N was deficient (30 g N kg^1).

In contrast to C_3 plants, high-[CO_2]-induced photosynthetic acclimation is not usually observed in C_4 plants grown under optimum conditions (Drake et al. 1997; Ghannoum et al. 2000). This is due to the fact that C_4 photosynthesis already operates in a high $[CO_2]$ environment. Generally, the changes in foliar nonstructural carbohydrates and N concentrations are small in C_4 plants at high $[CO_2]$ and, thus, are unlikely to entail significant photosynthetic acclimation as observed in C_3 photosynthesis (Ghannoum et al. 1997, 2000). Under low N supply, photosynthetic acclimation has been reported in some C_4 species (e.g., Wong, 1979; Morgan et al. 1994) but not in others (e.g., Ghannoum and Conroy 1998; von Caemmerer et al. 2001). Nitrogen deficiency precluded a growth response to elevated $[CO_2]$ in the tropical grasses, *Panicum coloratum* and *P. antidotale*, mainly as a result of reduced sink strength (Ghannoum and Conroy 1998). However, this did not occur for buffel grass (*Cenchrus ciliaris*).

An important question is whether high $[CO_2]$ alters the shoot nutrient concentration at which aboveground biomass stops growing. For N, this parameter plays a key role in modeling the impact of rising $[CO_2]$ on pasture productivity and animal-carrying capacity in Australia's tropical grasslands (Howden et al. 1999). Under current $[CO_2]$, the parameter has a strong influence on C_3 and C_4 competition, and values of 8.8 and 6.8 g N kg^{-1} shoot mass are used for C_3 and C_4 grasses, respectively (Howden et al. 1999). However, it is not known how rising $[CO_2]$ will affect these values. One could speculate that they would come closer together because of the increased efficiency of photosynthesis of C_3 plants at elevated $[CO_2]$, which is reflected in their lower critical N concentrations (see Table 3.1).

Experiments investigating the effect of $[CO_2]$ on nutrient concentrations required for plant survival are rare. Two-year-old pine trees (*Pinus radiata*) tended to die at a higher leaf P concentration at elevated than ambient $[CO_2]$ (Conroy 1989), and the reverse was true for the micronutrient Zn in rice (Defiani 1999). This may be related to the greater critical values for P at elevated $[CO_2]$ and the lower critical Zn concentration (see Table 3.1).

3.3.4 CLIMATE, RISING $[CO_2]$, AND PLANT NUTRITION

Published critical nutrient concentrations have generally been determined over a range of climatic conditions and temperatures that are typical of the regions where these crops are grown. A temperature rise of 1.4 to 5.8°C, predicted to accompany increases in greenhouse gas emissions, would seem to be of little importance to crops when diurnal oscillations commonly range over 25°C in a single day. However, a temperature rise of 1.4 to 5.8°C over a whole season is significant because the rate of plant development is dependent on the accumulation of average daily temperatures (Rawson 1992). The faster the temperature is accumulated, the faster the plant develops. For example, a rise of 4.5°C in the Australian wheat belt would increase the number of leaves on the main shoot by 60% after 50 days of growth (Rawson 1992). Crop duration would be reduced, with flowering occurring earlier. Although organs are initiated and develop faster at higher temperatures, they are smaller because there is insufficient time to produce photosynthate to maximize organ growth; that is, sink demand is increased more than source activity. The advantage of high $[CO_2]$ under these circumstances is that photosynthesis (source activity) is increased, providing that the temperature does not rise to levels that damage photosynthetic function. Consequently, the response to elevated $[CO_2]$ is generally expected to be greater at higher temperatures (Conroy et al. 1994), although this is not the case for wheat yield (Amthor 2001). Plants will also have to cope with episodes of extreme heat, which will occur more frequently as greenhouse gas concentrations rise. Elevated [CO_2] may ameliorate the damaging effect of these episodes on photosynthetic function (Taub et al. 2000).

How might climatic change affect nutrient demand? There have been no studies, as far we know investigating the effect of different climates and elevated [CO_2] on critical nutrient concentrations. We speculate that the requirement for P to support maximum C_3 photosynthetic rates may be less at high temperatures because the PCO cycle is favored over the PCR cycle at high temperatures, and this would lead to recycling of Pi within the chloroplast.

3.3.5 Implications of Rising [CO_2] and Climate Change for Mineral Nutrition in C_3 and C_4 Plants

The standard values that are published to aid farmers in fertilizer management have been obtained from experiments conducted under atmospheric [CO_2] ranging from approximately 320 to 350 ppm. For a number of nutrients, these standard values will be totally inappropriate for C_3 plants with rising atmospheric [CO_2] and, there-fore, will need to be reassessed. For example, a value of 37 g N kg^1 for wheat grown at a [CO_2] of 550 ppm (see Figure 3.2) would appear to be deficient based the published value of 45 g N kg^1 (Reuter and Robinson 1997). Farmers would conclude that N fertilizer application was required. This conclusion would be false because 37 g kg^1 is the correct critical concentration at a [CO_2] of 550 ppm. Increasing temperatures could further complicate the problem. Overfertilization with N not only leads to eutrophication of groundwater, but also causes soil acidity, which is a major problem in countries such as Australia. Similarly, the critical concentration of Zn is reduced for C_3 plants grown at elevated [CO_2]. In contrast to N and Zn, the critical P concentration is greater at elevated [CO_2] and, therefore, farmers may underesti-mate the amount of P fertilization required. Whether other nutrients are affected is not known. For C_4 plants, published critical concentrations are likely to be appro-priate as the atmospheric [CO_2] rises.

The total amount of N fertilizer required by C_3 crops to support maximum productivity at elevated [CO_2] is unlikely to increase despite an average enhancement of biomass production of 45% (Poorter and Navas 2003). The reason is that the greater productivity of C_3 crops at high [CO_2] is generally offset by a decrease in plant N concentration (Cotrufo et al. 1998). For C_4 crops, the quantity of N fertilizer needed to achieve maximum yield may be increased because the yield enhancement of 12 to 33% by elevated [CO_2] is not accompanied by a decrease in plant N concentration. In contrast to N, there is likely to be a larger demand for P fertilizer for both C_3 and C_4 crops because plant P concentrations are not substantially reduced by high [CO_2] (Rogers et al. 1993; Seneweera et al. 1994).

The shoot nutrient concentrations required to sustain growth of aboveground biomass and to ensure survival of plants is a crucial factor determining competition between C_3 and C_4 plants, and may be affected by elevated [CO_2]. In the temperate grasslands of North America and tropical grasslands of Australia, C_4 grasses tend to be more dominant when soil N levels are low because they require less N in their leaves to sustain maximum photosynthetic and growth rates (Sage and Kubien 2003). Consequently, C_4 species generally have a lower critical N concentration than C_3 species (see Table 3.1). As the atmospheric $[CO_2]$ rises, the critical N concentrations of C_3 plants will approach those of C_4 species (see Table 3.1) and, therefore, C_3 species are likely to be more competitive in low N sites. However, higher temperatures will favor C_4 species (Sage and Kubien 2003). There is anecdotal evidence from Australia that C_3 tropical grasses are less sensitive to P deficiency than C_4 grasses growing on the same sites. Rising $[CO_2]$ is likely to alter this because the P requirement of C_3 plants tends to increase at elevated $[CO_2]$. There have been few studies on shoot nutrient concentrations required for survival. Possible scenarios for the future are that lower N and Zn concentrations will be required to ensure survival of C_3 species, but that higher P concentrations will be needed.

Rising $[CO_2]$ must be taken into account when managing timing and amounts of fertilizer addition because sink development must match the higher photosynthetic rates if greater yields are to be achieved. For grain crops such as wheat and rice, N supply must be sufficiently high at critical stages, such as spikelet development, to ensure that the CO_2 response is realized. This was demonstrated in a rice FACE experiment in Japan, where grain yield was closely related to N uptake during panicle initiation. There was a 15% enhancement of grain yield at high $[CO_2]$ when N fertilizer supply and uptake were high during panicle initiation (Kim et al. 2001). This did not occur at low N supply. P deficiency precluded a vegetative growth response to elevated $[CO_2]$ in cotton, but not wheat in controlled environments (Rogers et al. 1993).

Changes in nutrient concentrations in leaves due to rising high $[CO_2]$ are likely to influence grain quality. N is particularly important for wheat because grain protein concentration determines its price and influences the suitability of flour as an ingredient for bread making. In a FACE (550 ppm $[CO_2]$) experiment with the spring wheat cultivar Yecora Roja, low N supply drastically reduced quality (protein concentration and loaf volume), which was exacerbated by high $[CO_2]$ (Kimball et al. 2001). A glasshouse experiment with the cultivar Hartog, grown with high N supplies at three $[CO_2]$ (280 ppm representing preindustrial $[CO_2]$, 350 ppm and 900 ppm) demonstrated that high $[CO_2]$ reduced protein concentration, which was associated with a change in dough strength (Rogers et al. 1998). Retrospective data for wheat grown in New South Wales, Australia, show that there was a steady decline in grain protein over the period between 1967 and 1990 (Figure 3.3). The explanations for this decline are that high-yielding varieties with lower grain-N concentrations have been introduced, and that soil N levels have declined due to poor management (Conroy and Hocking 1993). However, the rise in atmospheric $[CO_2]$ of approximately 15 ppm over this period should not be ignored as a contributing factor. In countries such as Australia and the U.S., where wheat is grown in areas where climatic conditions are already marginal for growth, temperature rises are likely to increase grain protein concentrations (Reyenga et al. 1999), possibly because starch synthesis is reduced. High $[CO_2]$ may ameliorate this problem to some extent, but higher temperatures are likely to have a deleterious effect on quality. For rice, protein is less important than amylose content, which affects the cooking quality. In a

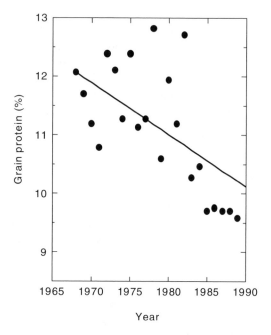

FIGURE 3.3 Retrospective data on average grain protein concentration in wheat grown in New South Wales, Australia from 1968 though 1990. (Adapted from Conroy, J. and Hocking, P. (1993), *Physiologia Plantarum*, 89, 570–576. With permission.) The solid line is drawn by inspection.

controlled environment experiment, high $[CO_2]$ reduced the amylose concentration suggesting that cooked rice from plants grown under rising $[CO_2]$ will be firmer in texture (Seneweera et al. 1996). Rice is the staple diet for a large portion of the world's population. Increasing problems for human health will arise if the decline in the Zn concentration in rice grain at high $[CO_2]$, observed for the cultivar Jarrah, is widespread among rice cultivars (Seneweera et al. 1996).

3.4 WATER DEMAND OF PLANTS UNDER GLOBAL CLIMATE CHANGE

Productivity of many natural and managed ecosystems is limited by water availability. Global climate change (in this section, changes in $[CO_2]$, temperature, and precipitation) will impact substantially on plant and canopy water balance, which will act to modulate the direct effects of climate change on productivity (Figure 3.4). It is often difficult to separate the direct impacts of climate change on the productivity of crops and pastures from those mediated by accompanying changes in the water budget of the agricultural system under study.

3.4.1 STOMATAL CONDUCTANCE AT ELEVATED $[CO_2]$

Stomata orchestrate the exchange of CO_2 and water vapor between the terrestrial vegetation and the atmosphere, thus exerting significant control on the global hydrological

Effects of Global Climate Change on Plant Water Demands

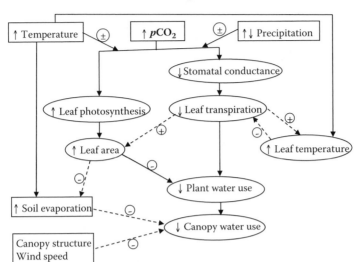

FIGURE 3.4 A schematic representation of the main routes by which global climate change impacts on the water use of leaves, plants, and canopies. The solid arrows indicate direct effects, while the dashed arrows indicate indirectly mediated effects. Plus and minus signs indicate positive or negative feedback effects.

cycle (Figure 3.4, Hetherington and Woodward 2003). There are variations in g_s among plant species and with leaf developmental stage and position in the canopy, as well as with biotic and abiotic factors (Farquhar and Sharkey 1982). Under most environmental conditions, leaf A and g_s are well coordinated such that carbon gain is maximized relative to water loss and a relatively constant intercellular-to-ambient $[CO_2]$ ratio (C_i/C_a) is maintained (Cowan and Farquhar 1977; Wong et al. 1979). Accordingly, elevated $[CO_2]$ increases the availability of the carbon substrate for photosynthesis, and provides the leaf with the opportunity to reduce its water loss by decreasing g_s without compromising CO_2 fixation. Elevated $[CO_2]$ causes a reduction in g_s in response to increasing C_i (Mott 1988). Two surveys of published data found that a doubling of current ambient $[CO_2]$ reduces g_s by 20% (Drake et al. 1997) to 40% (Morison 1987). In general, herbaceous species show a greater reduction in g_s (average of 36% for 16 crops and 40% in 8 C_4 grasses) than in tree species (average of 23% for 23 trees and 21% for European trees) (Figure 3.5; Morison and Gifford1984a, Field et al. 1995; Medlyn et al. 2001). Within the same functional group, plants show substantial variations in stomatal response to elevated $[CO_2]$. For example, cotton has a smaller reduction in g_s (10%) than other herbaceous crops (Samarakoon and Gifford 1995; Kimball et al. 2002), while g_s of some trees is unaffected by elevated $[CO_2]$ (Bunce 1992; Centritto et al. 2002). As argued by Morison (1985), stomatal response to elevated $[CO_2]$ depends not only on conditions during measurements, but also during growth (e.g., light, nutrition, soil moisture, temperature), which may explain part of the variations reported in the literature.

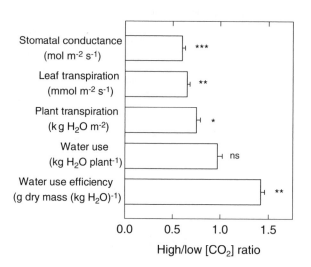

FIGURE 3.5 Stomatal conductance and leaf and plant water use of eight C_4 grasses grown at ambient (420 ppm) and elevated (680 ppm) $[CO_2]$ in naturally lit, temperature-controlled glasshouses for 6 weeks under ample water and nutrient supply. Statistical significance levels from one-way analysis of variance (ANOVA) are shown and they are: ns = not significant ($P > 0.05$), * = $P < 0.05$, ** < 0.01, *** = $P < 0.001$. (Adapted from Ghannoum, O., von Caemmerer, S., and Conroy, J.P. (2001) *Functional Plant Biology*, 28, 1207–1217. With permission.)

Growth at elevated $[CO_2]$ may cause stomatal acclimation (for example, Santrucek and Sage 1996) and induce lasting changes in stomatal frequency and pore size (Beerling and Woodward 1993; Maherali et al. 2002). Stomatal acclimation is defined as a change in stomatal sensitivity to $[CO_2]$ following long-term exposure to high $[CO_2]$. Due to the strong coupling between A and g_s, stomatal acclimation may be partly or largely related to the photosynthetic acclimation that is often observed under high $[CO_2]$ (Drake et al. 1997). Changes in stomatal sensitivity, frequency, or pore size can have significant implications for leaf E. However, as reviewed and argued by Drake et al. (1997) and Morison (1998), it is difficult to draw firm conclusions on these issues due to serious gaps in available data. Regardless of this contention, g_s is often significantly reduced at elevated $[CO_2]$, and this response represents the starting point of our discussion on the effects of rising atmospheric $[CO_2]$ on leaf, plant, and canopy E.

3.4.2 LEAF TRANSPIRATION AT ELEVATED $[CO_2]$

Leaf E can be described by the following equation:

$$E = g_s * VPD_l \tag{4.1}$$

where VPD_l is the leaf-to-air water vapor pressure difference. VPD_l depends on air temperature and humidity as well as T_l. A reduction in g_s at elevated $[CO_2]$ affects leaf E directly and causes feedback effects on T_l and, hence, VPD_l. Reduced g_s

decreases E (see Figure 3.4) according to the following equation (Jarvis and McNaughton 1986):

$$dE/E = (1 - \Omega) * dg_s/g_s \qquad (4.2)$$

where Ω, the decoupling coefficient, is a value between 0 and 1, describing the degree to which conditions at the leaf surface are decoupled from the atmosphere, and depends on the ratio g_s/g_b (g_b is the leaf boundary layer conductance) and temperature. When g_s/g_b is small, such as in moving air, high VPD_l, or low soil moisture, Ω tends to 0, and high-$[CO_2]$-induced reductions in g_s bring about large changes in E. When g_s/g_b is relatively large, such as in still air, low VPD_l, or high soil moisture, Ω tends to 1, and high-$[CO_2]$-induced reductions in g_s bring about small changes in E (Morison and Gifford 1984a; Jarvis and McNaughton 1986). When comparisons between $[CO_2]$ treatments are made under the same radiation, reductions in E at high $[CO_2]$ require adjustments in the sensible heat flux of the leaf, and, hence, its temperature (see Figure 3.4), according to the leaf energy balance equation (Nobel 1999):

$$Rn = H + \lambda E \qquad (4.3)$$

where Rn is the net flux of radiation (absorbed by the leaf minus that reflected and reradiated), H is the sensible heat flux between the leaf and its environment, λE is the latent heat flux associated with E, and λ is the latent heat of water vaporization. Consequently, a reduction in E at constant Rn entails an increase in H and, thus, T_l. The extent to which T_l will increase at high $[CO_2]$ depends on g_s and g_b, where the latter is not only influenced by wind speed, but also by leaf shape and structure (Jarvis and McNaughton 1986). In addition, as g_s increases and approaches g_b (i.e., g_s/g_b is large), T_l is more likely to increase at high $[CO_2]$ (Morison and Gifford 1984a). Increases in T_l at high $[CO_2]$ have been reported in some field and controlled environment experiments using C_3 and C_4 plants. For example, raising the $[CO_2]$ during growth increased the T_l of field-grown cotton (Idso et al. 1987), wheat (Kimball et al. 1995), and sorghum (Wall et al. 2001). High $[CO_2]$ also increased T_l of soybean grown in outdoor chambers (Allen et al., 2003), and two glasshouse-grown C_4 grasses (Siebke et al. 2002). However, increased T_l is not a universal response to high growth at $[CO_2]$, as has been observed in a nutrient-poor, water-limited grassland (Niklaus and Körner 2004). When T_l increases, it results in feedback on E for two main reasons. First, it increases the leaf water vapor pressure and, hence, VPD_l, which will in turn increase E. Secondly, increased T_l may increase A thus causing a draw down of C_i, which will tend to increase g_s, slightly counteracting the primary effect of high $[CO_2]$ (Field et al. 1995; Polley 2002). Most leaf-based measurements of E are made under standard conditions of temperature, humidity, and light in well-stirred gas exchange chambers. Under these conditions, elevated $[CO_2]$ invariably reduces E by a proportion close to the reductions in g_s. For example, E decreased by 35% for a 40% reduction in g_s in eight C_4 grasses grown at double ambient $[CO_2]$ (see Figure 3.5). According to Morison (1985), a 40% reduction in g_s for a doubling of ambient $[CO_2]$ will lead to a 25 to 35% reduction in leaf E.

This is close to the average of 34% reduction obtained by Kimball and Idso (1983) for 46 observations. These substantial decreases in leaf E lead to large increases (60 to 100%) in instantaneous leaf water use efficiency (A/E), whether or not leaf A is stimulated at high [CO_2] (Morison 1985; Eamus 1991; Drake et al., 1997; Wand et al. 1999).

3.4.2.1 Interaction of Water Availability and Temperature with Elevated [CO_2]-Induced Changes in Leaf Transpiration

The interaction of soil moisture (often a reduction) with the response of plants to elevated [CO_2] has been well studied (Morison 1993). One of the earliest responses of reduced soil moisture is a decrease in g_s, thus compounding the effects of elevated [CO_2] on leaf g_s, E, and T_l. However, elevated [CO_2] has a smaller effect on g_s, when g_s is made small due to environmental stresses such as low soil moisture or high VPD_l (Morison 1985; Bunce 1998b). The sensitivity of g_s to [CO_2] and VPD_l is linearly correlated with the absolute value of g_s for two C_3 and two C_4 species (Morison and Gifford 1983).

Warmer air temperatures will impact leaf E by affecting VPD_l in two main ways. First, warmer air has a greater water vapor holding capacity and, hence, will increase the evaporative gradient at the leaf surface. Secondly, higher air temperature will tend to increase T_l and, hence, VPD_l (Polley 2002). It is often impossible to separate the direct temperature effects on leaf g_s and E from those mediated by VPD_l. However, it is likely that VPD_l will rise with global warming, particularly where precipitation is expected to decrease. In addition, the leaf boundary layer is likely to become drier with more closed stomata. Therefore, projected increases in air temperature will counterbalance, to various extents, the high-[CO_2]-induced reductions of leaf g_s and E. For example, in soybean grown in outdoor chambers, elevated [CO_2] reduced g_s by 33% at an average growth temperature of 27°C, and by only 17% at 40°C (Allen et al. 2003). It is worth mentioning that temperature can have a direct effect on g_s or one that is mediated by C_i draw down. However, these responses are not well documented. In addition, the changes in g_s may be more accurately explained by variations in VPD_l than temperature, as has been observed for sweet gum trees (*Liquidambar styraciflua*) growing under FACE (Gunderson et al. 2002).

3.4.3 PLANT TRANSPIRATION AND WATER USE AT ELEVATED [CO_2]

Global climate change and rising atmospheric [CO_2] in particular will cause any number of feed forward and backward effects at the whole plant level (see Figure 3.4). Concerning the effects of elevated [CO_2] on plant E and water use, factors additional to those discussed in Section 3.4.1 for leaf E, must be taken into account. The main ones are leaf area, plant water relations, root distribution, phenology, and their interactions with diurnal and seasonal changes in environmental conditions. In this section we shall treat the plant as an isolated individual, uninfluenced by its neighbours. Issues related to upscaling will be discussed in Section 3.4.4. For transpirational purposes, the isolated plant may be considered as a "cluster of leaves," in keeping with Jarvis and McNaughton's (1986) analysis, such that plant g_s is equal to the average g_s of all transpiring leaves on the plant. In this context, g_b and Ω are not much different from those defined for an individual leaf, and Equation 3.2 can still be used to describe the influence of high-$[CO_2]$-induced reductions in g_s on plant E. However, for a cluster of leaves, average g_b is smaller and average Ω is larger than for a single leaf (Jarvis and McNaughton 1986). The difference between cluster and single leaf Ω depends on wind speed and leaf arrangement. The slower the wind and the closer the leaves are arranged together on the plant, the greater this difference will be. It is clear that in going from the single leaf to the individual plant, the fractional change in E due to high-$[CO_2]$-induced partial stomatal closure is reduced by the simple grouping of transpiring leaves. Consequently, the effect of elevated $[CO_2]$ on plant E is expected to be less than that on leaf E, by at least a fraction equal to the difference between leaf and plant Ω. This theoretical analysis is supported empirically. For example, in eight C_4 grasses, plant E averaged over the whole growth period was reduced by 25%, whereas instantaneous leaf E and g_s decreased by 35% and 40%, respectively, due to doubling ambient $[CO_2]$ (see Figure 3.5). In 11 crop species grown in wet and drying soil under controlled environmental conditions, average leaf g_s and whole plant daily E were reduced by 34% and 16%, respectively, at elevated $[CO_2]$ (Morison and Gifford 1984a). This damping effect on plant E operates in addition to any other direct or indirect effects that elevated $[CO_2]$ has on plant growth, of which the most pertinent is leaf area.

Elevated $[CO_2]$ stimulates leaf area production of most plants (Poorter 1993; Wand et al. 1999; Kimball et al. 2002). This stimulation varies among species and is highly dependent on growth conditions (for example, Morison and Gifford 1984a; Ghannoum et al. 2001). The increase in leaf area has the potential to offset the reductions in plant E at high $[CO_2]$ as is often observed. Relative to ambient $[CO_2]$, high-$[CO_2]$-grown maize, cotton, and wheat plants used less, more, and similar amounts of water, respectively. This was accompanied by a weak, strong, and intermediate increase in leaf area in response to elevated $[CO_2]$ in these three crops, respectively (Samarakoon and Gifford 1995). Figure 3.5 shows another example where on average, eight C_4 grasses grown under well-watered conditions consumed the same amount of water per plant at ambient and elevated $[CO_2]$. The reason was that increases in leaf area (Ghannoum et al. 2001) and temperature (Siebke et al. 2002) were sufficient to counterbalance reductions in leaf and plant E, and highlights the fact that C_4 species do not necessarily use less water at high $[CO_2]$. Similar results were reported for 16 crop species grown in wet and drying soil, at ambient and double ambient $[CO_2]$ (Morison and Gifford 1984a).

Elevated $[CO_2]$ can promote growth by improving shoot water relations of many plant species, particularly under conditions of atmospheric and soil water deficits (Knapp et al. 1993; Seneweera et al. 1998; Wall 2001; Kimball et al. 2002). Elevated $[CO_2]$ can also enhance root growth (Idso and Idso 1994; Wand et al. 1999; Kimball et al. 2002). Increased root length and volume may improve the plant's access to water and, hence, promote its growth and water use.

Whole plant WUE (water use efficiency) is often increased at elevated $[CO_2]$, regardless of the extent of the growth response or whether whole plant water use is decreased or not (Eamus 1991). The stimulation of plant WUE varies with species and growth conditions, and can reach a 100% with a doubling of $[CO_2]$ (Morison and Gifford 1984b; Centritto et al. 2002). In eight C_4 grasses, WUE increased by an average of 42% (see Figure 3.5).

3.4.3.1 Interaction of Water Availability and Temperature with Elevated $[CO_2]$-Induced Changes in Plant Transpiration and Water Use

Under water stress, the absolute and relative stomatal response to elevated $[CO_2]$ is smaller than in well-watered plants, as discussed in Section 3.4.2.1. Consequently, reductions in leaf and plant E are expected to be small or negligible during severe water stress (for example, Samarakoon and Gifford 1995; Seneweera et al. 1998). Water savings early in the drying cycle, when stress is mild to moderate, allow the high-$[CO_2]$-grown plants to photosynthesise and grow for longer under more favorable conditions. When the high-$[CO_2]$ response of leaf area is weak or g_s strong, plant water use decreases and water is saved or expended in extra growth (Morison and Gifford 1984a,b; Samarakoon and Gifford 1995; Owensby et al. 1997; De Luis et al. 1999; Seneweera et al. 2001; Centritto et al. 2002; Niklaus and Körner 2004). In both cases, elevated $[CO_2]$ increases WUE to a greater extent under water stress compared to well-watered conditions.

Increased air temperature can also interact with plant water use and E at elevated $[CO_2]$. As argued in Section 3.4.2.1, high temperatures will tend to offset some of the high-$[CO_2]$-induced reductions in leaf E. The same rationale will apply at the whole plant level, and for the same reasons discussed for a single leaf, high temperatures will also tend to diminish the beneficial effects of high $[CO_2]$ on plant E and water use.

3.4.4 CANOPY TRANSPIRATION AND WATER USE UNDER GLOBAL CLIMATE CHANGE

Predicting the effects of elevated $[CO_2]$ and the ensuing climate change on canopy E and water use is a daunting task without recourse to complex models. Canopy structure adds extra layers of complexity to the influence of changes in g_s on E (see Figure 3.4). Depending on canopy structure and wind speed and direction, substantial gradients of water vapor pressure and heat may develop vertically and horizontally within and across the canopy. Consequently, high-$[CO_2]$-induced partial stomatal closure will affect leaf E to different extents depending on the leaf position in the canopy. Scaling E from the plant to the canopy requires, at the simplest level, the introduction of an additional term, the canopy aerodynamic conductance (g_a), which describes the conductance of water vapor and heat fluxes between the canopy surface and the atmospheric layer above the canopy (Jarvis and McNaughton 1986). The term Ω (Equation 3.2) becomes the canopy decoupling coefficient, taking into account g_b and g_a. Aerodynamically rough canopies, with sparse vegetation and narrow leaves exposed to fast wind are better coupled to the free atmosphere above the canopy surface than are aerodynamically smooth, dense canopies with broad leaves exposed to slow wind. The outcome of these added complexities is that canopy E will be invariably less affected than leaf or plant E by a corresponding change in g_s. Increases in leaf area index (LAI) will further offset any changes that reduced canopy E has on canopy water use, while having the potential to reduce soil evaporation. Nevertheless, canopy E and evapotranspiration (ET) have been reported to decrease under elevated [CO_2] in field and OTC experiments. In field-grown wheat and sorghum crops exposed to FACE, ET decreased by 9% and 7%, respectively, relative to control plots (Conley et al. 2001; Grant et al. 2001). In a nutrient-poor, water-limited grassland system, there were small reductions in ET at elevated [CO_2]. These small but insignificant differences in ET accumulated over time, leading to marginal but significant water savings at elevated [CO_2] (Niklaus et al. 1998; Niklaus and Körner 2004). Similar results were reported in OTC experiments. In a C_4 tall-grass prairie grown in OTC, average ET over 34 days around the peak LAI was reduced by 22% at twice ambient compared to ambient [CO_2] (Ham et al. 1995; Owensby et al. 1997). Centritto et al. (1999a,b) found no water saving in cherry trees growing under well-watered conditions in OTC at high [CO_2] during two successive seasons. In the first season, decreases in g_s at high [CO_2] were offset by increases in leaf area. In the second season, although g_s decreased and leaf area was unchanged, water use did not differ between ambient- and elevated-[CO_2]-trees because Ω was high and the canopy poorly coupled to the atmosphere (Centritto et al. 1999a,b). Wilson et al. (1999) modeled ET for soybean and maize crops using simulations with increasing levels of complexity. At the simplest level, with no feedback operating, ET was reduced by 15.1% and 24.7% in soybean and maize, respectively, at double ambient [CO_2]. With maximum complexity, accounting for atmospheric feedbacks, soil fluxes, and physiological responses, ET decreased by 5.4% and 8.6%, in soybean and maize respectively, at high [CO_2]. Introducing water stress to the modeling scenarios increased the reductions in ET for both crops (Wilson et al. 1999). Indeed, environmental stresses interact strongly with the response of canopy ET and water use to elevated [CO_2]. In a wheat experiment, FACE decreased ET by 7% and 19% under high and low N supplies, respectively, most likely due to the weaker LAI response under low N (Grant et al. 2001). In contrast, soil water stress can enhance the reductions in ET under CO_2 enrichment as has been observed for sorghum (Conley et al. 2001). Warmer air temperatures have the potential to offset the effects of elevated [CO_2] on ET. In soybean growing in outdoor chambers, double ambient [CO_2] reduced ET by an average of 9% at 23°C, while high [CO_2] had no effect on ET above 35°C (Allen et al. 2003).

In summary, the available data suggests that elevated [CO_2] reduces ET of well-watered canopies. In poorly coupled canopies, E is mainly driven by Rn, and changes in g_s have little effect on the canopy water budget. Increases in air temperature will offset some of the reductions in ET, while moderate soil moisture deficits may promote high-[CO_2] reductions in ET. Whether reduced ET will translate into soil water saving in the field, will depend on a whole host factors, such as the response of LAI, water availability, and ambient temperature.

3.4.5 IMPLICATIONS OF CLIMATE CHANGE FOR PLANT WATER DEMANDS

Global climate change will bring about direct and indirect changes in plant and community water demands (see Figure 3.4).

Elevated $[CO_2]$ will tend always to reduce g_s more than leaf E, and reduce leaf E less than plant E and even less than canopy E.

At the plant level, reduction in water demands at elevated $[CO_2]$ will mainly depend on reductions in g_s, increases in leaf area and temperature, and how well the leaves on the plant are coupled to the surrounding atmosphere. In general, increased air temperature will tend to offset this response while reduced precipitation will tend to enhance it.

Scaling these responses to the canopy level is infinitely more complex because the feedback in operation are numerous and their outcome is seldom obvious. In general, it is true that the effect of elevated $[CO_2]$ and ensuing climate change on canopy E and water use is small in absolute terms, and smaller than the effects at the plant level. However, in many marginal natural and agricultural fields, even a small change in water demand can have significant impact on vegetative and reproductive yield, farmers' livelihood, and the long-term ecological health of the farm or field in question. Therefore, it is important for scientific and agricultural communities to make progress in predicting and ultimately factoring in the climate change impacts on canopy water demands. A reasonable approach that may provide a handle on this daunting task is to combine complex canopy modeling with long-term field experiments. It is the view of the authors of this chapter that the long-term indirect effects of climate change can be as or more important than the direct effects, and that future research efforts must be directed toward both aspects.

REFERENCES

Aben, S.K., Seneweera, S., Ghannoum, O., and Conroy, J. (1999) Nitrogen requirements for maximum growth and photosynthesis of rice *Oryza Sativa* L. cv. Jarrah grown at 36 and 70 Pa CO_2. *Australian Journal of Plant Physiology*, 26, 759–756.

Ainsworth, E.A., Davey, P.D., Hymus, G.J., Osborne, C.P., Rogers, A., Blum, H., Nosberger, J., and Long, S.P. (2003) Is stimulation of leaf photosynthesis by elevated carbon dioxide concentration maintained in the long term? A test with *Lolium perenne* grown for 10 years at two nitrogen fertilization levels under Free Air CO_2 Enrichment (FACE). *Plant, Cell and Environment*, 26, 705–714.

Allen, L.H., Pan, D., Boote, K.J., Pickering, N.B., and Jones, J.W. (2003) Carbon dioxide and temperature effects on evapotranspiration and water use efficiency of soybean. *Agronomy Journal*, 95, 1071–1081.

Amthor, J.S. (2001) Effects of atmospheric CO_2 concentration on wheat yield: review of results from experiments using various approaches to control CO_2 concentration. *Field Crops Research*, 73, 1–34.

Beerling, D.J. and Woodward, F.I., (1993) Ecophysiological responses of plants to global environmental change since Last Glacial Maximum. *New Phytologist,* 125, 641–648.

Berry, J.A. and Björkman, O. (1980) Photosynthetic response and adaptation to temperature in higher plants. *Annual Review of Plant Biology*, 31, 491–543.

Bunce, J.A. (1992) Stomatal conductance, photosynthesis and respiration of temperate deciduous tree seedlings grown outdoors at an elevated concentration of carbon dioxide.
Plant, Cell and Environment, 15, 541–549.

Bunce, J.A. (1998a) The temperature dependence of the stimulation of photosynthesis by
elevated carbon dioxide in wheat and barley. *Journal of Experimental Botany*, 49,
1555–1561.

Bunce, J.A. (1998b) Effects of humidity on short-term responses of stomatal conductance to
an increase in carbon dioxide concentration. *Plant, Cell and Environment*, 21,
115–120.

Cakmak, I. and Marschner, H. (1987) Mechanism of phosphorus-induced zinc-deficiency in
cotton. III. Changes in physiological availability of zinc in plants. *Physiologia Plantarum*, 70, 13–20.

Centritto, M., Lee, H.S.J., and Jarvis, P.G. (1999a) Interactive effects of elevated [CO_2] and
drought on cherry (*Prunus avium*) seedlings. I. Growth, whole-plant water use efficiency and water loss. *New Phytologist*, 141, 129–140.

Centritto, M., Lee, H.S.J., and Jarvis, P.G. (1999b) Interactive effects of elevated [CO_2] and
drought on cherry (*Prunus avium*) seedlings. II. Photosynthetic capacity and water
relations. *New Phytologist*, 141, 141–153.

Centritto, M., Lucas, M.E., and Jarvis, P.G. (2002) Gas exchange, biomass, whole-plant water-
use efficiency and water uptake of peach (*Prunus persica*) seedlings in response to
elevated carbon dioxide concentration and water availability. *Tree Physiology*, 22,
699–706.

Conley, M.M., Kimball, B.A., Brooks, T.J., et al. (2001) CO_2 enrichment increases water-use
efficiency in sorghum. *New Phytologist*, 151, 407–412.

Conroy, J.P. (1989) Influence of high CO_2 on *Pinus radiata*, Ph.D. thesis, Macquarie University, Sydney, Australia.

Conroy, J. and Hocking, P. (1993) Nitrogen nutrition of C-3 plants at elevated atmospheric
CO_2 concentrations. *Physiologia Plantarum*, 89, 570–576.

Conroy, J.P., Seneweera, S., Basra, A.S., Rogers, G., and Nissen-Wooller, B. (1994) Influence
of rising atmospheric CO_2 concentrations and temperature on growth, yield and grain
quality of cereal crops. *Australian Journal of Plant Physiology*, 21, 741–758.

Cotrufo, M.F., Ineson, P., and Scott, Y. (1998) Elevated CO_2 reduces the nitrogen concentration
of plant tissues. *Global Change Biology*, 4, 43–54.

Cowan, I.R. and Farquhar, G.D. (1977) Stomatal function in relation to leaf metabolism and
environment. *Society of Experimental Biology Symposium*, 31, 471–505.

Cure, J.D., Rufty, T.W., and Israel, D.W. (1988a) Phosphorus stress effects on growth and
seed yield responses to nonodulated soybean to elevated carbon dioxide. *Agronomy
Journal*, 80, 897–902.

Cure, J.D., Israel, D.W., and Rufty, T.W. (1988b) Nitrogen stress effects on growth and seed
yield of nodulated soybean exposed to elevated carbon dioxide. *Crop Science*, 28,
671–677.

Defiani, M.R. (1999) Zinc requirements of rice at elevated CO_2, MSc thesis, University of
Western Sydney, Sydney.

De Luis, J., Irigoyen, J.J., and Sanchez-Diaz, M. (1999) Elevated CO_2 enhances plant growth
in droughted N_2-fixing alfalfa without improving water status. *Physiologia Plantarum*,
197, 84–89.

Drake, B.G., Gonzàlez-Meler, M.A., and Long, S.P. (1997) More efficient plants: A consequence of rising atmospheric CO_2. *Annual Review of Plant Biology*, 47, 609–639.

Eamus, D. (1991) The interaction of rising CO_2 and temperature with water use efficiency.
Plant, Cell and Environment, 14, 843–852.

Evans, J.R. (1989) Photosynthesis and nitrogen relationships in leaves of C_3 plants. *Oecologia*, 78, 9–19.

FAO (Food and Agriculture Organization of the United Nations) (2002) *World Agriculture Towards 2015/2030*. Rome.

Farquhar, G.D. and von Caemmerer, S. (1982) A biochemical model of photosynthetic CO_2 assimilation in leaves of C_3 species. *Planta*, 149, 178–190.

Farquhar, G.D. and Sharkey, T.D. (1982) Stomatal conductance and photosynthesis. *Annual Review of Plant Physiology*, 33, 317–345.

Field, C.B., Jackson, R.B., and Mooney, H.A. (1995) Stomatal responses to increased CO_2: implications from the plant to the global scale. *Plant, Cell and Environment*, 18, 1214–1225.

Ghannoum, O. and Conroy, J.P. (1998) Nitrogen deficiency precludes a growth response to CO_2 enrichment in C_3 and C_4 *Panicum* grasses. *Australian Journal of Plant Physiology*, 25, 627–636.

Ghannoum, O., von Caemmerer, S., Ziska, L.H., and Conroy, J.P. (2000) The growth response of C_4 plants to rising atmospheric CO_2 partial pressure: a reassessment. *Plant, Cell and Environment*, 23, 931–942.

Ghannoum, O., von Caemmerer, S., and Conroy, J.P. (2001) Plant water use efficiency of 17 Australian NAD-ME and NADP-ME C_4 grasses at ambient and elevated CO_2 partial pressure. *Functional Plant Biology*, 28, 1207–1217.

Ghannoum, O., von Caemmerer, S., Barlow, E.W.R., and Conroy, J.P. (1997) The effect of CO_2 enrichment and irradiance on the growth, morphology and gas exchange of a C_3 (*Panicum laxum*) and a C_4 (*Panicum antidotale*) grass. *Functional Plant Biology*, 24, 227–237.

Grant, R.F., Kimball, B.A., Brook, T.J., et al. (2001) Modeling interactions among carbon dioxide, nitrogen and climate on energy exchange of wheat in a free air carbon dioxide experiment. *Agronomy Journal*, 93, 638–649.

Gunderson, C.A., Sholtis, J.D., Wullschleger, S.D., Tissue, D.T., Hanson, P.J., and Norby, R.J. (2002) Environmental and stomatal control of photosynthetic enhancement in the canopy of a sweetgum (*Liquidambar styraciflua* L.) plantation during 3 years of CO_2 enrichment. *Plant, Cell and Environment*, 25, 379–393.

Ham, J.M., Owensby, C.E., Coyne, P.I., and Bremer, D.J. (1995) Fluxes of carbon dioxide and water vapor from a prairie ecosystem exposed to ambient and elevated atmospheric carbon dioxide. *Agricultural and Forest Meteorology*, 77, 73–93.

Hatch, M.D. (1987) C_4 photosynthesis: a unique blend of modified biochemistry, anatomy and ultrastructure. *Biochimica et Biophysica Acta*, 895, 81–106.

Hetherington, A.M. and Woodward, F.I. (2003) The role of stomata in sensing and driving environmental change. *Nature*, 424, 901–908.

Hocking, P.J. and Meyer, C.P. (1991) Effects of CO_2 enrichment and nitrogen stress on growth, and partitioning of dry matter and nitrogen in wheat and maize. *Australian Journal of Plant Physiology*, 18, 339–356.

Howden, S.M., McKeon, G.M., Carter, J.O., and Beswick, A. (1999) Potential global change impacts on C_3–C_4 distributions in eastern Australian rangelands. In *Proceedings of the VI International Rangeland Congress*, (Eldridge, D. and Freudenberger, D., Eds.), International Rangeland Congress, Inc., Aitkinvale, 41–43..

Idso, K.E. and Idso, S.B. (1994) Plant responses to atmospheric CO_2 enrichment in the face of environmental constraints: a review of the past 10 years. *Agricultural and Forest Meteorology*, 69, 153–203.

Idso, S.B., Kimball, B.A., and Mauney, J.R. (1987) Atmospheric carbon dioxide enrichment effects on cotton midday foliage temperature: implications for plant water use and crop yield. *Agronomy Journal*, 79, 667–672.

IPCC (Intergovernmental Panel on Climate Change) (2001) *Climate Change 2001: The Scientific Basis*. Summary for Policymakers. Working group one assessment report. Cambridge University Press, Cambridge, U.K.

Jarvis, P.G. and McNaughton, K.G. (1986) Stomatal control of transpiration: Scaling up from leaf to region. *Advances in Ecological Research*, 15, 1–49.

Jitla, D.S., Rogers, G.S., Seneweera, S.P., Basra, A.S., Oldfield, R.J., and Conroy, J.P. (1997) Accelerated early growth of rice at elevated CO_2: is it related to developmental changes in the shoot apex. *Plant Physiology*, 115, 15–22.

Jordan, D.B. and Ogren, W.L. (1984) The CO_2/O_2 specificity of ribulose 1,5-bisphosphate carboxylase/oxygenase. *Planta*, 161, 308–313.

Kim, H-Y., Lieffering, M., Miura, S., Kobayashi, K. and Okada, M .(2001) Growth and nitrogen uptake of CO_2-enriched rice under filed conditions. *New Phytologist*, 150, 223–229.

Kimball, B.A. (1983) Carbon dioxide and agricultural yield: An assemblage and analysis of 430 prior observations. *Agronomy Journal*, 75, 779–788.

Kimball, B.A., Mauney, J.R., Nakayama, F.S., and Idso, S,B, (1993) Effects of increasing atmospheric CO_2 on vegetation. *Vegetatio*, 104/105, 65–75.

Kimball, B.A., Pinter, P.J., Garcia, R.L., et al. (1995) Productivity and water use of wheat under free-air CO_2 enrichment. *Global Change Biology*, 1, 429–442.

Kimball, B.A., Morris, C.F., Pinter, P.J., et al. (2001) Wheat grain quality as affected by elevated CO_2, drought and soil nitrogen. *New Phytologist*, 150, 295–303.

Kimball, B.A. and Idso, S.B. (1983) Increasing atmospheric CO_2: effects on crop yield, water use and climate. *Agricultural Water Management*, 7, 55–72.

Kimball, B.A., Kobayashi, K., and Bindi, M. (2002) Responses of agricultural crops to free-air CO_2 enrichment. *Advances in Agronomy*, 77, 293–368.

Kitagishi, K. and Obata, H. (1986) Effects of zinc deficiency on the nitrogen metabolism of meristematic tissues of rice plants with reference to protein synthesis. *Soil Science and Plant Nutrition*, 32, 397–406.

Knapp, A.K., Hamerlynck, E.P., and Owensby, C.E. (1993) Photosynthetic and water relations responses to elevated CO_2 in the C_4 grass *Andropogon gerardii*. *International Journal of Plant Science*, 154, 459–466.

Lawlor, D.W. and Cornic, G. (2002) Photosynthetic carbon assimilation and associated metabolism in relation to water deficits in higher plants. *Plant, Cell and Environment*, 25, 275–294.

Lloyd, J. and Farquhar, G.D. (1994) ^{13}C discrimination during CO_2 assimilation by the terrestrial biosphere. *Oecologia*, 99, 201–215.

Loladze, I. (2002) Rising atmospheric CO_2 and human nutrition: toward globally imbalanced plant stoichiometry? *Trends in Ecology and Evolution*, 17, 457–461.

Long, S.P. (1991) Modification of the response of photosynthesis productivity to rising temperature by atmospheric CO_2 concentrations: has its importance been underestimated? *Plant, Cell and Environment*, 14, 729–739.

Long, S.P. (1999) Environmental responses. In *C_4 Plant Biology* (Sage, R.F. and Monson, R.K., Eds.), Academic Press, San Diego, CA, 215–249.

Maherali, H., Reid, C.D., Polley, H.W., Johnson, H.B., and Jackson, R.B. (2002) Stomatal acclimation over a subambient to elevated CO_2 gradient in a C_3/C_4 grassland. *Plant, Cell Environment*, 25, 557–566.

Makino, A., Sakuma, H., Sudo, E., and Mae, T. (2003) Differences between maize and rice in N-use efficiency for photosynthesis and protein allocation. *Plant Cell Physiology*, 44, 952–956.

Marschner, H. (1995) *Mineral Nutrition of Higher Plants*, 2nd ed., Academic Press, San Diego, CA.

Medlyn, B.E., Barton, C.V.M., Broadmeadow, M.S.J., et al. (2001) Stomatal conductance of European forest species after long-term exposure to elevated [CO_2]: a synthesis of experimental data. *New Phytologist*, 149, 247–264.

Moore, B.D., Cheng, S.H., Sims, D., and Seemann, J.R. (1999) The biochemical and molecular basis for photosynthetic acclimation to elevated atmospheric CO_2. *Plant, Cell and Environment*, 22, 567–582.

Morgan, J.A., Hunt, H.W., Monz, C.A., and LeCain, D.R. (1994) Consequences of growth at two carbon dioxide concentrations and two temperatures for leaf gas exchange in *Pascopyrum smithii* (C_3) and *Bouteloua gracilis* (C_4). *Plant, Cell and Environment*, 17, 1023–1033.

Morison, J.I.L. (1987) Intercellular CO_2 concentration and stomatal response to CO_2. In *Stomatal Function*, Zeiger, E., Farquhar, G.D., and Cowan, I.R., Eds., Stanford University Press, Stanford, CA, 229–251.

Morison, J.I.L. (1985) Sensitivity of stomata and water use efficiency to high CO_2. *Plant, Cell and Environment*, 8, 467–474.

Morison, J.I.L. (1993) Response of plants to CO_2 under water limiting conditions. *Vegetatio*, 104/105, 193–209.

Morison, J.I.L. (1998) Stomatal response to increasing CO_2 concentration. *Journal of Experimental Botany*, 49, 443–452.

Morison, J.I.L. and Lawlor, D.W. (1999) Interactions between increasing CO_2 concentration and temperature on plant growth. *Plant, Cell and Environment*, 22, 659–682.

Morison, J.I.L. and Gifford, R.M. (1983) Stomatal sensitivity to carbon dioxide and humidity. A comparison of two C_3 and two C_4 grass species. *Plant Physiology*, 71, 789–796.

Morison, J.I.L. and Gifford, R.M. (1984a) Plant growth and water use with limited water supply in high CO_2 concentrations. II. Plant dry weight, partitioning and water use efficiency. *Australian Journal of Plant Physiology*, 11, 375–384.

Morison, J.I.L. and Gifford, R.M. (1984b) Plant growth and water use with limited water supply in high CO_2 concentrations. I. Leaf area, water use and transpiration. *Australian Journal of Plant Physiology*, 11, 361–374.

Mott, K.A. (1988) Do stomata respond to CO_2 concentrations other than intercellular? *Plant Physiology*, 86, 200–203.

Niklaus, P.A., Spinnler, D., and Körner, C. (1998) Soil moisture dynamics of calcareous grassland under elevated CO_2. *Oecologia*, 117, 201–208.

Niklaus, P.A. and Körner, C. (2004) Synthesis of a six-year study of calcareous grassland responses to *in situ* CO_2 enrichment. *Ecological Monographs*, 74, 491–511.

Nobel, S. (1999) *Physicochemical and Environmental Plant Physiology*, Academic Press, San Diego, CA.

Owensby, C.E., Ham, J.M., Knapp, A.K., Bremer, D., and Auen, L.M. (1997) Water vapour fluxes and their impact under elevated CO_2 in a C_4-tallgrass prairie. *Global Change Biology*, 3, 189–195.

Paul, M.J. and Pellny, T.K. (2003) Carbon metabolite feedback regulation of leaf photosynthesis and development. *Journal of Experimental Botany*, 54, 539–547.

Polley, H.W. (2002) Implications of atmospheric and climatic change for crop yield and water use efficiency. *Crop Science*, 42, 131–140.

Poorter, H. (1993) Interspecific variation in the growth response of plants to an elevated ambient CO_2 concentration. *Vegetatio*, 104/105, 77–97.

Poorter, H., Roumet, C., and Campbell, B.D. (1996) Interspecific variation in the growth response of plants to elevated CO_2: a search for functional types. In *Carbon Dioxide, Populations, and Communities*, Körner, C. and Bazzaz, F.A., Eds., Academic Press, New York, 375–412.

Poorter, H. and Navas, M-L. (2003) Plant growth and competition at elevated CO_2: on winners, losers and functional groups. *New Phytologist*, 157, 175–198.

Rawson, H.M. (1992) Plant responses to temperature under conditions of elevated CO_2. *Australian Journal of Botany*, 40, 473–490.

Reddy, K.R., Robana, R.R., Hodges, H.F., Liu, X.J., and McKinion, J.M. (1998) Interactions of CO_2 enrichment and temperature on cotton growth and leaf characteristics. *Environmental and Experimental Botany*, 39, 117–129.

Reuter, D.J. and Robinson, J.B. (1997) *Plant Analysis: An Interpretation Manual,* CSIRO Publishing, Melbourne, Australia.

Reyenga, R.J., Howden, S.M., Meinke, H., and McKeon, G.M. (1999) Modeling global change impacts on wheat cropping in south-east Queensland, Australia. *Environmental Modeling & Software*, 14, 297–306.

Rogers, G.S., Payne, L., Milham, P., and Conroy, J. (1993) Nitrogen and phosphorus requirements of cotton and wheat under changing atmospheric CO_2 concentrations. *Plant and Soil*, 156, 231–234.

Rogers, G.S., Milham, P.J., Thibaud, M.C., and Conroy, J.P. (1996a) Interactions between rising CO_2 concentration and N supply on cotton (*Gossypium hirsutum* L.). I. Leaf N concentration and growth. *Australian Journal of Plant Physiology*, 23, 119–125.

Rogers, G.S., Milham, P.J., Gillings, M., and Conroy, J.P. (1996b) Sink strength may be the key to growth and nitrogen responses in N-deficient wheat at elevated CO_2. *Australian Journal of Plant Physiology*, 23, 253–264.

Rogers, G.S. (1996) Influence of N and P nutrition on responses of wheat and cotton to elevated CO_2, Ph.D. thesis, University of Western Sydney, Australia.

Rogers, G., Gras, P., Payne, L., Milham, P., and Conroy, J., (1998) The influence of CO_2 concentration ranging from 280 to 900 L L1 on the protein starch and mixing properties of wheat flour (*Triticum aestivum* L. cv. Hartog and Rosella). *Australian Journal of Plant Physiology*, 25, 387–393.

Rudmann, S.G. (2000) Resource utilization of C_4 grasses at elevated CO_2, Ph.D. thesis, University of Western Sydney, Australia.

Rudmann, S.G., Milham, P.J., and Conroy, J.P. (2001) Influence of high CO_2 on nitrogen use efficiency of the C_4 grasses *Panicum coloratum* and *Cenchrus ciliaris* (NAD-ME and NADP-ME subtypes). *Annals of Botany*, 88, 571–577.

Sage, R.F. and Kubien, D.S. (2003) Quo vadis C_4? An ecophysiological perspective on global change and the future of C_4 plants. *Photosynthesis Res*earch, 77, 209–225.

Samarakoon, A.B. and Gifford, R.M. (1995) Soil water content under plants at high CO_2 concentration and interactions with the direct CO_2 effects: a species comparison. *Journal of Biogeography*, 22, 193–202.

Santrucek, J. and Sage, R.F. (1996) Acclimation of stomatal conductance to a CO_2-enriched atmosphere and elevated temperature in *Chenopodium album*. *Australian Journal of Plant Physiology*, 22, 467–478.

Seemann, J.R., Badger, M.R., and Berry, J.A. (1984) Variation in specific activity of ribulose-1,5-bisphosphate carboxylase between species using different photosynthetic pathways. *Plant Physiology*, 74, 791–794.

Seneweera, S.P., Milham, P., and Conroy, J.P. (1994) Influence of elevated CO_2 and phosphorus nutrition on the growth and yield of a short-duration rice (*Oryza sativa* L cv. Jarrah). *Australian Journal of Plant Physiology*, 21, 281–292.

Seneweera, S., Blakeney, A., Milham, P., Basra, A.S., Barlow, E.W.R., and Conroy, J. (1996) Influence of rising CO_2 and phosphorus nutrition on growth, yield and quality of rice (*Oryza sativa* L cv. Jarrah). *Cereal Chemistry*, 2, 239–243.

Seneweera, S.P., Ghannoum, O., and Conroy, J. (1998) High vapour pressure deficit and low soil water availability enhance shoot growth responses of a C_4 grass (*Panicum coloratum* cv. Bambatsi) to CO_2 enrichment. *Australian Journal of Plant Physiology*, 25, 287–292.

Seneweera, S.P., Ghannoum, O., and Conroy, J.P. (2001) Root and shoot factors contribute to the effect of drought on photosynthesis and growth of the C_4 grass *Panicum coloratum* at elevated CO_2 partial pressure. *Functional Plant Biology*, 28, 451–460.

Seneweera, S.P., Conroy, J.P., Ishimaru, K., Ghannoum, O., Okada, M., Lieffering, M., Kim, H-Y., and Kobayashi, K. (2002) Changes in source-sink relations during development influence photosynthetic acclimation of rice to free-air CO_2 enrichment (FACE). *Functional Plant Biology*, 29, 945–953.

Serraj, R., Allen, L.H., and Sinclair, T.R. (1999) Soybean leaf growth and gas exchange response to drought under carbon dioxide enrichment. *Global Change Biology*, 5, 283–291.

Sharkey, T.D. (1985) Photosynthesis in intact leaves of C_3 plants: physics, physiology and rate limitations. *Botanical Review*, 51, 53–105.

Siebke, K., Ghannoum, O., Conroy, J.P., and von Caemmerer, S. (2002) Elevated CO_2 increases the leaf temperature of two glasshouse-grown C_4 grasses. *Functional Plant Biology*, 29, 1377–1384.

Taub, D.R., Seeman, J.R., and Coleman, J.S. (2000) Growth in elevated CO_2 protects photosynthesis against high-temperature damage. *Plant, Cell and Environment*, 23, 649–656.

Thomas, J.F. and Harvey, C. (1983) Leaf anatomy of four species grown under continuous CO_2 enrichment. *Botanical Gazette*, 144, 303–309.

von Caemmerer, S., Ghannoum, O., Conroy, J.P., Clark, H., and Newton, P.C.D. (2001) Photosynthetic responses of temperate species to free air CO_2 enrichment (FACE) in a grazed New Zealand pasture. *Functional Plant Biology*, 28, 439–450.

Wall, G.W. (2001) Elevated atmospheric CO_2 alleviates drought stress in wheat. *Agriculture, Ecosystems and Environment*, 67, 261–271.

Wall, G.W., Brooks, T.J., Adam, N.R., et al. (2001) Elevated atmospheric CO_2 improved *Sorghum* plant water status by ameliorating the adverse effects of drought. *New Phytologist*, 152, 231–248.

Wand, S.J.E., Midgley, G.F., Jones, M.H., and Curtis, P.S. (1999) Responses of wild C_4 and C_3 grass (Poaceae) species to elevated atmospheric CO_2 concentration: a test of current theories and perceptions. *Global Change Biology*, 5, 723–741.

Wilson, K.B., Carlson, T.N., and Bunce, J.A. (1999) Feedback significantly influences the simulated effect of CO_2 on seasonal evapotranspiration from two agricultural species. *Global Change Biology*, 5, 903–917.

Wong, S.C. (1979) Elevated atmospheric partial pressure of CO_2 and plant growth. Interactions of nitrogen nutrition and photosynthetic capacity in C_3 and C_4 plants. *Oecologia*, 44, 68–74.

Wong, S.C., Cowan, I.R., and Farquhar, G.D. (1979) Stomatal conductance correlates with photosynthetic capacity. *Nature*, 282, 424–426.

Wong, S.C. (1990) Elevated atmospheric partial pressure of CO_2 and plant growth. II. Non-structural carbohydrate content in cotton plants and its effect on growth parameters. *Photosynthesis Research*, 23, 171–180.

Ziska, L.H. (1998) The influence of root zone temperature on photosynthetic acclimation to elevated carbon dioxide concentrations. *Annals of Botany*, 81, 717–721.

4 Climate Change and Symbiotic Nitrogen Fixation in Agroecosystems

Richard B. Thomas, Skip J. Van Bloem, and William H. Schlesinger

CONTENTS

4.1 INTRODUCTION

Nitrogen is the element most often limiting to net primary production in many terrestrial ecosystems (Lee et al., 1983; Binkley, 1986; Vitousek and Howarth, 1991). Symbiotic N_2 fixation, however, provides an important N input in many plant communities and has major impacts on N dynamics and ecosystem productivity, especially where N_2-fixing species are present during early successional stages (Boring et al., 1988; Vitousek and Howarth, 1991) or in agroecosystems where N_2-fixing plants are grown for products and green manure (Table 4.1; Danso et al., 1992; Unkovich and Pate, 2000). Forests with species capable of symbiotic N_2 fixation often have N_2 fixation rates that balance the annual uptake requirement for N (Binkley, 1986) and produce

TABLE 4.1
Estimated Symbiotic N_2 Fixation Rates for Diverse Terrestrial Ecosystems

Ecosystem	Species	N_2 Fixation Rate kg N ha^{-1} yr^{-1}	References
Agricultural and Pastoral Systems			
Soybean	Glycine max	15–142	Dakora & Keya, 1997; Unkovich & Pate, 2000
Common bean	Phaseolus vulgaris	17–85	Dakora & Keya, 1997; Unkovich & Pate, 2000
Peanut	Arachis hypogaea	32–175	Dakora & Keya, 1997; Unkovich & Pate, 2000
Chickpea	Cicer arietinum	57	Unkovich & Pate, 2000
Cowpea	Vigna unguiculata	24–201	Dakora & Keya, 1997
Field Pea	Pisum sativum	104–200	Unkovich & Pate, 2000
Lentil	Lens culinaris	80	Unkovich & Pate, 2000
Faba bean	Vicia faba	92	Unkovich & Pate, 2000
Bambara groundnut	Vigna subterranea	40–62	Dakora & Keya, 1997
Narrow-leaf lupin	Lupinus angustifolius	229	Unkovich & Pate, 2000
Agroforestry Species			
Leucaena	Leucaena leucocephala	110–548[a,b]	Danso et al., 1992; Dakora & Keya, 1997
Australian pine	Casuarina equisetifolia	43–60[a,b]	Danso et al., 1992
	Sesbania rostrata	505–581[a,b]	Danso et al., 1992
	Sesbania sesban	43–102[a,b]	Danso et al., 1992
	Albizia lebbeck	94[a,b]	Danso et al., 1992
	Acacia holosericia	36–108[a,b]	Dakora & Keya, 1997
Gliricidia	Gliricidia sepium	108[a,b]	Danso et al., 1992
Natural Systems			
Hawaii ashflow	Myrica faya	18	Vitousek et al., 1987
SE coastal plain	Myrica cerifera	<2–11	Permar & Fisher, 1983
Appalachian oak forest	Robinia pseudoacacia	30–75	Boring & Swank, 1984a; Boring & Swank, 1984b
Sonoran desert	Prosopis glandulosa	25–35	Rundel et al., 1982
Massachusetts peatland	Myrica gale	35	Schwintzer, 1983
Pacific northwest	Ceanothus velutinus	0–100	Tarrant, 1983
	Alnus rubra	40–160	Luken & Fonda, 1983
Alaskan boreal forest	Alnus incana	156–362	Van Cleve et al., 1971

[a] Above ground only.
[b] Using total N difference methods.

litter and plant tissues rich in N, greatly accelerating N cycling and availability for nonfixing plants (Mikola et al., 1983; Bernhard Reversat, 1996). Likewise, depending on the species, crop management, and climate, N_2 fixation can satisfy the majority of the N demand of a legume crop, or add N to soil reserves to meet a portion of the N requirements for nonlegume crops planted in tandem or in rotation (Hardarson and Atkins, 2003). Because of these benefits, N_2-fixing plants are important components in maintaining or improving soil fertility and exhibit great potential for use in soil stabilization, reforestation, and other silvicultural and agricultural practices.

Increasing atmospheric CO_2, as well as other components of climate change, has the potential to exert a strong influence on the productivity of symbiotic N_2-fixing organisms and the amounts of N contributed by these organisms to natural and agroecosystems. The combustion of fossil fuels and other human activities, including deforestation and other changes in land use, has caused the CO_2 concentration in the atmosphere to increase from the preindustrial level of ~270 µl l^{-1} to its current value of ~375 µl l^{-1}, and the concentration is expected to double from the preindustrial level later this century (IPCC, 2001). Carbon dioxide is a potent greenhouse gas and along with water vapor, methane, N_2O, and other gasses, it maintains the habitable temperatures on Earth, but its further accumulation in the atmosphere is the primary driver of global warming. Recent estimates suggest that a doubling of atmospheric CO_2 will force a 1 to 6°C increase in global mean temperature (IPCC, 2001), and this warming may be accompanied by shifts in the distribution of precipitation at regional and continental scales (Rind et al., 1990). Changes in these three linked variables (CO_2, temperature, and water) will alter plant growth, biomass, and plant community composition at local, regional, and global scales. Plant responses, through increased carbon sequestration, can affect the variables driving climate change. Predictions of future climate conditions require quantification of the effects of CO_2 enrichment on plants, the potential for the biosphere to offer feedback regulation on CO_2 in the atmosphere, and the factors that will modify feedback regulation.

Most biogeochemical models that incorporate future climate scenarios indicate that N limitations could exert considerable constraints on carbon sequestration by terrestrial ecosystems because N is frequently limiting to plant growth and decomposition (McMurtrie and Comins, 1996; Rastetter et al., 1997; Nadelhoffer et al., 1999; Hungate et al., 2003). In some cases, however, experimental increases in atmospheric CO_2 have been shown to stimulate symbiotic N_2 fixation, potentially providing an increased input of N into natural and agroecosystems. On the other hand, other factors such as soil phosphorus, water, and light availability, often limit the productivity of symbiotic N_2-fixing species (Vitousek and Howarth, 1991), and these factors may have important interactive effects on the responses of N_2-fixing species with respect to global environmental change. The objectives of this chapter are to examine the environmental constraints on the response of symbiotic N_2-fixing organisms to increasing CO_2, and to estimate the potential increase in N flux into natural and agroecosystems with a doubling of atmospheric CO_2 concentration.

4.2 N FLUXES BY N_2 FIXATION INTO TERRESTRIAL ECOSYSTEMS

The primary fluxes of N into terrestrial ecosystems can be separated into inputs by atmospheric wet and dry deposition, recycling from detritus, and either biological

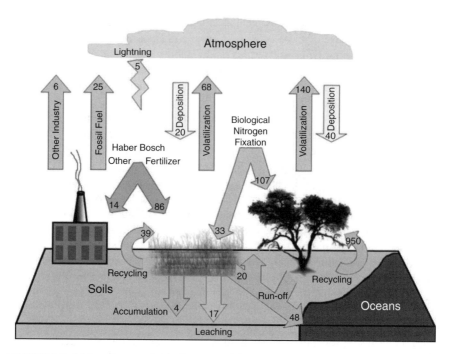

FIGURE 4.1 Major nitrogen fluxes for terrestrial systems with an emphasis on agricultural systems. Numbers are midpoints of estimates from various sources. Units are Tg yr^{-1} (1 Tg = 10^{12}g). The tree represents natural ecosystems and the pasture agroecosystems. The factory represents industrial processes. Volatilization includes denitrification as N$_2$, NO, and N$_2$O, as well as gaseous losses of NH$_3$ and NO$_3$. Sources of volatilization from natural ecosystems include: Biomass burning (40 Tg yr^{-1}), clearing of natural ecosystems (20 Tg yr^{-1}), draining wetlands (10 Tg yr^{-1}), and denitrification (70 Tg yr^{-1}). In agroecosystems, burning crop residue contributes approximately 5 Tg yr^{-1} to the volatile N pool, denitrification provides about 22 Tg yr^{-1}, loss of NH$_3$ from fertilizer about 11 Tg yr^{-1}, from animal, human, or pet waste about 26 Tg yr^{-1}, and direct gas loss from plants another 4 Tg yr^{-1}. Recycled N in agroecosystems includes 12 to 16 Tg yr^{-1} from crop residue and >20 Tg yr^{-1} from manure. Approximately 50 Tg yr^{-1} are removed from agricultural fields as harvest, going to human or animal food, seed, and postharvest processing losses. Some agricultural runoff is denitrified in wetlands and rivers; total N export to oceans by rivers is estimated at 36 Tg yr^{-1}. (From Schlesinger, 1997; Vitousek et al., 1997; Cleveland et al., 1999; Smil, 1999; Galloway and Cowling, 2002; Smil, 2002; Galloway et al., 2003; Galloway et al., 2004.)

or industrial N$_2$ fixation (Figure 4.1). Biological N$_2$ fixation by free-living and symbiotic bacteria occurs in almost every ecosystem, and accounts for ~140 Tg N yr^{-1} entering terrestrial systems, with ~76% of the N added to natural ecosystems, while the remaining goes into agroecosystems (Galloway et al., 2004). Most of this N is produced by symbiotic N$_2$-fixing bacteria because free-living bacteria have limited C supplies and, therefore, are thought to contribute relatively small amounts to the terrestrial N budget (Postgate, 1998). Humans increase N$_2$ fixation through energy and fertilizer production, supplementing N$_2$ fixation in agroecosystems by adding ~86 Tg N yr^{-1}.

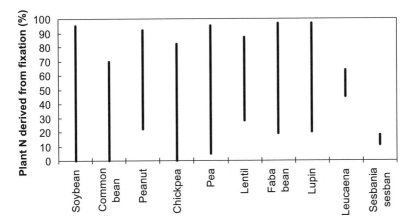

FIGURE 4.2 Range of the percent of whole plant N derived from N_2 fixation measured in field-grown legumes commonly used as crops or in agroforestry (Danso et al., 1992; Unkovich and Pate, 2000).

Legume crops, including dry beans, peanuts, peas, and forage (e.g., alfalfa, clover), fix 25 to 53 Tg N yr^{-1} (Galloway et al., 1995; Smil, 1999), although a large amount of this is removed from agroecosystems through harvests. Legume crops provide approximately 20% of food protein worldwide and are found in all agricultural regions, totaling over 140 million ha in 1997. In addition to food production, legumes are used for green manure cover crops, soil remediation in disturbed sites, and production forestry. Approximately 21 Tg N yr^{-1} of the input into agroecosystems is removed for human dietary consumption and ultimately returns to the environment as losses to soil, water, or atmosphere (Smil, 1999; Smil, 2002).

Different legume species and cultivars vary in their ability to fix N_2 (Table 4.1), ranging from 15 to 325 kg N ha^{-1} yr^{-1}, although environmental and management factors heavily influence rates of N_2 fixation (Hardarson et al., 1987; Hardarson and Atkins, 2003). These factors also strongly influence the amount of N that is supplied to the plant from N_2-fixing bacteria, and the amount of N that plants remove from the soil with their roots. The percentage of plant N supplied from N_2 fixation often averages ~50 to 60% in grain legumes and N_2-fixing trees, but is as high as 70 to 80% in forage legumes (Figure 4.2) (Danso, 1995).

Analysis of changing land cover and N fluxes from 1860 to the early 1990s suggests the conversion of natural systems to agroecosystems reduced N inputs from biological N_2 fixation by ~13 Tg N yr^{-1}, but these losses have been replaced by a net gain of ~3.5 Tg N yr^{-1} from increased use of N_2-fixing crops over the same period (Galloway et al., 2004). In addition, the total amount of N contributed to agroecosystems by legumes is probably increasing because the land area devoted to N_2-fixing species is increasing. For example, soybean production in Brazil has increased from about 1.5 M ha in 1970 to 13 M ha in 2002; in the process, soybean has become Brazil's leading crop in production and made Brazil the second largest soy producer in the world (Alvez et al., 2003).

There are few estimates of the total global N flux contributed by symbiotic N_2 fixation in managed forests and agroforestry. Based on several studies, however, tree species used in agroforestry fix as much or more N_2 than crops, and some species, such as *Gliricidia sepium* or *Leucaena leucocephala*, can surpass the total amount of N_2 fixed by tree species in natural ecosystems (see Table 4.1).

4.3   REGULATION OF BIOLOGICAL N_2 FIXATION

The atmosphere contains the largest global pool of N, but it exists primarily as N_2 gas and is not useful to most organisms until it has been *fixed* or converted into a reduced form either biologically by bacteria or abiotically by lightening or industrial processes. Biological N_2 fixation is a complex process and highly regulated. The nitrogenase enzyme catalyzes the reduction of N_2 to NH_4^+. This is an energetically expensive conversion that becomes less efficient as O_2 concentrations increase in bacterial cells. In symbiotic systems, plants provide carbohydrates to N_2-fixing bacteria and reduce the O_2 in the near vicinity of nitrogenase by forming nodules in root cortex cells to house bacteria. Gutschick (1981) estimated that the process of N_2 fixation requires approximately 22.8 g glucose per g N_2 reduced, divided in roughly equal amounts between the direct energetic costs of N_2 fixation and the costs of construction and maintenance of nodules. Thus, the carbohydrate costs of this symbiosis are substantial for the plant. It is difficult to determine these costs under field conditions because it is hard to separate all of the belowground costs associated with nutrient uptake and carbohydrate supply to mycorrhizal fungi and N_2-fixing bacteria (Chapin et al., 2002). However, several experiments have estimated that as much as 25 to 33% of the carbon fixed in daily photosynthesis by legumes is used by N_2-fixing bacteria in the root nodules (Pate, 1986; Ta et al., 1987; Lambers et al., 1998). In return for the carbohydrate contributed to the N_2-fixing bacteria, the plant receives an N supply for photosynthesis and growth. Leaves of N_2-fixers are rich in N and it is often not fully resorbed by the plant. When leaves are shed during cold or dry seasons, they provide significant N inputs into the ecosystem as they decompose.

Vitousek and Howarth (1991) have suggested that symbiotic N_2-fixing plants pose a paradox that asks why, in a plant world limited by N availability, symbiotic N_2-fixing plants do not have a strong competitive advantage over other plants until soil N becomes nonlimiting. They proposed that the reason for this is that N_2-fixers are strongly limited by abiotic and biotic factors other than N availability. These factors include the availability of resources involved with energy production, such as light and phosphorus (Vitousek and Howarth, 1991), as well as other resources, such as water supply (Serraj et al., 1999a). Pollutants, like tropospheric ozone, also have a negative impact on N_2-fixing plants, particularly agricultural legumes (Morgan et al., 2003). In addition, N_2-fixing plants senesce tissues rich in N and increased N in the soil inhibits N_2 fixation (Hunt and Layzell, 1993). Although N_2-fixing plants usually have abundant N provided by bacteria, they are not good at competing for soil N with nonfixing plants. The abiotic and biotic factors that affect the productivity of symbiotic N_2-fixing plants form the same list of factors that will affect the response of symbiotic N_2-fixing plants as climate change occurs.

Both nodulation and specific nitrogenase activity (nitrogenase activity per unit weight of nodule) respond to many environmental variables, yet there is no consensus as to the exact mechanism of these responses. The source control hypothesis states that the environment limits resource supply for N_2 fixation. For example, light availability affects photosynthesis which in turn affects the amount of carbohydrate available for N_2 fixation in nodules, which in turn affects nodulation and nitrogenase activity. Regulation could stem from low energy supply for N_2 fixation or through a change in the O_2 permeability of nodules because O_2 is required for energy production in the nodule, but must be maintained at low concentrations to prevent inhibition of nitrogenase (Minchin, 1997). The source control hypothesis is the classical ecophysiological perspective, but the observation that current carbohydrate supply has little effect on N_2 fixation (Hartwig et al., 1990; Weisbach et al., 1996) has led to a reassessment of the source control paradigm. Alternately, one of the important recent breakthroughs in understanding the physiology of symbiotic N_2 fixation is the recognition that nodulation and nitrogenase activity are regulated in response to many factors in basically the same way — by a feedback mechanism mediated by the plant's demand for N (Parsons et al., 1993; Hartwig, 1998). This hypothesis states that nodulation and nitrogenase activity are regulated by *sink control*, indicating that N_2 fixation responds to changes in the N demand by plant organs. Evidence points to this feedback being controlled by the soluble N pool in the plant (Serraj et al., 1999a). The N feedback hypothesis was first proposed as the mechanism by which nodulation and nitrogenase activity respond to soil NO_3 supply (Parsons et al., 1993). It has now been hypothesized to be the mechanism by which N_2-fixing plants respond to drought (Serraj et al., 1999a), soil P deficiency (Almeida et al., 2000), changes in plant source-sink balance during seed production (Schulze, 2003), defoliation (Hartwig and Nosberger, 1994; Hartwig and Trommler, 2001), and elevated CO_2 (Luscher et al., 2000; Serraj and Sinclair, 2003).

4.4 EFFECTS OF INCREASED CO_2 ON SYMBIOTIC N_2-FIXING SYSTEMS

Photosynthesis does not operate at maximum efficiency in C_3 photosynthesis because the primary carboxylating enzyme, ribulose bisphoshate carboxylase-oxygenase (rubisco), is not saturated at current CO_2 concentrations, and because the reaction catalyzed by this enzyme is competitively inhibited by O_2 (Zelitch, 1973). As CO_2 partial pressure increases, the ratio of carboxylation to oxygenation by rubisco increases and photorespiration decreases, thus increasing photosynthesis and bio-mass production of plants. Quantitative analyses using experiments conducted with plant species from a wide variety of functional groups and over a wide range of treatments have indicated that the average growth response to a doubling of CO_2 is between 30 and 60% (Cure and Acock, 1986; Poorter, 1993; Curtis, 1996; Koch and Mooney, 1996; Poorter et al., 1996). There is considerable variation in the magnitude of the growth response both between species and among species when environmental factors, such as nutrient supply, are varied (Curtis and Wang, 1998; Poorter, 1998;

Luo and Mooney, 1999). Several possible mechanisms exist, however, whereby CO_2 enrichment may alleviate nutrient limitations to growth:

- Elevated CO_2 may stimulate belowground processes of plants leading to an enhanced ability of plants to acquire nutrients from the soil (Berntson, 1994; Norby, 1994). Elevated CO_2 can alter the architecture, morphology, and size of plant root systems (Berntson and Woodward, 1992; Rogers et al., 1992; Rogers et al., 1994; Berntson and Bazzaz, 1996) as well as the nutrient uptake kinetics of fine roots (Bassirirad et al., 1996a; Bassirirad et al., 1996b; Jackson and Reynolds, 1996).
- Reallocation of nutrients within the plant or increased efficiency of metabolic processes could lower nutrient limitations (Drake et al., 1997).
- Symbiotic root associations with mycorrhizal fungi or N_2-fixing bacteria may be stimulated by CO_2 enrichment and result in increased nutrient supply to the plant (Norby, 1987; Arnone and Gordon, 1990; Thomas et al., 1991; O'Neill, 1994; Godbold and Berntson, 1997; Luscher et al., 2000; Thomas et al., 2000).

The growth response of symbiotic N_2-fixing plants to elevated CO_2 appears to be larger than that of other functional groups (Hunt et al., 1991; Hunt et al., 1993; Poorter, 1993; Luscher et al., 1998; Warwick et al., 1998; Navas et al., 1999; Edwards et al., 2001; Lee et al., 2003b), reflecting the potentially unlimited N source from the atmosphere provided by the N_2-fixing bacteria. The ability of N_2-fixing plants to respond to elevated CO_2 under conditions of low soil N availability increases their competitive capability with nonfixing plants (Poorter and Navas, 2003). A recent analysis of 165 studies that examined the responses of pastures and rangeland to global change indicates that content of legume increases by about 10% in grass-legume swards with a doubling of atmospheric CO_2 concentration (Campbell et al., 2000).

The beneficial effects of CO_2 enrichment are primarily due to increased mass and numbers of nodules, and in a lesser number of studies, stimulation of specific nodule activity, together leading to greater amounts of whole-plant N contents. Elevated CO_2 stimulates carbon allocation to roots and nodules of symbiotic N_2-fixing plants, increasing sugar concentrations in nodules as well as root starch concentrations (Hardy and Havelka, 1976; Phillips et al., 1976; Masterson and Sherwood, 1978; Williams et al., 1981; Finn and Brun, 1982; Allen et al., 1988; Thomas et al., 2000; Cabrerizo et al., 2001). Using $^{14}CO_2$ as a tracer, Tissue et al. (1997) found that as elevated CO_2 stimulates leaf photosynthesis, there was an increase in the rates of C transport to the nodules, and C use by nodules of *Gliricidia sepium,* a common tree species used in agroforestry systems. Elevated CO_2 has also been found to increase the size of bacterial populations in the rhizosphere of N_2-fixing plants (Schortemeyer et al., 1996; Marilley et al., 1999) and stimulate the N_2-fixing activity of rhizosphere bacteria around nonfixing plants in a temperate wetland (Dakora and Keya, 1997). In addition, elevated CO_2 reduced the accumulation of soluble N compounds in soybean plants, thereby reducing the negative impact of these N compounds on N_2-fixing activity (Serraj and Sinclair, 2003).

Thus, CO_2 enrichment positively affects N_2-fixing plants in ways that positively affect N_2 fixation through either of the hypothesized regulatory pathways of N_2 fixation (source control or sink control). Elevated CO_2 increases plant mass thereby increasing the N demand by plant tissues and decreasing the buildup of soluble N compounds that may lead to inhibition of N_2-fixing activity. At the same time, elevated CO_2 stimulates carbohydrate production and allocation to belowground processes, thereby increasing the carbon supply for nodules. Overall, elevated atmospheric CO_2 may be expected to stimulate the growth and N_2 fixation of most symbiotic N_2-fixing plants when grown under optimum environmental conditions. Despite 30 years of research on the effects of elevated CO_2 on plants, however, relatively few studies have attempted to assess interactions between elevated CO_2 and other environmental factors on roots, nodules, and nodule activity of N_2-fixing plants.

4.5 ENVIRONMENTAL FACTORS INFLUENCING CO_2 EFFECTS

4.5.1 Soil N Availability

Agroecosystems and other managed ecosystems are often subjected to additional N inputs through fertilization, even when symbiotic N_2-fixing plants are a component of the system. Soil N availability has a large influence on symbiotic N_2 fixation. N_2-fixing plants often grow in soils that are N limited (Boring et al., 1988). The greatest nodule activity occurs when N in the soil is available in amounts that are sufficient for vigorous growth of young plants before nodules are able to produce N (Minchin et al., 1981). High-N sites usually do not support N_2 fixation because NO_3 or NH_4^+ uptake and assimilation may cost less in terms of carbon. High levels of N may inhibit nodulation, nodule size, or specific nodule activity (reviewed in Streeter, 1988, and see N-sink hypothesis covered previously). Thus, the ability of symbiotic N_2-fixing species to contribute N inputs to ecosystems, as well as the potential for CO_2 stimulation of N inputs, will be dependent on soil N that is already available for plant productivity.

The inhibitory effects of N fertilization on symbiotic N_2 fixation may be offset at least partially by elevated CO_2, and this could have important implications at the plant, community, and ecosystem levels by increasing the duration of N_2-fixing species in successional systems. Several studies report that elevated CO_2 stimulates nodule production (Sa and Israel, 1991; De Luis et al., 1999; Thomas et al., 2000; Lee et al., 2003a; Temperton et al., 2003), and sometimes specific nitrogenase activity (Thomas et al., 2000), even under conditions of high soil N availability. Stable isotope analyses with symbiotic N_2-fixing plants have indicated that the percentage of plant N derived from N_2 fixation increases under CO_2 enrichment, while the percentage of plant N derived from soil N decreases (Soussana and Hartwig, 1996; Zanetti et al., 1996; Zanetti and Hartwig, 1997; Luscher et al., 2000; Thomas et al., 2000; Lee et al., 2003a). This response, however, has not been demonstrated consistently (Vogel et al., 1997; Zanetti et al., 1998; Schortemeyer et al., 1999), even when including field studies conducted in very low nutrient ecosystems (Niklaus et al., 1998).

The addition of N fertilizer typically inhibits N_2 fixation through a reduction in nodule initiation, nodule development, or specific nodule activity. All of these processes could be affected by elevated CO_2. Bacterial infection of root hairs may be affected by CO_2 enrichment by the stimulation of root mass production and numbers of root hairs available for infection. CO_2 enrichment may also affect NO_3 inhibition of specific nitrogenase activity by increasing the carbohydrates available to meet the energy requirements for nodule activity or for reducing the O_2 diffusion resistance into the nodules. If the effect of elevated CO_2 is to increase N demand in a legume due to the large stimulation of biomass production, then N_2 fixation rates would be regulated to meet the N demand by maintaining nodule growth and activity despite conditions such as high substrate N from fertilizer.

4.5.2 SOIL PHOSPHORUS AVAILABILITY

In many ecosystems, growth of symbiotic N_2-fixing plants is phosphorus limited (Ae et al., 1990), and because N availability also has an important effect on symbiotic N_2 fixation, it is currently thought that what controls N_2 fixation is not the overall level of N or P in the soil, but the total N to available P ratio (nutrient supply ratio Guofan and Tingxiu, 1991; Smith, 1992; Crews, 1993). N_2 fixation is known to require large amounts of P (Gutschick, 1980; Dixon and Wheeler, 1983; Hogberg, 1989) and *Rhizobium* nodules often contain two to three times more P per unit of dry weight than the root on which they are formed (Barrea et al., 1988). It has been suggested that the high energetic costs of supporting the rhizobial symbiosis require the uptake of large amounts of P to meet the need for adenine triphosphate (ATP) (Tjepkema and Winship, 1980; Sa and Israel, 1991; Tang et al., 2001). In soybean plants, N_2 fixation is stimulated to a greater degree than host plant growth by P fertilization (Israel, 1987). Phosphorus fertilization often increases N_2 fixation, nodule weight, nitrogenase activity, nodule N and starch content, and plant N content (Gates, 1974; Graham and Rosas, 1979; Singleton et al., 1985; Israel, 1987; Pereira and Bliss, 1987; Reinsvold and Pope, 1987; Sa and Israel, 1991). It is still not clear, however, whether the effect of soil P availability on N_2 fixation is a direct effect on the bacterial capacity to fix N_2 or whether it affects photosynthetic capacity of the host plant, which in turn affects the bacterial symbiont (Hardy and Havelka, 1976).

Soil P will probably severely limit the responses of symbiotic N_2-fixing plants to CO_2 enrichment, and experimental evidence indicates that adequate soil P is required for elevated CO_2 to positively affect symbiotic N_2 fixation. For example, Sa and Israel (1991) found that while soybean plants grown with adequate P responded to elevated CO_2 with greater nodule mass, specific nodule activity, and total plant N, soil P deficiency eliminated the CO_2 responses of N_2 fixation. Similar results have been found in clover grown where elevated CO_2 stimulated photosynthesis, but had no effect on nodulation or nitrogenase activity under severe soil P limitations (Almeida et al., 2000).

The interaction between elevated CO_2 and soil phosphorus supply on symbiotic N_2 fixation may be particularly important in acid tropical soils, in grasslands, and pastures grown on calcareous soils that bind P at high pH (7–8) (e.g., Niklaus et al., 1998), or in areas where large amounts of P are lost as fertilizer runoff (e.g., South Florida). Perhaps the best evidence of the importance of soil P availability in the response of N_2-fixing plants to CO_2 enrichment comes from a study conducted on species from a calcareous grassland in Basel, Switzerland (Leadley and Stocklin, 1996; Leadley et al., 1999), where symbiotic N_2 fixation is highly restricted by low P availability; P fertilization greatly increased the response of N_2-fixing plants to CO_2 enrichment, stimulating total plant N (Niklaus et al., 1998; Stocklin and Korner, 1999).

The effects of elevated CO_2 on mycorrhizal colonization of N_2-fixing plants is also important in the consideration of how P limitations affect the responses on symbiotic N_2 fixation to climate change because mycorrhizal fungi typically stimulate the uptake of phosphorus. For example, it has been shown that the infection of *Robinia pseudoacacia* roots by mycorrhizal fungi was increased under elevated CO_2, concurrent with increases in nodule mass and total plant N and P content (Olesniewicz and Thomas, 1999).

4.5.3 INCREASED TEMPERATURE

In leaves, the specificity of rubisco (its relative affinity for CO_2 vs. O_2) is temperature dependent, decreasing strongly with rising temperature (Long and Drake, 1992). A consequence of this decline in specificity is an increase in photorespiration of C_3 plants with increasing temperature. As the CO_2 concentration increases, however, the stimulation of photosynthesis progressively increases with temperature, and thus it has been suggested that the stimulation of photosynthesis and perhaps growth will be greatest at high temperatures (Long, 1991; Long and Drake, 1992; Drake et al., 1997).

Oxidative respiration used to maintain cellular integrity, build ion gradients, and transport materials is also highly sensitive to temperature, and this sensitivity results in a strong relationship between short-term variation in temperature and the rates of CO_2 evolution by plant tissues (Atkin and Tjoelker, 2003). In addition, the rate of oxidative respiration varies with the supply of carbohydrates, resulting in a linkage to the rate of photosynthesis and to the utilization of carbohydrates by sink tissues (Dewar et al., 1999; Atkin and Tjoelker, 2003).

N_2-fixing plants and their associated strains of N_2-fixing bacteria display a remarkable resilience to temperature regimes of their natural environment and occur in both cold and warm climates (Dixon and Wheeler, 1983). For example, *Alnus incana* as well as other N_2-fixing species grow within the Arctic Circle. At the opposite extreme, *Casuarina* spp. tolerate temperatures of 35°C or more in Australia and Indonesia. The temperature range for host growth is typically broader than the range for optimal N_2 fixation (Mulder et al., 1977). The current success of many legume crops in the warm tropics suggests that future warming may have little effect on N_2 fixation.

Our understanding of the effects of temperature on nodule development comes primarily from studies of a small subset of agricultural legumes. Upper temperature limits for nodulation are higher in tropical than temperate species, although the precise effects vary with species of host plant and with *Rhizobium* strain (Gibson, 1971). High temperatures may affect different stages of nodule development. At 30°C, there was reduced nodulation of peas and beans due to inhibition of root hair function (Lie, 1974). In addition, *Trifolium subterraneum* grown at 30°C showed accelerated degeneration of nodules, thereby reducing the period of N_2-fixing activity (Pankhurst and Gibson, 1973). However, adaptation to higher soil temperatures by developing nodules deeper on the root system has also been observed (Munns et al., 1977).

Nitrogenase activity of legumes shows a broad temperature optima of maximum activity between 20 and 30°C with temperate legumes having optima at the lower end of this range and tropical species at the upper end (Waughman, 1977). *Trifolium repens* showed a nearly linear increase in nitrogenase activity with temperature between 5 and 25°C, but there was no change in temperature response between 25 and 33°C (Ryle et al., 1989). Nitrogenase activity has also been shown to increase with a temperature from 18 to 28°C in *T. repens* and *Medicago sativa* (Crush, 1993). In actinorhizal species, nitrogenase activity may be somewhat more sensitive to temperature. In temperate species (*Alnus glutinosa*, *Myrica gale*, and *Hippophae rhamnoides*), the optima is around 20 to 25°C (Wheeler, 1971; Waughman, 1977; Hensley and Carpenter, 1979). For subtropical *Casuarina* species, optima of 35°C or more have been found (Bond and Mackintosh, 1975). Temperature effects on nitrogenase activity, however, vary with species (Hardy et al., 1968; Waughman, 1977) and with the temperature conditions under which the plants have been grown (Gibson, 1976), a factor often ignored.

A few cursory studies have addressed the interaction of elevated CO_2 and higher temperatures on N_2 fixation, and these generally support the hypothesis that the effects of elevated CO_2 will be greatest at high temperatures (Crush, 1993; Uselman et al., 1999; Lilley et al., 2001). However, more studies are needed that address CO_2 and temperature interactions on N_2 fixation because changes in global temperatures are not independent of increases in atmospheric CO_2.

4.5.4 WATER STRESS

Water stress is a major environmental factor limiting growth and symbiotic N_2 fixation and has been extensively reviewed (Serraj et al., 1999b). N_2 fixation is very sensitive to drought, decreasing with decreasing leaf and nodule water potentials and stomatal conductance (Pankhurst and Gibson, 1973; Patterson et al., 1979; Finn and Brun, 1982). Water deficits reduce nitrogenase activity through reductions in the formation of nodules (Sprent, 1972), and specific nitrogenase activity (Bennett and Albrecht, 1984; Durand et al., 1987; Djekoun and Planchon, 1991). Several mechanisms have been proposed for reductions in N_2-fixing activity with increasing water stress, including reduced photosynthate production (Huang et al., 1975; McNabb et al., 1977; Seiler and Johnson, 1984; Harrington and Seiler, 1988) and increased resistance of O_2 diffusion into nodules (Sprent, 1981; Witty et al., 1986). Current evidence, however, suggests that reductions in N_2 fixation due to water stress may be controlled through a negative feedback by the accumulation of soluble N in the plant (Serraj et al., 1999a).

In experiments combining CO_2 enrichment with drought conditions, it has generally been shown that elevated CO_2 ameliorates the effects of water stress because high CO_2 concentrations cause partial stomatal closure while stimulating photosynthesis, increasing the photosynthetic water use efficiency. Relatively few studies have been conducted that examine the interactive effects of elevated CO_2 and water stress on N_2 fixation. In the few studies that have been conducted with N_2-fixing species, elevated CO_2 at least partially compensated for drought-induced reductions in N_2 fixation by stimulating nodule mass and specific nodule activity (Serraj et al., 1998; De Luis et al., 1999; Serraj et al., 1999a). In soybean, elevated CO_2 stimulated the accumulation of nonstructural carbohydrates while reducing concentrations of soluble nitrogen compounds that might reduce N_2 fixation through negative feedback (Serraj et al., 1998; Serraj and Sinclair, 2003), suggesting that both N_2-fixing plants and the process of N_2 fixation could become more drought tolerant as atmospheric CO_2 continues to increase.

4.5.5 ELEVATED TROPOSPHERIC OZONE

Tropospheric ozone levels have risen more rapidly than atmospheric CO_2 over the last few decades and average summer concentrations in the Northern Hemisphere are expected to continue to increase by 0.5 to 2.5% per year over the next 2 to 3 decades (Runeckles and Krupa, 1994). Ozone negatively impacts plant growth, development, economic yield, biodiversity, and ecosystem function (Bishop and Cook, 1981; Skelly et al., 1983; Materna, 1984; Treshow, 1984; Reich and Amundson, 1985; Winner and Atkinson, 1986; Lefohn et al., 1988; Heagle, 1989; Davison and Barnes, 1998; Turcsanyi et al., 2000). Agricultural loss caused by ozone-induced reduction of plant growth and yield has amounted to an estimated \$3 billion per year in the United States and a 20% decrease in crop productivity in Europe (Adams et al., 1989; IPCC Working Group I, 2001).

Ozone is highly reactive, binding to plasma membranes in plants and altering metabolism. Ozone reactions produce toxic oxygen species harmful to cells including peroxide, superoxide, singlet oxygen, and hydroxyl radicals. The exact mechanism of O_3 on the photosynthetic apparatus is not well understood. However, the effective dose of O_3 causing damage to plants is a function of stomatal opening as well as the O_3 concentration (Rich et al., 1970; Martin et al., 1996; Assmann, 1999). Common responses for many plants subjected to elevated O_3 levels is the induction of accelerated foliar senescence accompanied by a decrease in photosynthesis, chlorophyll, activity and content of rubisco, photosystem II activity, specific leaf mass, total leaf area, and N content per unit leaf area (Lehnherr et al., 1987; Dann and Pell, 1989; Pell et al., 1992; Farage and Long, 1995; Farage, 1996; Reid and Fiscus, 1998; Reid et al., 1998; Reid et al., 1999; Morgan et al., 2003). In addition, O_3 inhibits phloem loading and restricts allocation of newly fixed C to developing organs such as roots and fruits (Oshima et al., 1979; Grantz and Yang, 1996). Pausch et al. (1996a), using stable isotope and leaf gas exchange techniques on soybean, demonstrated that chronic O_3 exposure inhibited carbohydrate translocation from leaves to other organs by reducing phloem loading, possibly without a concomitant reduction in the amount of C fixed. Miller et al. (1998) found that elevated O_3 suppressed leaf and root weight ratios, increased pod weight ratios, and decreased specific leaf weight in soybeans grown in open-top chambers.

Despite a large body of knowledge of the influence of O_3 on the productivity of crop and native plant species, there are few studies that examine the effect of this pollutant or the interaction of O_3 and CO_2 on the process of symbiotic N_2 fixation (Li and Gupta, 1995; Pausch et al., 1996b; Morgan et al., 2003). In a review of O_3 effects on soybean, Morgan et al. (2003) concluded that elevated CO_2 reversed or diminished the detrimental effects of the pollutant on photosynthesis and shoot weight, but only about one-fourth of the studies included addressed root mass, and these showed no improvement with the addition of CO_2. None of the studies addressed nodulation or nitrogenase activity (Morgan et al., 2003). One fairly consistent response of plants to exposure to elevated O_3 is the reduction in biomass allocation to roots and root function (Cooley and Manning, 1987; Reiling and Davison, 1992; Grantz and Yang, 1996). Thus, one might expect a deleterious effect of elevated O_3 on symbiotic N_2 fixation, particularly nodule production and specific nodule activity. If this is true, then reduced N inputs into ecosystems by symbiotic N_2 fixers might be predicted as O_3 continues to increase in the troposphere. Elevated CO_2, on the other hand, may offset this negative impact of O_3 by reducing conductance of stomata, the entry point of O_3 into the plant, and by stimulating carbon transport to roots and nodules. Even if emissions are stabilized, year-to-year variability in weather would still result in high O_3 years, resulting in substantial crop yield loss.

Even though O_3 does not enter the soil and acts exclusively on the shoot, root development and function are also negatively affected by elevated O_3 concentrations. Reduction of root-to-shoot ratio was observed in 17 out of 20 species subjected to elevated levels of O_3 (Cooley and Manning, 1987), and root biomass decreased by an average of 21% in 48 studies of soybean (Morgan et al., 2003). Treatments that restrict carbohydrate supply to nodules, such as enzymatic NO_3 reduction (Vessey et al., 1988a; Vessey et al., 1988b; Walsh, 1990), light deprivation (Minchin et al., 1985), nodule excision (Hunt et al., 1987), defoliation (Hartwig et al., 1987), and stem girdling (Vessey et al., 1988b), increase O_2 diffusion resistance into nodules and subsequently lower nitrogenase activity. Li and Gupta (1995) showed that specific nodule activity was reduced with exposure to elevated O_3 levels. Similarly, reduced photosynthate translocation in soybeans was sufficient to maintain moderate rates of soil N uptake, but not adequate to maintain high rates of energy-expensive N_2 fixation (Pausch et al., 1996b).

A few studies have shown that elevated O_3 produces a short-term enhancement of fine root mass and mycorrhizal infection of some tree species, indicating that some belowground activity might be stimulated instead of inhibited (Gorissen et al., 1991; Qiu et al., 1993; Rantanen et al., 1994). However, these potentially beneficial responses by belowground plant parts to O_3 are probably transient, and long-term effects of elevated O_3 on root growth, mycorrhizal formation, and soil microbial processes are typically negative (Meier et al., 1990; Dighton and Jansen, 1991; Edwards and Kelly, 1992; Morgan et al., 2003).

As single factors, CO_2 and O_3 have opposing effects on carbon dynamics of plants. Elevated CO_2 stimulates photosynthesis and growth of plants. O_3 inhibits photosynthesis and growth. Research to date, however, indicates that there are potentially strong interactions between O_3 and CO_2 levels where increased CO_2 concentrations have resulted in the reduction of O_3 damage. For example, Heagle et al. (1998) and Fiscus et al. (1997) found that elevated CO_2 ameliorated growth and yield suppression of soybean due to O_3. In some cases, the deleterious O_3 effects are completely offset by CO_2 enrichment, but in most cases only partial amelioration of O_3 effects by CO_2 occurs (Mulholland et al., 1997; Miller et al., 1998). The view that elevated CO_2 should reduce the deleterious effects of O_3 is based on the expected decrease in stomatal conductance that CO_2 enrichment produces, which should reduce the flux of O_3 into the leaf (Allen, 1990; Fiscus et al., 1997). In addition, a stimulation of photosynthesis by CO_2 enrichment increases substrate availability for repair and detoxification, and this may offset damage by O_3 (Rao et al., 1995).

The effects of elevated CO_2 and O_3 on roots is not well understood, but Reinert and Ho (1995) found that exposure of soybean plants to elevated O_3 (~7.5 times ambient) suppressed root mass by 37% when compared to an ozone-free environment, but O_3 in combination with a doubling of CO_2 concentration reduced root weight by only 12%. When comparing root mass at ambient O_3 to ~4 times ambient O_3, CO_2 levels between 480 and 600 μL L^{-1} prevented decreases in root mass (Miller et al., 1998). At higher levels of CO_2, root mass increased by 50% even with elevated O_3. On the other hand, the negative effects of O_3 on fine root production and mycorrhizal infection of Scots pine, a nonfixer, were not ameliorated by elevated CO_2 (Perez Soba et al., 1995; Kasurinen et al., 1999). However, if elevated CO_2 offsets the negative impacts on photosynthesis and carbohydrate dynamics of N_2-fixing plants, then elevated CO_2 could offset the deleterious effects of elevated O_3 on symbiotic N_2 fixation.

Interactive effects of elevated CO_2 with elevated O_3 will undoubtedly be mediated by both genetic and environmental factors that influence O_3 impacts on legumes. For example, soybean cultivars vary in O_3 tolerance (Foy et al., 1995; Hardarson and Atkins, 2003), and models suggest that the observed reduction of O_3 damage to water-stressed soybeans results from reduced rates of O_3 uptake (Kobayashi et al., 1993). Still, soybean is among the most sensitive crops to O_3 (Figure 4.3) and benefits of elevated CO_2 may be more obvious in species that are more resistant to O_3 damage.

4.6 EXPECTED EFFECTS ON ECOSYSTEM N AVAILABILITY

In natural ecosystems, elevated CO_2 will have its greatest influence on ecosystem N cycling by affecting immobilization and mineralization fluxes (Thornley and Cannell, 2000). The amount of N in plant, soil organic matter, and microbial pools vastly outweighs the amount of N input by fixation and export by leaching and volatilization (see Figure 4.1). In agroecosystems, N_2 fixation comprises a larger proportion of N flux than natural ecosystems (see Figure 4.1), so changes in N_2 fixation caused by atmospheric CO_2 may have a larger impact on N cycling. Perennial, pasture, and forestry systems with their capacity to store greater amounts of N in biomass and soil organic matter than agroecosystems probably represent an

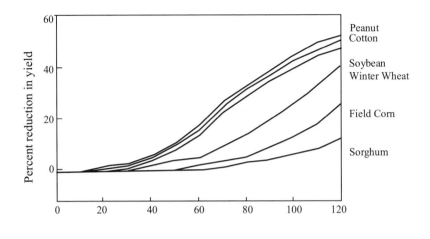

FIGURE 4.3 Estimated decrease in yield from growing season exposure to ozone. (From U.S. Environmental Protection Agency, 1986.)

intermediate level on a gradient between agricultural monocultures and natural systems in regard to the influence of N_2 fixation on N cycling.

Leguminous plants add large quantities of N to many ecosystems, fixing 15 to 200 kg N ha^{-1} yr^{-1} (see Table 4.1), which is an important reason for their use in crop rotation and intercropping agricultural systems. Our previous discussion suggested that in many cases elevated CO_2 and other components of climate change will stimulate N inputs through symbiotic N_2 fixation. In agroecosystems, much of this N remains in plant matter that is harvested, but some N is exported either directly or indirectly to soils (Hardarson et al., 1987). Many studies have examined the physiological responses of individual symbiotic N_2-fixing plants to climate change, but until recently, few experiments have been conducted at an appropriate spatial and temporal scale to examine how elevated CO_2 affects ecosystem processes regulating the N cycle. With the development of FACE (free air CO_2 enrichment) technology (Hendrey, 1992; Miglietta et al., 2001; Okada et al., 2001), it has become possible to increase the concentration of atmospheric carbon dioxide in large plots in intact ecosystems without altering other microclimatic variables and without restricting the movement of animals, including important herbivores. Several FACE experiments contain N_2-fixing plants and are examining how components of climate change affect N inputs into ecosystems by symbiotic N_2-fixing plants. These experiments include SoyFACE (in Urbana, IL), where CO_2 and tropospheric O_3 are used as treatments in a corn–soybean rotation system, and SwissFACE in Eschicon, Switzerland, where CO_2 and fertilizer N are added to a pasture of clover and ryegrass. Perhaps the most complete information of the effects of elevated CO_2 on ecosystem N dynamics comes from SwissFACE because this has been a primary focus since the experiment began in 1993. Care must be used in extrapolating SwissFACE results to other agricultural systems, however, because this is a pasture grown with perennial species at high density, mixed swards have closely spaced plants, and there is no crop rotation. In addition, components of N cycling, such as N mineralization and microbial biomass, can have highly variable responses to elevated CO_2 (Zak et al., 2000). Nevertheless, the clover monocultures at the SwissFACE experiment may be indicative of patterns for other N_2-fixing crops, and mixed swards may represent other intercropped or weedy fields (see below).

N is added to ecosystems through symbiotic N_2 fixation primarily through decomposition of litter and root turnover. One predicted consequence of elevated CO_2 on ecosystems is a decline in litter quality and rates of decomposition, resulting from increased nutrient translocation from mobile pools in plant tissues and higher C:N ratio of litter (Strain and Bazzaz, 1983). Experimental tests of this prediction with a number of species, including N_2-fixing and nonfixing plants, have not con-sistently supported this prediction (Norby et al., 2001). In grass-clover pastures in New Zealand, decomposition rates were increased by elevated CO_2, based on CO_2 production from soils, but there were no consistent effects on soil N (Ross et al., 1995; Ross et al., 1996). At the SwissFACE experiment, N concentration of litter remained steady in grasslands grown with elevated CO_2 because of increased N_2 fixation in the plant community (Soussana and Hartwig, 1996; Zanetti et al., 1996; Zanetti and Hartwig, 1997; Niklaus et al., 1998; Hartwig et al., 2000).

Direct loss of N from roots and nodules of symbiotic N_2-fixing plants has rarely been observed in the field, but these losses may be an important component in N inputs to soil. Direct transfer of fixed N_2 was not measured in alfalfa swards (Russelle et al., 1994) and very little N was found in soil from root exudates of *Alnus incana* (Huss Danell, 1986). Uselman et al. (1999) found that approximately 1% of the N fixed in nodules was excreted by nodules of *Robinia pseudoacacia*, and while there was no effect of elevated CO_2 on exudation, there was increased N exudation from plants grown at higher temperatures. On the other hand, in hydroponic systems, ryegrass has been found to take up NH_4^+ exuded from clover (Paynel et al., 2001; Paynel and Cliquet, 2003), but in natural systems exudates are more likely to be accessed first by microbes and subsequently by plants as microbial populations turn over (Hodge et al., 2000). At the SwissFACE experiment, ryegrass grown in mixtures with clover received 10 to 50% of its N, depending on plant age and seasonal climate, from N_2 fixation by clover (Soussana and Hartwig, 1996; Zanetti and Hartwig, 1997). Mixed swards growing in elevated CO_2 had significantly higher proportions of N transfer (20 to 60%) while addition of N fertilizer decreased N transfer (Zanetti and Hartwig, 1997). The exact source of the N transferred (e.g., exudates, decomposition of nodule, root, or litter) was not determined.

Microbes play an important role in determining the fate of N that enters the soil from fixation (see Figure 4.1). Some N is stored in microbial communities and accumulating humus, some is recycled for plant use and stored in plant tissues, and some is lost to volatilization and leaching. Decomposition of leaves, root, and nodule material in soil, resulting almost entirely from microbial activity, becomes available for plant growth. Annual rates of N_2O volatilization by denitrification increased in SwissFACE fertilizer-enriched ryegrass and mixed ryegrass and clover pastures with elevated CO_2, but there was no change in denitrification in clover monocultures with elevated CO_2 (Baggs et al., 2003a; Baggs et al., 2003b). For all swards, added fertilizer increased denitrification. N_2O loss increased with soil moisture content, but the amount of increase was greater under elevated CO_2 and was also related to the amount of available NO_3 in the system (Baggs et al., 2003b). Complete denitrification to N_2 gas also increased at elevated CO_2 and appears to be a significant sink for fixed N_2 (Baggs et al., 2003b).

Future N cycling in ecosystems will be highly dependent on the effect of elevated CO_2 on soil N pools and microbial communities. When CO_2 was increased experimentally in SwissFACE pastures, soil NO_3 and NH_4^+ pools decreased while soil microbes immobilized more N without decreasing plant N pools (Niklaus et al., 1998). Excess N produced in these systems was sequestered by microbial pools below ground in addition to being transferred to aboveground grass biomass, increasing the size of the N pool in the ecosystem. Microbial population sizes, activity, and community composition increased in clover or mixed pastures under elevated CO_2, but not in ryegrass monocultures (Montealegre et al., 2002), indicating that CO_2 will not only stimulate N_2 fixation, but will also stimulate microbial mediation of N cycling.

The ability of agroecosystems to hold increased levels of N has important implications for future site fertility and habitat quality for aquatic systems that would receive N through leaching and runoff (Vitousek et al., 1997). In some systems, only small amounts of additional N have been leached as N_2 fixation is stimulated (Niklaus et al., 1998). At SwissFACE, leaching was reduced under elevated CO_2 in pure grass or mixed clover-grass fields (Soussana and Hartwig, 1996), but significant increases in N leaching from pure clover fields were measured among all CO_2 treatments (Soussana and Hartwig, 1996). These results suggest that nonfixing plants are required as an N sink to absorb the additional N that will be fixed under future atmospheric CO_2 conditions, and that without this sink, monocultures of N_2-fixing crops may export the additional N from the system. Of course, soil management is important in accumulation rates because agricultural soils frequently incur high rates of erosion and leaching, which export N to waterways (Howarth, 1998). The bottom line, therefore, in the effects of elevated CO_2 and other components of climate change on N cycling in agroecosystems, is that N_2 fixation will generally be stimulated, but without a biotic sink to store additional N (nonfixing plants, higher N content in harvested material, and microbes), much of the added N may be lost to the system by leaching or denitrification.

4.7 SUMMARY

As a single factor, increased CO_2 in the atmosphere stimulates photosynthesis and biomass production of N_2-fixing plants. Thus, elevated CO_2 stimulates N_2 fixation in response to the greater N demand of the growing plant tissues by increasing nodulation, and to a lesser extent, specific nitrogenase activity. Increased N inputs from N_2-fixing plants could stimulate ecosystem productivity and potentially carbon sequestration. However, assumptions that atmospheric CO_2 acts as a single factor are too simplistic because nutrient availability, temperature, moisture, and pollutants will interact with atmospheric CO_2 to influence the responses of N_2-fixing plants. Elevated CO_2 may ameliorate some of the negative effects of some of these environmental factors, such as drought, thereby increasing N inputs into ecosystems. At the other extreme, environmental factors such as soil phosphorus deficiencies may eliminate much of the positive response of N_2-fixing plants to elevated CO_2.

Leguminous plants add large quantities of N to many ecosystems, and the effects of elevated CO_2 on the subsequent flux of added N within a system depends on the plant community present, management regime, and size of preexisting N pools and fluxes. Although elevated CO_2 may stimulate N_2 fixation in natural and agroecosystems, depending on the interaction of other environmental factors, the relative importance in altering the N cycle will vary greatly. In ecosystems with relatively large pools of N in soil organic matter, perennial biomass, and litter, such as natural or production forests, the effects of CO_2 on decomposition and uptake of N may be more important than changes in symbiotic N_2 fixation. Systems at the other end of the spectrum, such as conventionally tilled, single-crop agroecosystems with low pools of biomass N, will see notable increases of N inputs by N_2 fixation. Pastures, perennial crops, and no-till systems will fall somewhere in between. Whether N added by symbiotic N_2 fixation will remain within the ecosystem will depend on the presence of a long-term N sink, such as nonfixing plants.

REFERENCES

Adams, R.M., Glyer, J.D., Johnson, S.L., and McCarl, B.A. (1989) A reassessment of the economic effects of ozone on United States agriculture. *Japca—The Journal of the Air and Waste Management Association*, 39, 960–968.

Ae, N., Arihara, J., Okada, K., Yoshihara, T., and Johansen, C. (1990) Phosphorus uptake by pigeon pea and its role in cropping systems of the Indian subcontinent. *Science*, 248, 477–480.

Allen, L.H. (1990) Plant responses to rising carbon-dioxide and potential interactions with air-pollutants. *Journal of Environmental Quality*, 19, 15–34.

Allen, L.H., Vu, J.C.V., Valle, R.R., Boote, K.J., and Jones, P.H. (1988) Nonstructural carbohydrates and nitrogen of soybean grown under carbon-dioxide enrichment. *Crop Science*, 28, 84–94.

Almeida, J.P.F., Hartwig, U.A., Frehner, M., Nosberger, J., and Luscher, A. (2000) Evidence that P deficiency induces N feedback regulation of symbiotic N_2 fixation in white clover (*Trifolium repens* L.). *Journal of Experimental Botany*, 51, 1289–1297.

Alvez, B.J.R., Boddey, R.M., and Urquiaga, S. (2003) The success of BNF in soybeans in Brazil. *Plant and Soil*, 252, 1–9.

Arnone, J.A. and Gordon, J.C. (1990) Effect of nodulation, nitrogen-fixation and CO_2 enrichment on the physiology, growth and dry mass allocation of seedlings of *Alnus rubra* Bong. *New Phytologist*, 116, 55–66.

Assmann, S.M. (1999) The cellular basis of guard cell sensing of rising CO_2. *Plant Cell and Environment*, 22, 629–637.

Atkin, O.K. and Tjoelker, M.G. (2003) Thermal acclimation and the dynamic response of plant respiration to temperature. *Trends in Plant Science*, 8, 343–351.

Baggs, E.M., Richter, M., Cadisch, G., and Hartwig, U.A. (2003a) Denitrification in grass swards is increased under elevated atmospheric CO_2. *Journal Soil Biology and Biochemistry*, 35, 729–732.

Baggs, E.M., Richter, M., Hartwig, U.A., and Cadisch, G. (2003b) Nitrous oxide emissions from grass swards during the eighth year of elevated atmospheric pCO_2 (SwissFACE). *Global Change Biology*, 9, 1214–1222.

Barrea, J.M., Azcon Aguilar, C., and Azcon, R. (1988). The role of mycorrhiza in improving the establishment and function of the *Rhizobium*-legume system under field conditions. In *Nitrogen Fixation by Legumes in Mediterranean Agriculture*, Beck, D.P. and Materon, L.A., Eds., Dordrecht, Boston, 153–161.

Bassirirad, H., Griffin, K.L., Strain, B.R., and Reynolds, J.F. (1996a) Effects of CO_2 enrichment on growth and root $(NH_4^+)N^{15}$ uptake rate of loblolly pine and ponderosa pine seedlings. *Tree Physiology*, 16, 957–962.

Bassirirad, H., Thomas, R.B., Reynolds, J.F., and Strain, B.R. (1996b) Differential responses of root uptake kinetics of NH_4^+ and NO_3^- to enriched atmospheric CO_2 concentration in field-grown loblolly pine. *Plant Cell and Environment*, 19, 367–371.

Bennett, J.M. and Albrecht, S.L. (1984) Drought and flooding effects on N_2 fixation, water relations, and diffusive resistance of soybean. *Agronomy Journal*, 76, 735–740.

Bernhard Reversat, F. (1996) Nitrogen cycling in tree plantations grown on a poor sandy savanna soil in the Congo. *Applied Soil Ecology*, 4, 161–172.

Berntson, G.M. (1994) Modeling root architecture — Are there tradeoffs between efficiency and potential of resource acquisition. *New Phytologist*, 127, 483–493.

Berntson, G.M. and Bazzaz, F.A. (1996) The allometry of root production and loss in seedlings of *Acer rubrum* (Aceraceae) and *Betula papyrifera* (Betulaceae): implications for root dynamics in elevated CO_2. *American Journal of Botany*, 83, 608–616.

Berntson, G.M. and Woodward, F.I. (1992) The root-system architecture and development of *Senecio vulgaris* in elevated CO_2 and drought. *Functional Ecology*, 6, 324–333.

Binkley, D. (1986) *Forest Nutrition Management*, John Wiley & Sons, New York.

Bishop, J.A. and Cook, L.M., eds. (1981) *Genetic Consequences of Man Made Change*, Academic Press, London.

Bond, G. and Mackintosh, A.H. (1975) Diurnal changes in nitrogen-fixation in root nodules of *Casuarina*. *Proceedings of the Royal Society of London Series B-Biological Sciences*, 192, 1–12.

Boring, L.R. and Swank, W.T. (1984a) Symbiotic nitrogen-fixation in regenerating black locust (*Robinia pseudoacacia* L) stands. *Forest Science*, 30, 528–537.

Boring, L.R. and Swank, W.T. (1984b) The role of black locust (*Robinia pseudoacacia*) in forest succession. *Journal of Ecology*, 72, 749–766.

Boring, L.R., Swank, W.T., Waide, J.B., and Henderson, G.S. (1988) Sources, fates, and impacts of nitrogen inputs to terrestrial ecosystems: review and synthesis. *Biogeochemistry*, 6, 119–159.

Cabrerizo, P.M., Gonzalez, E.M., Aparicio-Tejo, P.M., and Arrese-Igor, C. (2001) Continuous CO_2 enrichment leads to increased nodule biomass, carbon availability to nodules and activity of carbon-metabolising enzymes but does not enhance specific nitrogen fixation in pea. *Physiologia Plantarum*, 113, 33–40.

Campbell, B.D., Stafford Smith, D.M., and GCTE Pastures and Rangelands Network Members (2000) A synthesis of recent global change research on pasture and rangeland production: reduced uncertainties and their management implications. *Agriculture Ecosystems and Environment*, 82, 39–55.

Chapin, F.S., Matson, P.A., and Mooney, H.A. (2002) *Principles of Terrestrial Ecosystem Ecology*, Springer, New York.

Cleveland, C.C., Townsend, A.R., Schimel, D.S., Fisher, H., Howarth, R.W., Hedin, L.O., Perakis, S.S., Latty, E.F., Von Fischer, J.C., Elseroad, A., and Wasson, M.F. (1999) Global patterns of terrestrial biological nitrogen (N_2) fixation in natural ecosystems. *Global Biogeochemical Cycles*, 13, 623–645.

Cooley, D.R. and Manning, W.J. (1987) The impact of ozone on assimilate partitioning in plants — a review. *Environmental Pollution*, 47, 95–113.

Crews, T.E. (1993) Phosphorus regulation of nitrogen-fixation in a traditional Mexican agro-ecosystem. *Biogeochemistry*, 21, 141–166.

Crush, J.R. (1993) Hydrogen evolution from root-nodules of *Trifolium repens* and *Medicago sativa* plants grown under elevated atmospheric CO_2. *New Zealand Journal of Agricultural Research*, 36, 177–183.

Cure, J.D. and Acock, B. (1986) Crop responses to carbon-dioxide doubling — a literature survey. *Agricultural and Forest Meteorology*, 38, 127–145.

Curtis, P.S. (1996) A meta-analysis of leaf gas exchange and nitrogen in trees grown under elevated carbon dioxide. *Plant Cell and Environment*, 19, 127–137.

Curtis, P.S. and Wang, X.Z. (1998) A meta-analysis of elevated CO_2 effects on woody plant mass, form, and physiology. *Oecologia*, 113, 299–313.

Dakora, F.D. and Keya, S.O. (1997) Contribution of legume nitrogen fixation to sustainable agriculture in Sub-Saharan Africa. *Journal Soil Biology and Biochemistry*, 29, 809–817.

Dann, M.S. and Pell, E.J. (1989) Decline of activity and quantity of ribulose bisphosphate carboxylase oxygenase and net photosynthesis in ozone-treated potato foliage. *Plant Physiology*, 91, 427–432.

Danso, S.K.A. (1995). Sustainable agriculture: the role of biological nitrogen fixing plants. In *Nuclear Techniques in Soil-Plant Studies for Sustainable Agriculture and Environmental Preservation,* (International Atomic Energy Agency, Ed., IAEA, Vienna, 205–224.

Danso, S.K.A., Bowen, G.D., and Sanginga, N. (1992) Biological nitrogen-fixation in trees in agroecosystems. *Plant and Soil*, 141, 177–196.

Davison, A.W. and Barnes, J.D. (1998) Effects of ozone on wild plants. *New Phytologist*, 139, 135–151.

De Luis, I., Irigoyen, J.J., and Sánchez-Díaz, M. (1999) Elevated CO_2 enhances plant growth in droughted N_2-fixing alfalfa without improving water status. *Physiologia Plantarum*, 107, 84–89.

Dewar, R.C., Medlyn, B.E., and McMurtrie, R.E. (1999) Acclimation of the respiration photosynthesis ratio to temperature: insights from a model. *Global Change Biology*, 5, 615–622.

Dighton, J. and Jansen, A.E. (1991) Atmospheric pollutants and ectomycorrhizae — More questions than answers. *Environmental Pollution*, 73, 179–204.

Dixon, R.O.D. and Wheeler, C.T. (1983). Biochemical, physiological and environmental aspects of symbiotic nitrogen fixation. In *Biological Nitrogen Fixation in Forest Ecosystems: Foundations and Applications,* Gordon, J.C. and Wheeler, C.T., Eds., Nijhoff/Junk Publishers, The Hague, 107–171.

Djekoun, A. and Planchon, C. (1991) Water status effect on dinitrogen fixation and photosynthesis in soybean. *Agronomy Journal*, 83, 316–322.

Drake, B.G., Gonzàlez-Meler, M.A., and Long, S.P. (1997) More efficient plants: a consequence of rising atmospheric CO_2? *Annual Review of Plant Physiology and Plant Molecular Biology*, 48, 609–639.

Durand, J.L., Sheehy, J.E., and Minchin, F.R. (1987) Nitrogenase activity, photosynthesis and nodule water potential in soybean plants experiencing water-deprivation. *Journal of Experimental Botany*, 38, 311–321.

Edwards, G.S. and Kelly, J.M. (1992) Ectomycorrhizal colonization of loblolly-pine seedlings during 3 growing seasons in response to ozone, acidic precipitation, and soil Mg status. *Environmental Pollution*, 76, 71–77.

Edwards, G.R., Newton, P.C.D., Tilbrook, J.C., and Clark, H. (2001) Seedling performance of pasture species under elevated CO_2. *New Phytologist*, 150, 359–369.

Farage, P.K. (1996) The effect of ozone fumigation over one season on photosynthetic processes of *Quercus robur* seedlings. *New Phytologist*, 134, 279–285.

Farage, P.K. and Long, S.P. (1995) An in-vivo analysis of photosynthesis during short-term O_3 exposure in 3 contrasting species. *Photosynthesis Research*, 43, 11–18.

Finn, G.A. and Brun, W.A. (1982) Effect of atmospheric CO_2 enrichment on growth, non-structural carbohydrate content, and root nodule activity in soybean. *Plant Physiology*, 69, 327–331.

Fiscus, E.L., Reid, C.D., Miller, J.E., and Heagle, A.S. (1997) Elevated CO_2 reduces O_3 flux and O_3-induced yield losses in soybeans: Possible implications for elevated CO_2 studies. *Journal of Experimental Botany*, 48, 307–313.

Foy, C.D., Lee, E.H., Rowland, R.A., and Devine, T.E. (1995) Ozone tolerances of soybean cultivars and near-isogenic lines in a fumigation chamber. *Journal of Plant Nutrition*, 18, 649–667.

Galloway, J.N., Aber, J.D., Erisman, J.W., Seitzinger, S.P., Howarth, R.W., Cowling, E.B., and Cosby, B.J. (2003) The nitrogen cascade. *Bioscience*, 53, 341–356.

Galloway, J.N. and Cowling, E.B. (2002) Reactive nitrogen and the world: 200 years of change. *Ambio*, 31, 64–71.

Galloway, J.N., Dentener, F.J., Capone, D.G., Boyer, E.W., Howarth, R.W., Sietzinger, S.P., Asner, G.P., Cleveland, C.C., Green, P., Holland, E., Karl, D.M., Michaels, A.F., Porter, J.H., Townsend, A., and Vorosmarty, C. (2004) Nitrogen cycles: Past, present and future. *Biogeochemistry*, 70, 153–226.

Galloway, J.N., Schlesinger, W.H., Levy, H., Michaels, A., and Schnoor, J.L. (1995) Nitrogen fixation: Anthropogenic enhancement-environmental response. *Global Biogeochemical Cycles*, 9, 235–252.

Gates, C.T. (1974) Nodule and plant development in *Stylosanthes humilis* Hbk — Symbiotic response to phosphorus and sulfur. *Australian Journal of Botany*, 22, 45–55.

Gibson, A.H. (1971) Factors in physical and biological environment affecting nodulation and nitrogen fixation by legumes. *Plant and Soil*, special volume, 139–152.

Gibson, A.H. (1976). Recovery and compensation by nodulated legumes to environmental stress. In *Symbiotic Nitrogen Fixation in Plants,* (Nutman, P.S., Ed.), Cambridge University Press, London.

Godbold, D.L. and Berntson, G.M. (1997) Elevated atmospheric CO2 concentration changes ectomycorrhizal morphotype assemblages in Betula papyrifera. *Tree Physiology*, 17, 347–350.

Gorissen, A., Joosten, N.N., and Jansen, A.E. (1991) Effects of ozone and ammonium-sulfate on carbon partitioning to mycorrhizal roots of juvenile douglas-fir. *New Phytologist*, 119, 243–250.

Graham, P.H. and Rosas, J.C. (1979) Phosphorus fertilization and symbiotic nitrogen-fixation in common bean. *Agronomy Journal*, 71, 925–926.

Grantz, D.A. and Yang, S.D. (1996) Effect of O_3 on hydraulic architecture in pima cotton. *Plant Physiology*, 112, 1649–1657.

Guofan, L. and Tingxiu, D. (1991) Mathematical model of the relationship between nitrogen fixation by the black locust and soil conditions. *Journal Soil Biology and Biochemistry*, 23, 1–7.

Gutschick, V.P. (1980). Energy flow in the nitrogen cycle, especially in fixation. In *Nitrogen Fixation,* Newton, W.E. and Orme-Johnson, W.H., Eds., Vol. 1, University Park Press, Baltimore, MD, 17–27.

Gutschick, V.P. (1981) Evolved strategies in nitrogen acquisition by plants. *American Naturalist*, 188, 607–637.

Hardarson, G. and Atkins, C. (2003) Optimising biological N$_2$ fixation by legumes in farming systems. *Plant and Soil*, 252, 41–54.

Hardarson, G., Danso, S.K.A., and Zapata, F. (1987). Biological nitrogen fixation in field crops. In *Handbook of Plant Science in Agriculture,* (Christie, B.R., Ed.), CRC Press, Boca Raton, FL, 165–192.

Hardy, R.W.F. and Havelka, U.D. (1976). Photosynthate as a major factor limiting nitrogen fixation by field-grown legumes with emphasis on soybeans. In *Symbiotic Nitrogen Fixation in Plants,* Nutman, P.S., Ed., Cambridge University Press, Cambridge, 421–439.

Hardy, R.W.F., Holsten, R.D., Jackson, E.K., and Burns, R.C. (1968) The acetylene-ethylene assay for N$_2$ fixation: Laboratory and field evaluation. *Plant Physiology*, 43, 1185–1207.

Harrington, J.T. and Seiler, J.R. (1988) Acetylene-reduction in black alder seedlings as affected by direct and indirect moisture deficits using a split-pot growing system. *Environmental and Experimental Botany*, 28, 225–230.

Hartwig, U.A. (1998) The regulation of symbiotic N$_2$ fixation: a conceptual model of N feedback from the ecosystem to the gene expression level. *Perspectives in Plant Ecology, Evolution, and Systematics*, 1, 92–120.

Hartwig, U., Boller, B., and Nosberger, J. (1987) Oxygen-supply limits nitrogenase activity of clover nodules after defoliation. *Annals of Botany*, 59, 285–291.

Hartwig, U., Boller, B.C., Baurhoch, B., and Nosberger, J. (1990) The influence of carbohydrate reserves on the response of nodulated white clover to defoliation. *Annals of Botany*, 65, 97–105.

Hartwig, U.A., Luscher, A., Daepp, M., Blum, H., Soussana, J.F., and Nosberger, J. (2000) Due to symbiotic N$_2$ fixation, five years of elevated atmospheric pCO$_2$ had no effect on the N concentration of plant litter in fertile, mixed grassland. *Plant and Soil*, 224, 43–50.

Hartwig, U.A. and Nosberger, J. (1994) What triggers the regulation of nitrogenase activity in forage legume nodules after defoliation. *Plant and Soil*, 161, 109–114.

Hartwig, U.A. and Trommler, J. (2001) Increase in the concentrations of amino acids in the vascular tissue of white clover and white lupin after defoliation: an indication of a N feedback regulation of symbiotic N$_2$ fixation. *Agronomie*, 21, 615–620.

Heagle, A.S. (1989) Ozone and crop yield. *Annual Review of Phytopathology*, 27, 397–423.

Heagle, A.S., Miller, J.E., and Pursley, W.A. (1998) Influence of ozone stress on soybean response to carbon dioxide enrichment: III. Yield and seed quality. *Crop Science*, 38, 128–134.

Hendrey, G.R. (1992) FACE — free-air CO$_2$ enrichment for plant research in the field — Introduction. *Critical Reviews in Plant Sciences*, 11, 59–60.

Hensley, D.L. and Carpenter, P.L. (1979) Effect of temperature on N$_2$ fixation (C$_2$H$_2$ reduction) by nodules of legume and actinomycete-nodulated woody species. *Botanical Gazette*, 140, S58-S64.

Hodge, A., Robinson, D., and Fitter, A. (2000) Are microorganisms more effective than plants at competing for nitrogen? *Trends in Plant Science*, 5, 304–308.

Hogberg, P. (1989) Soil nutrient availability, root symbioses and tree species composition in tropical Africa: a review. *Journal of Tropical Ecology*, 2, 359–372.

Howarth, R.W. (1998) An assessment of human influences on fluxes of nitrogen from the terrestrial landscape to the estuaries and continental shelves of the North Atlantic Ocean. *Nutrient Cycling in Agroecosystems*, 52, 213–223.

Huang, C.Y., Boyer, J.S., and Vanderhoef, L.N. (1975) Limitation of acetylene-reduction (nitrogen-fixation) by photosynthesis in soybean having low water potentials. *Plant Physiology*, 56, 228–232.

Hungate, B.A., Dukes, J.S., Shaw, M.R., Luo, Y., and Field, C.B. (2003) Nitrogen and climate change. *Science*, 302, 1512–1513.

Hunt, R., Hand, D.W., Hannah, M.A., and Neal, A.M. (1991) Response to CO_2 enrichment in 27 herbaceous species. *Functional Ecology*, 5, 410–421.

Hunt, R., Hand, D.W., Hannah, M.A., and Neal, A.M. (1993) Further responses to CO_2 enrichment in British herbaceous species. *Functional Ecology*, 7, 661–668.

Hunt, S., King, B.J., Canvin, D.T., and Layzell, D.B. (1987) Steady and nonsteady state gas-exchange characteristics of soybean nodules in relation to the oxygen diffusion barrier. *Plant Physiology*, 84, 164–172.

Hunt, S. and Layzell, D.B. (1993) Gas-Exchange of Legume Nodules and the Regulation of Nitrogenase Activity. *Annual Review of Plant Physiology and Plant Molecular Biology*, 44, 483–511.

Huss Danell, K. (1986) Nitrogen in shoot litter, root litter and root exudates from nitrogen-fixing *Alnus incana*. *Plant and Soil*, 91, 43–49.

IPCC (2001) *Climate Change 2001*, Cambridge University Press, Cambridge, U.K.

IPCC Working Group I (2001). Atmospheric chemistry and greenhouse gases. In *Climate Change 2001: The Scientific Basis,* Houghton, J.T., Ding, Y., Griggs, D.J., Noguer, M., van der Linden, P.J., Da, X., Maskell, K. and Johnson, C.A., Eds., Cambridge University Press, Cambridge, U.K., 239–288.

Israel, D.W. (1987) Investigation of the role of phosphorus in symbiotic dinitrogen fixation. *Plant Physiology*, 84, 835–840.

Jackson, R.B. and Reynolds, H.L. (1996) Nitrate and ammonium uptake for single- and mixed-species communities grown at elevated CO_2. *Oecologia*, 105, 74–80.

Kasurinen, A., Helmisaari, H.S., and Holopainen, T. (1999) The influence of elevated CO_2 and O_3 on fine roots and mycorrhizas of naturally growing young Scots pine trees during three exposure years. *Global Change Biology*, 5, 771–780.

Kobayashi, K., Miller, J.E., Flagler, R.B., and Heck, W.W. (1993) Model analysis of interactive effects of ozone and water-stress on the yield of soybean. *Environmental Pollution*, 82, 39–45.

Koch, G.W. and Mooney, H.A. (1996). The response of terrestrial ecosystems to elevated CO_2: a synthesis and summary. In *Carbon Dioxide and Terrestrial Ecosystems,* Mooney, H.A. and Koch, G.W., Eds., Academic Press, San Diego, 415–429.

Lambers, H., Chapin, F.S., and Pons, T.L. (1998) *Plant Physiological Ecology* Springer, New York.

Leadley, P.W., Niklaus, P.A., Stocker, R., and Korner, C. (1999) A field study of the effects of elevated CO_2 on plant biomass and community structure in a calcareous grassland. *Oecologia*, 118, 39–49.

Leadley, P.W. and Stocklin, J. (1996) Effects of elevated CO_2 on model calcareous grasslands: Community, species, and genotype level responses. *Global Change Biology*, 2, 389–397.

Lee, J.A., Harmer, R., and Ignaciuk, R. (1983). Nitrogen as a limiting factor in plant communities. In *Nitrogen as an Ecological Factor,* Lee, J.A., McNeill, S. and Rorison, S., Eds., Blackwell Scientific, Oxford, U.K., 95–112.

Lee, T.D., Reich, P.B., and Tjoelker, M.G. (2003a) Legume presence increases photosynthesis and N concentrations of co-occurring non-fixers but does not modulate their responsiveness to carbon dioxide enrichment. *Oecologia,* 137, 22–31.

Lee, T.D., Tjoelker, M.G., Reich, P.B., and Russelle, M.P. (2003b) Contrasting growth response of an N_2 fixing and non-fixing forb to elevated CO_2: dependence on soil N supply. *Plant and Soil,* 255, 475–486.

Lefohn, A.S., Laurence, J.A., and Kohut, R.J. (1988) A comparison of indexes that describe the relationship between exposure to ozone and reduction in the yield of agricultural crops. *Atmospheric Environment,* 22, 1229–1240.

Lehnherr, B., Grandjean, A., Machler, F., and Fuhrer, J. (1987) The effect of ozone in ambient air on ribulosebisphosphate carboxylase oxygenase activity decreases photosynthesis and grain-yield in wheat. *Journal of Plant Physiology,* 130, 189–200.

Li, Y.C. and Gupta, G. (1995) Physiological-changes in soybean treated with ozone and molybdenum. *Communications in Soil Science and Plant Analysis,* 26, 1649–1658.

Lie, T.A. (1974). Environmental effects on nodulation and symbiotic nitrogen fixation. In *The Biology of Nitrogen Fixation,* Quispel, A., Ed., Amsterdam Press, Amsterdam, 555–582.

Lilley, J.M., Bolger, T.P., Peoples, M.B., and Gifford, R.M. (2001) Nutritive value and the nitrogen dynamics of *Trifolium subterraneum* and *Phalaris aquatica* under warmer, high CO_2 conditions. *New Phytologist,* 150, 385–395.

Long, S.P. (1991) Modification of the response of photosynthetic productivity to rising temperature by atmospheric CO_2 concentrations — Has its importance been underestimated. *Plant Cell and Environment,* 14, 729–739.

Long, S.P. and Drake, B.G. (1992). Photosynthetic CO_2 assimilation and rising atmospheric CO_2 concentrations. In *Topics in Photosynthesis. Crop Photosynthesis: Spatial and Temporal Determinants,* Thomas, H. and Baker, N.R., Eds., Vol. 2, Elsevier, Amsterdam, 69–107.

Luken, J.O. and Fonda, R.W. (1983) Nitrogen accumulation in a chronosequence of red alder communities along the Hoh River, Olympic National Park, Washington. *Canadian Journal of Forest Research — Revue Canadienne De Recherche Forestiere,* 13, 1228–1237.

Luo, Y. and Mooney, H.A. (1999) *Carbon Dioxide and Environmental Stress* Academic Press, San Diego.

Luscher, A., Hartwig, U.A., Suter, D., and Nosberger, J. (2000) Direct evidence that symbiotic N_2 fixation in fertile grassland is an important trait for a strong response of plants to elevated atmospheric CO_2. *Global Change Biology,* 6, 655–662.

Luscher, A., Hendrey, G.R., and Nosberger, J. (1998) Long-term responsiveness to free air CO_2 enrichment of functional types, species and genotypes of plants from fertile permanent grassland. *Oecologia,* 113, 37–45.

Marilley, L., Hartwig, U.A., and Aragno, M. (1999) Influence of an elevated atmospheric CO_2 content on soil and rhizosphere bacterial communities beneath *Lolium perenne* and *Trifolium repens* under field conditions. *Microbial Ecology,* 38, 39–49.

Martin, M.J., Humphries, S.W., Farage, P.K., McKee, I.F., and Long, S.P. (1996). A mechanistic model for the prediction of the effects of rising tropospheric ozone concentration on wheat photosynthesis. In *Photosynthesis: from Light to Biosphere,* (Mathis, P., Ed.), Vol. 5, Kluwer Academic Publishers, New York, 829–832.

Masterson, C.L. and Sherwood, M.T. (1978) Some effects of increased atmospheric carbon-dioxide on white clover (*Trifolium repens*) and pea (*Pisum sativum*). *Plant and Soil*, 49, 421–426.

Materna, J. (1984). Impact of atmospheric pollution on natural ecosystems. In *Air Pollution and Plant Life*, Treshow, M., Ed., John Wiley & Sons, Chichester, U.K., 397–416.

McMurtrie, R.E. and Comins, H.N. (1996) The temporal response of forest ecosystems to doubled atmospheric CO_2 concentration. *Global Change Biology*, 2, 49–57.

McNabb, D.H., Geist, J.M., and Youngberg, C.T. (1977) Nitrogen fixation by *Ceanothus velutinus* in northeastern Oregon. *Agronomy Abstracts*, 127–128.

Meier, S., Grand, L.F., Schoeneberger, M.M., Reinert, R.A., and Bruck, R.I. (1990) Growth, ectomycorrhizae and nonstructural carbohydrates of loblolly-pine seedlings exposed to ozone and soil-water deficit. *Environmental Pollution*, 64, 11–27.

Miglietta, F., Peressotti, A., Vaccari, F.P., Zaldei, A., deAngelis, P., and Scarascia-Mugnozza, G. (2001) Free-air CO_2 enrichment (FACE) of a poplar plantation: the POPFACE fumigation system. *New Phytologist*, 150, 465–476.

Mikola, P., Uomala, P., and Malkonen, E. (1983). Application of biological nitrogen fixation in European silviculture. In *Biological Nitrogen Fixation in Forest Ecosystems: Foundations and Applications,* Gordon, J.C. and Wheeler, C.T., Eds., Nijhoff/Junk, The Hague, 279–294.

Miller, J.E., Heagle, A.S., and Pursley, W.A. (1998) Influence of ozone stress on soybean response to carbon dioxide enrichment: II. Biomass and development. *Crop Science*, 38, 122–128.

Minchin, F.R. (1997) Regulation of oxygen diffusion in legume nodules. *Journal Soil Biology and Biochemistry*, 29, 881–888.

Minchin, F.R., Sheehy, J.E., Minguez, M.I., and Witty, J.F. (1985) Characterization of the resistance to oxygen diffusion in legume nodules. *Annals of Botany*, 55, 53–60.

Minchin, F.R., Summerfield, R.J., and Neves, M.C.P. (1981) Nitrogen nutrition of cowpeas (*Vigna unguiculata*): Effects of timing of inorganic nitrogen applications on nodulation, plant-growth and seed yield. *Tropical Agriculture*, 58, 1–12.

Montealegre, C.M., van Kessel, C., Russelle, M.P., and Sadowsky, M.J. (2002) Changes in microbial activity and composition in a pasture ecosystem exposed to elevated atmospheric carbon dioxide. *Plant and Soil*, 243, 197–207.

Morgan, P.B., Ainsworth, E.A., and Long, S.P. (2003) How does elevated ozone impact soybean? A meta-analysis of photosynthesis, growth, and yield. *Plant, Cell and Environment*, 26, 1317–1328.

Mulder, E.G., Lie, T.A., and Houwers, A. (1977). The importance of legumes under temperate conditions. In *A Treatise on Dinitrogen Fixation. IV. Agronomy and Ecology,* Hardy, R.W.F. and Gibson, A.H., Eds., John Wiley & Sons, New York, 221–242.

Mulholland, B.J., Craigon, J., Black, C.R., Colls, J.J., Atherton, J., and Landon, G. (1997) Effects of elevated carbon dioxide and ozone on the growth and yield of spring wheat (*Triticum aestivum* L). *Journal of Experimental Botany*, 48, 113–122.

Munns, D.N., Fogle, V.W., and Hallock, B.G. (1977) Alfalfa root nodule distribution and inhibition of nitrogen-fixation by heat. *Agronomy Journal*, 69, 377–380.

Nadelhoffer, K.J., Emmett, B.A., Gunderson, P., Kjønaas, O.J., Koopmans, C.J., Schleppi, P., Tietema, A., and Wright, R.F. (1999) Nitrogen deposition makes a minor contribution to carbon sequestration in temperate forests. *Nature*, 398, 145–148.

Navas, M.L., Garnier, E., Austin, M.P., and Gifford, R.M. (1999) Effect of competition on the responses of grasses and legumes to elevated atmospheric CO_2 along a nitrogen gradient: differences between isolated plants, monocultures and multi-species mixtures. *New Phytologist*, 143, 323–331.

Niklaus, P.A., Leadley, P.W., Stocklin, J., and Korner, C. (1998) Nutrient relations in calcareous grassland under elevated CO_2. *Oecologia*, 116, 67–75.

Norby, R.J. (1987) Nodulation and nitrogenase activity in nitrogen-fixing woody-plants stimulated by CO_2 enrichment of the atmosphere. *Physiologia Plantarum*, 71, 77–82.

Norby, R.J. (1994) Issues and perspectives for investigating root responses to elevated atmospheric carbon-dioxide. *Plant and Soil*, 165, 9–20.

Norby, R.J., Cotrufo, M.F., Ineson, P., O'Neill, E.G., and Canadell, J.G. (2001) Elevated CO_2, litter chemistry, and decomposition: a synthesis. *Oecologia*, 127, 153–165.

O'Neill, E.G. (1994) Responses of soil biota to elevated atmospheric carbon-dioxide. *Plant and Soil*, 165, 55–65.

Okada, M., Lieffering, M., Nakamura, H., Yoshimoto, M., Kim, H.Y., and Kobayashi, K. (2001) Free-air CO_2 enrichment (FACE) using pure CO_2 injection: system description. *New Phytologist*, 150, 251–260.

Olesniewicz, K.S. and Thomas, R.B. (1999) Effects of mycorrhizal colonization on biomass production and nitrogen fixation of black locust (Robinia pseudoacacia) seedlings grown under elevated atmospheric carbon dioxide. *New Phytologist*, 142, 133–140.

Oshima, R.J., Braegelmann, P.K., Flagler, R.B., and Teso, R.R. (1979) Effects of ozone on the growth, yield, and partitioning of dry-matter in cotton. *Journal of Environmental Quality*, 8, 474–479.

Pankhurst, C.E. and Gibson, A.H. (1973) Rhizobium strain influence on disruption of clover nodule development at high root temperature. *Journal of General Microbiology*, 74, 219–231.

Parsons, R., Stanforth, A., Raven, J.A., and Sprent, J.I. (1993) Nodule growth and activity may be regulated by a feedback mechanism involving phloem nitrogen. *Plant Cell and Environment*, 16, 125–136.

Pate, J.S. (1986). Economy of symbiotic N fixation. In *On the Economy of Plant Form and Function,* (Givnish, T.J., Ed., Cambridge University Press, Cambridge, U.K., 299–325.

Patterson, R.P., Raper, C.D., and Gross, H.D. (1979) Growth and specific nodule activity of soybean during application and recovery of a leaf moisture stress. *Plant Physiology*, 64, 551–556.

Pausch, R.C., Mulchi, C.L., Lee, E.H., Forseth, I.N., and Slaughter, L.H. (1996a) Use of ^{13}C and ^{15}N isotopes to investigate O_3 effects on C and N metabolism in soybeans .1. C fixation and translocation. *Agriculture Ecosystems and Environment*, 59, 69–80.

Pausch, R.C., Mulchi, C.L., Lee, E.H., and Meisinger, J.J. (1996b) Use of ^{13}C and ^{15}N isotopes to investigate O_3 effects on C and N metabolism in soybeans .2. Nitrogen uptake, fixation, and partitioning. *Agriculture Ecosystems and Environment*, 60, 61–69.

Paynel, F. and Cliquet, J.B. (2003) N transfer from white clover to perennial ryegrass, via exudation of nitrogenous compounds. *Agronomie*, 23, 503–510.

Paynel, F., Murray, P.J., and Cliquet, J.B. (2001) Root exudates: a pathway for short-term N transfer from clover and ryegrass. *Plant and Soil*, 229, 235–243.

Pell, E.J., Eckardt, N., and Enyedi, A.J. (1992) Timing of ozone stress and resulting status of ribulose bisphosphate carboxylase oxygenase and associated net photosynthesis. *New Phytologist*, 120, 397–405.

Pereira, P.A.A. and Bliss, F.A. (1987) Nitrogen-fixation and plant-growth of common bean (*Phaseolus vulgaris* L) at different levels of phosphorus availability. *Plant and Soil*, 104, 79–84.

Perez Soba, M., Dueck, T.A., Puppi, G., and Kuiper, P.J.C. (1995) Interactions of elevated CO_2, NH_3 and O_3 on mycorrhizal infection, gas-exchange and N-metabolism in saplings of Scots pine. *Plant and Soil*, 176, 107–116.

Permar, T.A. and Fisher, R.F. (1983) Nitrogen-fixation and accretion by wax myrtle (*Myrica cerifera*) in slash pine (*Pinus elliottii*) plantations. *Forest Ecology and Management*, 5, 39–46.

Phillips, D.A., Newell, K.D., Hassell, S.A., and Felling, C.E. (1976) Effect of CO_2 enrichment on root nodule development and symbiotic N_2 reduction in *Pisum sativum* L. *American Journal of Botany*, 63, 356–362.

Poorter, H. (1993) Interspecific Variation in the Growth-Response of Plants to an Elevated Ambient CO_2 Concentration. *Vegetatio*, 104, 77–97.

Poorter, H. (1998) Do slow-growing species and nutrient-stressed plants respond relatively strongly to elevated CO2? *Global Change Biology*, 4, 693–697.

Poorter, H. and Navas, M.L. (2003) Plant growth and competition at elevated CO_2: on winners, losers and functional groups. *New Phytologist*, 157, 175–198.

Poorter, H., Roumet, C., and Campbell, B.D. (1996). Interspecific variation in the growth response of plants to elevated CO_2: A search for functional types. In *Carbon Dioxide, Populations, and Communities,* Korner, C. and Bazzaz, F.A., Eds., Academic Press, San Diego, 375–412.

Postgate, J. (1998) *Nitrogen Fixation*, 3rd edn. Cambridge University Press, Cambridge, U.K.

Qiu, Z., Chappelka, A.H., Somers, G.L., Lockaby, B.G., and Meldahl, R.S. (1993) Effects of ozone and simulated acidic precipitation on ectomycorrhizal formation on loblolly-pine seedlings. *Environmental and Experimental Botany*, 33, 423–431.

Rantanen, L., Palomaki, V., and Holopainen, T. (1994) Interactions between exposure to O_3 and nutrient status of trees — Effects on nutrient content and uptake, growth, mycorrhiza and needle ultrastructure. *New Phytologist*, 128, 679–687.

Rao, M.V., Hale, B.A., and Ormrod, D.P. (1995) Amelioration of ozone-induced oxidative damage in wheat plants grown under high-carbon dioxide — Role of antioxidant enzymes. *Plant Physiology*, 109, 421–432.

Rastetter, E.B., Agren, G.I., and Shaver, G.R. (1997) Responses of N-limited ecosystems to increased CO_2: A balanced-nutrition, coupled-element-cycles model. *Ecological Applications*, 7, 444–460.

Reich, P.B. and Amundson, R.G. (1985) Ambient levels of ozone reduce net photosynthesis in tree and crop species. *Science*, 230, 566–570.

Reid, C.D. and Fiscus, E.L. (1998) Effects of elevated $[CO_2]$ and/or ozone on limitations to CO_2 assimilation in soybean (*Glycine max*). *Journal of Experimental Botany*, 49, 885–895.

Reid, C.D., Fiscus, E.L., and Burkey, K.O. (1998) Combined effects of chronic ozone and elevated CO_2 on Rubisco activity and leaf components in soybean (*Glycine max*). *Journal of Experimental Botany*, 49, 1999–2011.

Reid, C.D., Fiscus, E.L., and Burkey, K.O. (1999) Effects of chronic ozone and elevated atmospheric CO_2 concentrations on ribulose-1,5-bisphosphate in soybean (*Glycine max*). *Physiologia Plantarum*, 106, 378–385.

Reiling, K. and Davison, A.W. (1992) The response of native, herbaceous species to ozone — growth and fluorescence screening. *New Phytologist*, 120, 29–37.

Reinert, R.A. and Ho, M.C. (1995) Vegetative growth of soybean as affected by elevated carbon dioxide and ozone. *Environmental Pollution*, 89, 89–96.

Reinsvold, R.J. and Pope, P.E. (1987) Combined effect of soil-nitrogen and phosphorus on nodulation and growth of *Robinia pseudoacacia*. *Canadian Journal of Forest Research — Revue Canadienne De Recherche Forestiere*, 17, 964–969.

Rich, S., Waggoner, P.E., and Tomlinson, H. (1970) Ozone uptake by bean leaves. *Science*, 169, 79–80.

Rind, D., Suozzo, R., Balachandran, N.K., and Prather, M.J. (1990) Climate change and the middle atmosphere .1. The doubled CO_2 climate. *Journal of the Atmospheric Sciences*, 47, 475–494.

Rogers, H.H., Peterson, C.M., McCrimmon, J.N., and Cure, J.D. (1992) Response of plant-roots to elevated atmospheric carbon-dioxide. *Plant Cell and Environment*, 15, 749–752.

Rogers, H.H., Runion, G.B., and Krupa, S.V. (1994) Plant-responses to atmospheric CO_2 enrichment with emphasis on roots and the rhizosphere. *Environmental Pollution*, 83, 155–189.

Ross, D.J., Saggar, S., Tate, K.R., Feltham, C.W., and Newton, P.C.D. (1996) Elevated CO_2 effects on carbon and nitrogen cycling in grass/clover turves of a Psammaquent soil. *Plant and Soil*, 182, 185–198.

Ross, D.J., Tate, K.R., and Newton, P.C.D. (1995) Elevated CO_2 and temperature effects on soil carbon and nitrogen cycling in ryegrass/white clover turves of an Endoaquept soil. *Plant and Soil*, 176, 37–49.

Rundel, P.W., Nilsen, E.T., Sharifi, M.R., Virginia, R.A., Jarrell, W.M., Kohl, D.H., and Shearer, G.B. (1982) Seasonal dynamics of nitrogen cycling for a prosopis woodland in the Sonoran desert. *Plant and Soil*, 67, 343–353.

Runeckles, V.C. and Krupa, S.V. (1994) The impact of UV-B radiation and ozone on terrestrial vegetation. *Environmental Pollution*, 83, 191–213.

Russelle, M.P., Allan, D.L., and Gourley, C.J.P. (1994) Direct assessment of symbiotically fixed nitrogen in the rhizosphere of alfalfa. *Plant and Soil*, 159, 233–243.

Ryle, G.J.A., Powell, C.E., Timbrell, M.K., and Gordon, A.J. (1989) Effect of temperature on nitrogenase activity in white clover. *Journal of Experimental Botany*, 40, 733–739.

Sa, T.M. and Israel, D.W. (1991) Energy status and functioning of phosphorus-deficient soybean nodules. *Plant Physiology*, 97, 928–935.

Schlesinger, W.H. (1997) *Biogeochemisty: An Analysis of Global Change*, 2nd edn. Academic Press, San Diego.

Schortemeyer, M., Atkin, O.K., McFarlane, N., and Evans, J.R. (1999) The impact of elevated atmospheric CO_2 and nitrate supply on growth, biomass allocation, nitrogen partitioning and N_2 fixation of *Acacia melanoxylon*. *Australian Journal of Plant Physiology*, 26, 737–747.

Schortemeyer, M., Hartwig, U.A., Hendrey, G.R., and Sadowsky, M.J. (1996) Microbial community changes in the rhizospheres of white clover and perennial ryegrass exposed to free air carbon dioxide enrichment (FACE). *Journal Soil Biology and Biochemistry*, 28, 1717–1724.

Schulze, J. (2003) Source-sink manipulations suggest an N-feedback mechanism for the drop in N_2 fixation during pod-filling in pea and broad bean. *Journal of Plant Physiology*, 160, 531–537.

Schwintzer, C.R. (1983) Non-symbiotic and symbiotic nitrogen-fixation in a weakly minerotrophic peatland. *American Journal of Botany*, 70, 1071–1078.

Seiler, J.R. and Johnson, J.D. (1984) Growth and acetylene-reduction of black alder seedlings in response to water-stress. *Canadian Journal of Forest Research — Revue Canadienne De Recherche Forestiere*, 14, 477–480.

Serraj, R., Allen, L.H., and Sinclair, T.R. (1999a) Soybean leaf growth and gas exchange response to drought under carbon dioxide enrichment. *Global Change Biology*, 5, 283–291.

Serraj, R. and Sinclair, T.R. (2003) Evidence that carbon dioxide enrichment alleviates ureide-induced decline of nodule nitrogenase activity. *Annals of Botany*, 91, 85–89.

Serraj, R., Sinclair, T.R., and Allen, L.H. (1998) Soybean nodulation and N_2 fixation response to drought under carbon dioxide enrichment. *Plant Cell and Environment*, 21, 491–500.

Serraj, R., Sinclair, T.R., and Purcell, L.C. (1999b) Symbiotic N_2 fixation response to drought. *Journal of Experimental Botany*, 50, 143–155.

Singleton, P.W., Abdel Magid, H.M., and Tavares, J.W. (1985) Effect of phosphorus on the effectiveness of strains of *Rhizobium japonicum*. *Soil Science of America Journal*, 49, 613–616.

Skelly, J.M., Yand, Y.-S., Chevonne, B.I., Long, S.P., Nellessen, J.E., and Winner, W.E. (1983). Ozone concentrations and their influence on forest species in the Blue Ridge Mountains of Virginia. In *Air Pollution and the Productivity of the Forest,* David, D.D., Miller, A.A. and Dochinger, L., Eds., Isaac Walton League of America, Arlington, VA, 143–160.

Smil, V. (1999) Nitrogen in crop production: An account of global flows. *Global Biogeochemical Cycles*, 13, 647–662.

Smil, V. (2002) Nitrogen for food production: Proteins for human diets. *Ambio*, 31, 126–131.

Smith, V.H. (1992) Effects of nitrogen — phosphorus supply ratios on nitrogen-fixation in agricultural and pastoral ecosystems. *Biogeochemistry*, 18, 19–35.

Soussana, J.F. and Hartwig, U.A. (1996) The effects of elevated CO_2 on symbiotic N_2 fixation: A link between the carbon and nitrogen cycles in grassland ecosystems. *Plant and Soil*, 187, 321–332.

Sprent, J.I. (1972) The effects of water stress on nitrogen fixing root nodules III. Effects of osmotically applied stress. *New Phytologist*, 71, 451–460.

Sprent, J.I. (1981). Nitrogen fixation. In *The Physiology and Biochemistry of Drought Resistance in Plants,* Paleg, L.G. and Aspinall, D., Eds., Academic Press, New York, 131–143.

Stocklin, J. and Korner, C. (1999) Interactive effects of elevated CO2, P availability and legume presence on calcareous grassland: results of a glasshouse experiment. *Functional Ecology*, 13, 200–209.

Strain, B.R. and Bazzaz, F.A. (1983). Terrestrial plant communities. In *CO_2 and Plants,* (Lemon, E.R., Ed.), Westview, Boulder, CO, 177–222.

Streeter, J. (1988) Inhibition of legume nodule formation and N_2 fixation by nitrate. *Crc Critical Reviews in Plant Sciences*, 7, 1–23.

Ta, T.C., Macdowall, F.D.H., and Faris, M.A. (1987) Utilization of carbon from shoot photosynthesis and nodule CO2 fixation in the fixation and assimilation of nitrogen by alfalfa root-nodules. *Canadian Journal of Botany — Revue Canadienne De Botanique*, 65, 2537–2541.

Tang, C.X., Hinsinger, P., Jaillard, B., Rengel, Z., and Drevon, J.J. (2001) Effect of phosphorus deficiency on the growth, symbiotic N_2 fixation and proton release by two bean (*Phaseolus vulgaris*) genotypes. *Agronomie*, 21, 683–689.

Tarrant, R.F. (1983). Nitrogen fixation in American forestry: Research and application. In *Biological Nitrogen Fixation in Forest Ecosystems: Foundations and Applications,* Gordon, J.C. and Wheeler, C.T., Eds., M. Nijhoff/W. Junk, The Hague, 261–278.

Temperton, V.M., Grayston, S.J., Jackson, G., Barton, C.V.M., Millard, P., and Jarvis, P.G. (2003) Effects of elevated carbon dioxide concentration on growth and nitrogen fixation in *Alnus glutinosa* in a long-term field experiment. *Tree Physiology*, 23, 1051–1059.

Thomas, R.B., Bashkin, M.A., and Richter, D.D. (2000) Nitrogen inhibition of nodulation and N_2 fixation of a tropical N_2-fixing tree (*Gliricidia sepium*) grown in elevated atmospheric CO_2. *New Phytologist*, 145, 233–243.

Thomas, R.B., Richter, D.D., Ye, H., Heine, P.R., and Strain, B.R. (1991) Nitrogen dynamics and growth of seedlings of an N-fixing tree (*Gliricidia sepium* (Jacq.) Walp.) exposed to elevated atmospheric carbon dioxide. *Oecologia*, 88, 415–421.

Thornley, J.H.M. and Cannell, M.G.R. (2000) Dynamics of mineral N availability in grassland ecosystems under increased [CO_2]: Hypotheses evaluated using the Hurley Pasture Model. *Plant and Soil*, 224, 153–170.

Tissue, D.T., Megonigal, J.P., and Thomas, R.B. (1997) Nitrogenase activity and N_2 fixation are stimulated by elevated CO_2 in a tropical N_2-fixing tree. *Oecologia*, 109, 28–33.

Tjepkema, J.D. and Winship, L.J. (1980) Energy requirement for nitrogen-fixation in actinorhizal and legume root-nodules. *Science*, 209, 279–281.

Treshow, M. (1984) *Air Pollution and Plant Life,* John Wiley & Sons, New York.

Turcsanyi, E., Cardoso-Vilhena, J., Daymond, J., Gillespie, J., Balaguer, L., Ollerenshaw, J., and Barnes, J.D. (2000). Impacts of tropospheric ozone: past, present, and likely future. In *Trace Gas Emissions and Plants,* (Singh, S.N., Ed., Kluwer, Dordrecht, The Netherlands, 249–272.

U.S. Environmental Protection Agency (1986). Air quality criteria for ozone and other photochemical oxidants, Rep. No. EPA-600/8-84-020aF-eF. Office of Health and Environmental Assessment, Environmental Criteria and Assessment Office, Research Triangle Park, NC.

Unkovich, M.J. and Pate, J.S. (2000) An appraisal of recent field measurements of symbiotic N_2 fixation by annual legumes. *Field Crops Research*, 65, 211–228.

Uselman, S.M., Qualls, R.G., and Thomas, R.B. (1999) A test of a potential short cut in the nitrogen cycle: The role of exudation of symbiotically fixed nitrogen from the roots of a N-fixing tree and the effects of increased atmospheric CO_2 and temperature. *Plant and Soil*, 210, 21–32.

Van Cleve, K., Viereck, L.A., and Schlentor, R.L. (1971) Accumulation of N in alder (*Alnus*) ecosystems near Fairbanks, Alaska. *Arctic and Alpine Research*, 3, 101–114.

Vessey, J.K., Walsh, K.B., and Layzell, D.B. (1988a) Can a limitation in phloem supply to nodules account for the inhibitory effect of nitrate on nitrogenase activity in soybean. *Physiologia Plantarum*, 74, 137–146.

Vessey, J.K., Walsh, K.B., and Layzell, D.B. (1988b) Oxygen limitation of N_2 fixation in stem-girdled and nitrate-treated soybean. *Physiologia Plantarum*, 73, 113–121.

Vitousek, P.M., Aber, J.D., Howarth, R.W., Likens, G.E., Matson, P.A., Schindler, D.W., Schlesinger, W.H., and Tilman, D.G. (1997) Human alteration of the global nitrogen cycle: Sources and consequences. *Ecological Applications*, 7, 737–750.

Vitousek, P.M. and Howarth, R.W. (1991) Nitrogen limitation on land and in the sea — How can it occur? *Biogeochemistry*, 13, 87–115.

Vitousek, P.M., Walker, L.R., Whiteaker, L.D., Mueller-Dombois, D., and Matson, P.A. (1987) Biological invasion by *Myrica faya* alters ecosystem development in Hawaii. *Science*, 238, 802–804.

Vogel, C.S., Curtis, P.S., and Thomas, R.B. (1997) Growth and nitrogen accretion of dinitrogen-fixing *Alnus glutinosa* (L) Gaertn under elevated carbon dioxide. *Plant Ecology*, 130, 63–70.

Walsh, K.B. (1990) Vascular transport and soybean nodule function 3. Implications of a continual phloem supply of carbon and water. *Plant Cell and Environment*, 13, 893–901.

Warwick, K.R., Taylor, G., and Blum, H. (1998) Biomass and compositional changes occur in chalk grassland turves exposed to elevated CO_2 for two seasons in FACE. *Global Change Biology*, 4, 375–385.

Waughman, G.J. (1977) Effect of temperature on nitrogenase activity. *Journal of Experimental Botany*, 28, 949–960.

Weisbach, C., Hartwig, U.A., Heim, I., and Nosberger, J. (1996) Whole-nodule carbon metabolites are not involved in the regulation of the oxygen permeability and nitrogenase activity in white clover nodules. *Plant Physiology*, 110, 539–545.

Wheeler, C.T. (1971) Causation of diurnal changes in nitrogen fixation in nodules of *Alnus glutinosa*. *New Phytologist*, 70, 487–495.

Williams, L.E., Dejong, T.M., and Phillips, D.A. (1981) Carbon and nitrogen limitations on soybean seedling development. *Plant Physiology*, 68, 1206–1209.

Winner, W.E. and Atkinson, C.J. (1986) Absorption of air-pollution by plants, and consequences for growth. *Trends in Ecology and Evolution*, 1, 15–18.

Witty, J.F., Minchin, F.R., Skot, L., and Sheehy, J.E. (1986). Nitrogen fixation and oxygen in legume root nodules. In *Oxford Surveys of Plant Molecular and Cellular Biology*, (Miflin, B.J., Ed.), Vol. 3, Oxford University Press, Oxford, U.K., 275–314.

Zak, D.R., Pregitzer, K.S., Curtis, P.S., Vogel, C.S., Holmes, W.E., and Lussenhop, J. (2000) Atmospheric CO_2, soil-N availability, and allocation of biomass and nitrogen by *Populus tremuloides*. *Ecological Applications*, 10, 34–46.

Zanetti, S. and Hartwig, U.A. (1997) Symbiotic N_2 fixation increases under elevated atmospheric pCO_2 in the field. *Acta Oecologica — International Journal of Ecology*, 18, 285–290.

Zanetti, S., Hartwig, U.A., Luscher, A., Hebeisen, T., Frehner, M., Fischer, B.U., Hendrey, G.R., Blum, H., and Nosberger, J. (1996) Stimulation of symbiotic N_2 fixation in *Trifolium repens* L under elevated atmospheric pCO_2 in a grassland ecosystem. *Plant Physiology*, 112, 575–583.

Zanetti, S., Hartwig, U.A., and Nosberger, J. (1998) Elevated atmospheric CO_2 does not affect per se the preference for symbiotic nitrogen as opposed to mineral nitrogen of *Trifolium repens* L. *Plant Cell and Environment*, 21, 623–630.

Zelitch, I. (1973) Plant productivity and control of photorespiration. *Proceedings of the National Academy of Sciences of the United States of America*, 70, 579–584.

5 Belowground Food Webs in a Changing Climate

Joseph C. Blankinship and Bruce A. Hungate

CONTENTS

5.1 INTRODUCTION

Amidst a network of tunnels and pores, soil organisms recycle carbon and nutrients. They mix plant litter and detritus, causing changes in soil structure that can ultimately influence the amount of water and oxygen available for plant roots. They are also eaten by aboveground predators, providing direct connections between below- and aboveground food webs. Root herbivory is very difficult to quantify, but likely just as important as leaf herbivory. As we gain an appreciation for the importance of mycorrhizal fungi, an appreciation of organisms that graze on these fungi seems to follow.

In spite of the importance of these activities in contributing to ecosystem functioning, soil organisms — and the feeding relationships between soil organisms collectively referred to as the *belowground food web* — have received much less attention than their aboveground counterparts. After all, it's dark down there. This makes wildlife viewing a bit problematic. Ecological processes at this microscale are difficult to mimic in the laboratory and difficult to monitor in the field, and sampling inevitably alters habitats. Belowground food webs have also received less attention because of their often overwhelming concentration of biodiversity and complex feeding interactions among organisms resulting from the strong prevalence of omnivory. Life-history traits, pathogens, lateral gene transfer between microorganisms, and symbioses also introduce interesting possibilities of complicating our understanding of belowground food webs.

Climatic change experiments provide an important avenue for understanding belowground food webs. Anthropogenic climatic change is a perturbation through which we can investigate how organisms collectively influence soil resource availability, and whether the trophic structure and species composition of the food web modifies these influences. Adjustments may occur in the activity, abundance, diversity, and distribution of organisms in the soil profile, allowing the belowground community to tolerate novel conditions. Can a species adapted to cold Arctic tundra tolerate warmer summer temperatures and increased soil moisture due to the melting of permafrost? Can a drought-adapted species tolerate even longer droughts? Can a community take advantage of increased plant root production and exudation by harvesting more energy under elevated atmospheric CO_2? Changes at the level of individual species within food webs may, in some cases, have negligible effects of trophic interactions. In other cases, such changes may be important, altering belowground trophic structure and plant resource availability. Trophic interactions between belowground organisms determine the amount of energy available for carbon and nutrient cycling, litter mixing, and aboveground food webs. How will current rates of anthropogenic climatic change affect these trophic interactions?

5.2   CURRENT PARTICIPANTS AND MECHANISMS

Although most belowground food web analyses have been performed in intensively managed arable soils (Scheu 2002), it is expected that many of the same participants (organisms) and mechanisms (processes) operate in relatively unmanaged soils. For example, it is expected that soil structure has a similar influence on belowground food webs in both managed and unmanaged ecosystems, such that fine-textured soils provide greater physical protection for bacteria against protozoan grazing than coarse-textured soils (Rutherford and Juma 1992). Spatial heterogeneity in soil resources is assumed to influence all belowground food webs. Similarly, indirect effects mediated by plants, such as changes in microclimate and litter production, are expected to influence all belowground food webs. However, it is also expected that management practices and greater perturbation in agroecosystems have accumulated differences in soil structure, resource availability, and plant interactions that could significantly modify the activity, abundance, diversity, and distribution of organisms in the soil profile. In order to identify differences between intensively managed and relatively unmanaged ecosystems, we first need a basis for comparison.

Food web diagrams display trophic interactions between organisms in one of three forms (Scheu et al. 2002). In simplest form, a food web diagram displays *connectivity* between different species (e.g., a protozoan and a bacterium) or different feeding guilds (e.g., herbivores and carnivores). Connectivity food webs display who eats whom and nothing more. Improved understanding can be displayed in the second type of food web diagram, which includes *energy flow*. While connectivity diagrams indicate how energy is transferred between organisms, energy flow diagrams indicate how much energy is transferred. This additional information can help identify the dominant pathways of energy transfer within a food web. The third and most detailed food web diagram includes *interaction strengths* between organisms. Interaction strength food webs display the per capita effect a predator exerts on its prey or vice versa. This form of food web diagram is truly functional and can be applied to field conditions in order to understand possible food web responses to perturbation. In a majority of natural ecosystems, our understanding of belowground food webs remains at the stage of connectivity diagrams.

The trophic structure of a food web depends on the abundance, distribution, and feeding preferences of individual organisms. In order to study trophic interactions between these organisms, ecologists organize individuals into populations of species, and species into feeding groups. A *feeding group* (or "guild") consists of populations that eat a similar substrate. For example, *detritivores* are a feeding group that eats the common substrate of detritus (unrecognizable dead plant and animal material). Likewise, *herbivores* eat live plant material, *bacterivores* eat bacteria, *fungivores* eat fungi, and *predators* eat live animals. The feeding group is the basal entity of any food web and ideally includes all species that perform a functionally equivalent role in the food web. When different species obtain their energy through the same number of feeding transfers from primary production, they belong to the same *trophic level*. Low trophic levels are closely connected to primary producers and eat organisms that have participated in a small number of feeding transfers. High trophic levels (or predators) eat organisms that have participated in a larger number of feeding transfers. *Trophic structure* refers to the distribution of biomass and energy in different trophic levels in a food web. Biomass distribution can take the form of an upright or inverted trophic pyramid depending on instantaneous rates of bottom–up and top–down forces. Energy distribution always forms an upright pyramid with more energy processed at lower trophic levels.

Species probably do not aggregate into discrete, homogeneous trophic levels (Polis and Strong 1996). In reality, omnivory, diets shifts, and spatial heterogeneity in resources blur the boundaries between discrete trophic levels. *Omnivores* (organisms that eat multiple types of substrates) complicate the concept of trophic levels. Omnivores are food generalists and cannot be placed into a single trophic level. Similarly, ontogenetic (e.g., age) and environmentally induced changes (e.g., wildfire) in the diet of species may confuse the concept of discrete trophic levels (Polis 1984). Spatial heterogeneity in soil structure and the distribution of populations means that just because two populations are in the same general location, feeding transfers are not necessarily occurring. For example, a bacterium may be able to escape predation by hiding in soil pores that are too small for protozoa to enter (Rutherford and Juma 1992). Rather than discrete levels, trophic structure is more accurately described as a continuum of feeding transfers or a "trophic spectrum" (Darnell 1961).

For some food webs, the collection of trophic transfers can be usefully organized into linear food chains. But such food chains likely oversimplify trophic dynamics in soil, where a web pattern more accurately reflects the complex connections between different trophic levels. Trophic systems are often described as either plant-based (live plant material) or detritus-based (i.e., dead plant and animal material), depending on what is viewed as the ultimate source of energy (after the sun!). In reality, these two systems form one food web, as live plant material senesces and joins the detritus-based trophic system. Or a carnivore connects the two trophic systems by eating root herbivores in the plant-based system and detritivores (e.g., earthworms) in the detritus-based system.

While plant- and detritus-based trophic systems emphasize bottom–up effects (i.e., lower to higher trophic levels) of biomass and energy transfer within food webs, trophic cascades emphasize top–down effects (i.e., higher to lower trophic levels). Much like the overlapping boundaries and connections between trophic levels and food chains, bottom–up and top–down effects are intertwined and operate simultaneously in most food webs. A population may be limited by resource availability on one soil particle (i.e., bottom–up), predation on an adjacent soil particle (i.e., top–down), or by both on yet another soil particle (i.e., bottom–up and top–down forces can affect populations simultaneously). A *trophic cascade* occurs when changes in abundances at one trophic level alters abundances at a lower trophic level across more than one link in a food web (Pace et al. 1999). The green world hypothesis (Hairston et al. 1960) postulates strong top–down control in aboveground food webs. If a food chain has an odd number of trophic levels, the world is *green* because (in the case of a three-level system) predators reduce herbivore abundances, releasing plants from grazing pressure. If a food chain has an even number of trophic levels (for example, four), the world is *barren*, because secondary predators reduce abundances of primary predators, which then reduces predation on herbivores, and more plants are ultimately eaten. There is evidence that trophic cascades also occur in belowground food webs. For example, in a desert plant litter, the removal of predatory mites from a model system increased the abundance of bacterivorous nematodes, and decreased the abundance of bacteria (Santos et al., 1981). In this case, the presence or absence of the fourth trophic level cascaded down the food chain to affect the second trophic level. However, when applied to more complex field systems, there are reasons to believe that trophic cascades are less common belowground than aboveground (McLaren and Peterson 1994; Rypstra and Carter 1995). First, omnivory rates are probably higher belowground, which makes it rare for a predator species to significantly reduce abundances of one prey species before switching to a different prey species, among the wide array of soil biodiversity. Second, spatial heterogeneity in the soil profile may prevent trophic cascades by providing refugia for prey species. And third, trophic cascades may be absent because of high microbial turnover rates in detritus-based trophic systems. Detritivores are dominated by bacteria and fungi, which have high reproductive rates that may compensate for the effects of top–down grazing. For example, bacterial abundances can be unaffected by food chain length, and fungal abundances can actually be higher in the presence of predators (Mikola and Setälä 1998).

A *trophic transfer* refers to the feeding of one organism on another, or on detritus. These individual trophic transfers collectively form the food web. Although simple in theory, trophic transfers are difficult to quantify, especially in a spatially heterogeneous and opaque environment such as soil. A single trophic transfer consists of three components — consumption, assimilation, and production — that determine the amount of energy and biomass that are transferred during a feeding event. The more energy or biomass that is transferred, the higher the *trophic efficiency*.

Consumption efficiencies in belowground food webs are generally lower than in aboveground food webs because deterministic biological interactions, such as predation, are restricted to small spatial scales with inherent restrictions in perception and locomotion (Ekschmitt and Griffiths 1998). This is expected to reduce consumption efficiencies and lead to overestimations of the overall energy and biomass cycled through belowground food webs. On the other hand, detritivores have extremely high consumption efficiencies because they repeatedly ingest, defecate, and reingest the same detritus. Generally, about 1% of available energy consumed at one trophic level successfully contributes to production at the next highest trophic level (Colinvaux and Barnett 1979).

Size is often used to categorize soil organisms. Organisms with a body width less than 0.1 mm are referred to as microflora or microfauna. Larger organisms between 0.1 and 2 mm are referred to as mesofauna, and the largest organisms greater than 2 mm are macrofauna. Size can be a useful means of characterizing a belowground food web, much like characterizing soil texture by sieving. An organism must generally have a large enough mouth to physically consume another organism. A small organism may be able to escape predation by having exclusive access to small pores within a soil particle, and has a larger surface-to-volume ratio to absorb soluble substrates more rapidly. On the other hand, a larger organism has a smaller surface-to-volume ratio and is more resistant to desiccation. Therefore, the distribution of different body sizes in a belowground food web provides information about how trophic transfers occur in a particular soil.

Food web diagrams are models of what ecologists suspect — either through experiments or observations — to be happening. The two main types of food web models are those that are organism oriented or process oriented (Paustian 1994). *Organism-oriented food web models* describe the flow of matter or energy through different groups of organisms that are organized by taxonomy or trophic levels (Figure 5.1). *Process-oriented food web models* describe the activities mediating the transformations and storage of matter or energy (Figure 5.2).

With such a wide variety of organisms participating in the belowground food web — "if [only] we are willing to sweep our vision down from the world lined by the horizon to include the world an arm's length away" (Wilson 1994) — it is understandable that the subject of soil biodiversity is of considerable interest. Biological diversity within litter and soils may be orders of magnitude greater than aboveground diversity, but it has not been fully documented in any ecosystem (Adams and Wall 2000). Accurate descriptions of biodiversity are often hampered by extraction methods, choice of sampling depth, selective sorting, and insufficient taxonomic resolution (André et al. 2002). These factors introduce bias and tend to underestimate soil biodiversity. For example, the most commonly used extraction methods have efficiencies around 40%, and most studies focus on a selected half of microarthropods (mites or springtails) (André et al. 2002). Taxonomic resolution seems to be related to organism size, with smaller organisms, such as bacteria and fungi, being the least resolved.

One of the central motivations for studying biodiversity has been to identify connections with ecosystem function. Biodiversity may be important in maintaining current ecosystem function and in providing insurance for future ecosystem function

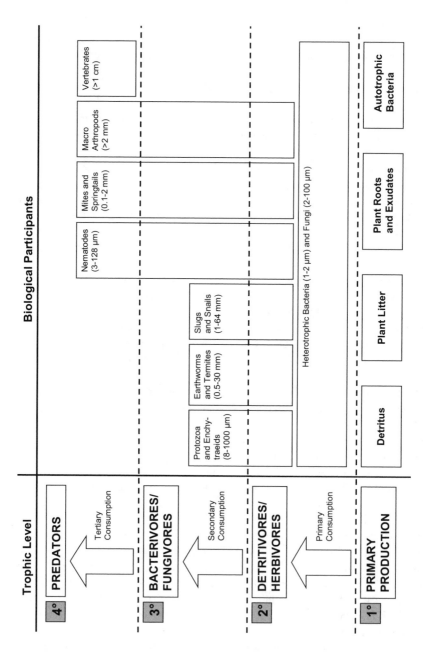

FIGURE 5.1 Organism-oriented belowground trophic level model. Size ranges refer to body widths.

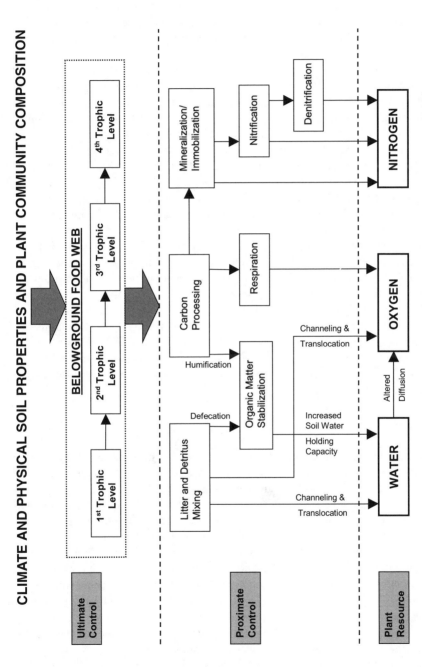

FIGURE 5.2 Process-oriented belowground food web model showing controls on soil water, oxygen, and nitrogen availability.

during perturbation (e.g., climatic change) (Bengtsson et al. 1997). *Functional redundancy* describes a hypothesis in ecology in which different species perform the same function in nature, so that changes in species diversity do not affect ecosystem function (Loreau 2004). In the case of food webs, *function* refers to the trophic level in which the organism participates. If two species are functionally redundant, then in theory, one species can go extinct without any changes in net energy transferred between trophic levels. In model systems, there is strong evidence that belowground trophic structure can affect ecosystem function. For example, in the absence of predatory mites, fungivorous nematodes decreased fungal biomass, which leads to a 40% reduction in decomposition rates (Santos et al. 1981). In more complex systems, however, functional redundancy within trophic levels is expected to compensate for any losses of individual species. Another study showed that the presence of bacterivorous nematodes on C and N mineralization seemed to vary with time (with an initial increase followed by a longer-term decrease) and litter quality (Bouwman et al. 1994). The third trophic level appeared to indirectly affect C mineralization via changes in bacterial activity, and directly affect N mineralization via digestion and excretion.

While taxonomic descriptions of biodiversity are one means of organizing soil organisms for food web analyses, trophic levels are the preferred categories by providing additional information about how different species function within the soil environment. Food web stability depends more on the number of trophic levels than on species richness (de Ruiter et al. 1998) and changes in trophic structure are known to affect ecosystem function (Chapin III et al. 1997). The functional importance of a soil organism is hypothesized to be inversely related to the trophic position of the organism (Laakso and Setälä 1999). In other words, feeding interactions are most likely to influence ecosystem-level processes when they occur near the bottom of the food web (Setälä 2002).

5.3   EFFECTS OF CLIMATIC CHANGE

Our knowledge of climatic change impacts on soil organisms lags behind other branches of terrestrial ecology (Ingram and Freckman 1998). As noted earlier, the entire belowground food web is rarely investigated at the same time, and rarely in relation to climatic change. Little is known about how changes in belowground trophic structure currently influence ecosystem function, and we cannot yet rule out the possibility that functional redundancy within trophic levels increases the resistance and resilience of belowground food webs to anthropogenic climatic change. Through studying belowground food webs in the context of climatic change, we hope to improve our predictions of future conditions. Climatic change experiments can also help improve our understanding of current interactions between community structure and activity. Specific mechanisms for changes in ecosystem processes may become clearer as we begin to identify the responses of individual trophic groups. Because the entire belowground food web has never been studied in a single climatic change experiment (the closest attempts include: Klironomos et al. 1996; Cotrufo and Gorissen 1997; Yeates et al. 1997; Lussenhop et al. 1998; Rillig et al. 1999b;

Taylor et al. 2004), it is necessary to combine the results of separate studies in order to understand how climatic change affects trophic structure.

Climatic change can affect soil carbon and nitrogen cycling (see Chapter 2), and the organisms in charge of the cycling are living within a complex food web. Thus, documenting food web responses may help us explain observed changes in soil biogeochemistry. Why is a particular ecosystem process not affected by climatic change? Through measuring the abundance, diversity, distribution, and efficiency of trophic transfers, we may be able to account for adjustments made by the belowground food web that allow the community to resist climatic change. If the process does change, is it simply because of changes in abiotic factors, or are changes in trophic interactions partly responsible? Are there any common global patterns of response to climatic change factors, or do results appear to be context dependent?

5.3.1 DIRECT VS. INDIRECT EFFECTS OF CLIMATIC CHANGE ON BELOWGROUND FOOD WEBS

Changes in precipitation, temperature, and atmospheric CO_2 can affect the belowground food web through both direct and indirect mechanisms (Table 5.1). Direct effects occur when a climatic change factor alters the abiotic environment and immediately elicits a physiological or behavioral response from soil organisms. For example, many organisms increase rates of respiration when temperature increases. Indirect effects occur when a climatic change factor (or two or three) interacts with an ecosystem property to affect soil organisms. For example, enhancing carbon input to soil can stimulate ammonium consumption by soil heterotrophs, thereby reducing ammonium availability to nitrifying bacteria. Accumulating experimental evidence indicates that effects of climatic change on soil food webs are primarily indirect responses to changes in vegetation (Wardle et al. 1998).

Altered abundance and annual distribution of precipitation can directly affect soil organisms through three types of mechanisms. First, altered precipitation may affect metabolic rates. Second, some soil organisms (such as protozoa and nematodes) depend on water films between soil particles for mobility. If precipitation increases, it will be easier for these organisms to move around and perhaps consume previously inaccessible resources. If precipitation decreases, less mobility is expected. And third, altered precipitation may change the vertical distribution of populations in the soil profile, allowing a population to move down in the soil profile where it is wetter (to escape desiccation) or up in the soil profile where it is drier (to escape drowning).

Altered precipitation can also have indirect effects. Repeated drying cycles have been shown to decrease soil aggregate stability (Soulides and Allison 1961; Young et al. 1998), indicating that reduced precipitation can indirectly affect organisms through changes in soil structure. Altered precipitation can indirectly affect soil aeration by influencing oxygen consumption rates through respiration, and positively affect plant root production and turnover (Norby and Jackson 2000), which represents a change in available surface area for microbial colonization and herbivory in the rhizosphere. Altered precipitation can also influence plant community composition (Lloret et al. 2004) and phenology, which may indirectly affect belowground communities through changes in litter abundance, quality, and host species.

TABLE 5.1
Direct and Indirect Effects of Climatic Change on Belowground Food Webs

	ALTERED PRECIPITATION	ELEVATED TEMPERATURE	ELEVATED CARBON DIOXIDE
DIRECT EFFECTS	Δ metabolism Δ mobility Δ vertical distribution of populations	Δ metabolism Δ vertical distribution of populations Δ latitudinal distribution of pops.	none
INDIRECT EFFECTS	Δ soil structure Δ O$_2$ availability Δ plant root production/turnover Δ plant species composition Δ length/timing of growing season	Δ soil moisture Δ soil structure Δ plant root production/turnover Δ plant species composition Δ length/timing of growing season	↑ plant litter and root production Δ plant species composition Δ soil structure ↑ soil moisture Δ length/timing of growing season

Elevated temperature can directly affect soil organisms by influencing metabolic rates or shifting spatial distributions. Changes in distribution may occur within the soil profile (e.g., as organisms seek deeper, cooler temperatures) or between latitudes and altitudes.

Warming can also have indirect effects. Elevated temperatures can alter soil water content, reducing it directly by enhancing evapotranspiration (Goyal 2004) or altering it indirectly through changes in plant senescence (e.g., Zavaleta et al. 2003a). Freeze-thaw cycles (which are expected to become more common during the winter in colder ecosystems and less common in warmer ecosystems) tend to break apart soil particles and decrease aggregate stability (Soulides and Allison 1961; Oztas and Fayetorbay 2003). This could indirectly affect soil organisms through changes in soil structure (e.g., porosity for diffusion of oxygen) and habitat stability. Like altered precipitation, elevated temperature can change plant root production and turnover, often increasing both (Pregitzer et al. 2000a), and change plant species composition. Experiments suggest that warming reduces plant species richness through the loss of less abundant species (Chapin III et al. 1995; Lloret et al. 2004), and it is unknown whether belowground biodiversity follows this change in aboveground biodiversity. And finally, growing seasons begin earlier, measured either through date of budbreak (Farnsworth et al. 1995; Norby et al. 2003) or snowmelt (Price and Waser 1998), when average conditions are warmer. In colder ecosystems, this provides a longer growing season for plants, and probably for higher belowground activity too.

Soil organisms are not expected to directly respond to elevated atmospheric CO_2 because soil CO_2 concentrations are already quite high and are unlikely to significantly change with atmospheric CO_2 (Lavelle et al. 1997). Plant responses to elevated CO_2, however, are likely to indirectly affect belowground food webs. First, elevated CO_2 has been shown to increase plant litter abundance (Owensby et al. 1999; Ferris et al. 2001; Dijkstra et al. 2002), plant root production (King et al. 1996; Owensby et al. 1999; Pregitzer et al. 2000b; Arnone et al. 2000), and plant root exudation (Paterson et al. 1997). Changes in root production and exudation are expected to directly affect belowground herbivores and microbes that can metabolize labile carbon substrates. Second, elevated CO_2 can affect plant species composition (Zavaleta et al. 2003b), which may modify belowground communities. Third, elevated CO_2 can influence soil structure by altering the size and stability of soil aggregates (Rillig et al. 1999a). Elevated CO_2 tends to increase mycorrhizal production of glomalin (Rillig et al. 1999a), a protein that is known to play a role in binding soil aggregates, but net effects on aggregate size (and soil porosity?) may be ecosystem specific. For example, in a sandstone grassland, elevated CO_2 increased aggregate size (Rillig et al. 1999a), while in a calcareous grassland, elevated CO_2 reduced aggregate size (Niklaus et al. 2003). Elevated CO_2 has been shown to increase soil moisture (Field et al. 1997; Niklaus et al. 1998; Hungate et al. 2002; Nelson et al. 2004), and influence the length and timing of the growing season (Badeck et al. 2004), which are certain to affect biological activity.

5.3.2 EFFECTS OF ALTERED PRECIPITATION

A number of experiments have investigated effects of moisture on soil organisms (Table 5.2), but field manipulations of precipitation are rare. Some soils are treated in the field, others are treated in the laboratory, and others are treated through latitudinal or altitudinal transplantation (e.g., Sohlenius and Boström 1999). Some soils are treated for 24 hours (e.g., Kang et al. 2003) while others are treated for 13 years (e.g., Lindberg and Persson 2004). But at this point in our understanding, it would be premature to ignore any type of experimental design.

The second trophic level of the belowground food web tends to respond positively to elevated precipitation and negatively to reduced precipitation. Wetter conditions increase the abundance (Clarholm 1981) and biomass (Lundgren and Söderström 1983; Gallardo and Schlesinger 1995; Fierer and Schimel 2002) of bacteria and fungi. It is hypothesized that elevated precipitation favors bacterial-based food chains over fungal-based food chains because of differences in physiology (Wardle et al. 1998). Drier conditions have been shown to decrease bacterial and fungal biomass, but only in the more exposed surface litter layer (Salamanca et al. 2003), which is most vulnerable to drying. While individual bacterial cells may not be migrating deeper, this does illustrate that population distributions within a soil profile can be shifted by changes in precipitation. Repeated wetting events also tend to decrease bacterial and fungal biomass (Taylor et al. 2004), which is probably caused by cell death upon wetting and subsequent consumption by other detritivores. Altered microbial substrate utilization under elevated precipitation (Papatheodorou et al. 2004) indicates changes in metabolism, but there is no evidence of changes in microbial community structure (Wilkinson et al. 2002). Similarly, snail and slug community structure is unaffected by altered soil moisture (Sternberg 2000). Earthworms (Presley et al. 1996) and herbivorous nematodes (Todd et al. 1999) tend to prefer wetter conditions. Collectively, experiments provide evidence that elevated precipitation tends to increase production at the second trophic level, and reduced precipitation has the opposite effect.

The third trophic level tends to be unresponsive to elevated precipitation and to respond negatively to reduced precipitation. Despite increased production by the second trophic level under wetter conditions, there is little evidence of population explosions in the third trophic level. While protozoa do increase in abundance (Clarholm 1981), perhaps in response to a favored bacterial-based food web, there are no changes in abundances of enchytraeids (Lindberg et al. 2002), nematodes (Sohlenius and Wasilewska 1984; Todd et al. 1999; Sohlenius and Boström 1999; Papatheodorou et al. 2004), or microarthropods (Lindberg and Persson 2004). When a positive effect of elevated precipitation on nematodes was found, it only occurred in surface litter (Freckman et al. 1987), suggesting that surface nematodes are most likely to be limited (either via mobility or physiology) by water availability. Nematode diversity appears to be unaffected by wetter conditions (Sohlenius and Wasilewska 1984; Papatheodorou et al. 2004), and microarthropod diversity can be positively affected (Tsiafouli et al. 2005) or unaffected (Lindberg and Persson 2004). The third trophic level tends to be more responsive to reduced precipitation. In most cases, drier conditions decrease the biomass of organisms in the third trophic level. For example, drying

TABLE 5.2
Effects of Altered Precipitation on Belowground Trophic Structure

Trophic Level	Ecosystem Type	Type of Manipulation	Duration of Manipulation	Results	Reference
II. Detritivores/Herbivores	warm desert	wetting	10 days	increase in microbial biomass-N	Gallardo and Schlesinger, 1995
	boreal forest	wetting	17 days	increase in abundance of bacteria	Clarholm, 1981
	boreal forest	*ambient precipitation*	2 years	increase in abundance of bacteria with increase in organic soil water content, no change in abundance of bacteria in mineral soil	Lundgren and Söderström, 1983
	grassland	wetting	6 months	altered microbial substrate utilization	Papatheodorou et al., 2004
	coniferous forest	wetting	4 months	no change in microbial community structure using PLFA	Wilkinson et al., 2002
	deciduous forest	drying	1 year	no change in microbial biomass	Salamanca et al., 2003
	grassland	drying-rewetting, *laboratory incubation*	10 days	decrease in microbial biomass-C with rewetting	Kieft et al., 1987
	grassland, oak woodland	drying-rewetting, *laboratory incubation*	2 months	increase in microbial biomass with increased drying-rewetting frequency	Fierer and Schimel, 2002
	coniferous forest	wetting *constant vs. fluctuating*	5 months	decrease in microbial biomass with fluctuating irrigation	Taylor et al., 2004
	grassland	wetting	4 years	increase in abundance of herbivorous nematodes	Todd et al., 1999
	wetland	drying and wetting, *laboratory incubation*	1 year	increase in survivorship of earthworms at intermediate soil water contents, increase in fecundity of earthworms at high soil water contents	Presley et al., 1996
	grassland	drying and wetting, *laboratory incubation*	2 years	no change in species composition of slugs and snails	Sternberg, 2000
	deciduous forest	drying and wetting, *laboratory incubation*	24 hours	decrease in soil respiration with extreme drying and wetting	Kang et al., 2003
	grassland	drying and wetting, *laboratory incubation*	16 days	decrease in soil respiration with drying	Fierer et al., 2003
	deciduous forest	drying and wetting, *laboratory incubation*	4 months	no change in soil respiration	Zak et al., 1999
	polar tundra, taiga	drying-rewetting, *laboratory incubation*	24 hours	decrease in soil respiration with drying, increase in respiration with rewetting	Gulledge and Schimel, 1998
	grassland, oak woodland	drying-rewetting, *laboratory incubation*	2 months	increase in soil respiration with increased drying-rewetting frequency in grassland, decrease in soil respiration with increased drying-rewetting frequency in oak woodland	Fierer and Schimel, 2002

TABLE 5.2 (CONTINUED)
Effects of Altered Precipitation on Belowground Trophic Structure

Trophic Level	Ecosystem Type	Type of Manipulation	Duration of Manipulation	Results	Reference
III. Bacterivores/Fungivores					
	boreal forest	wetting	17 days	increase in abundance of amoebae	Ciarholm, 1981
	boreal forest	drying and wetting	8 and 10 years	decrease in abundance of enchytraeids with drying	Lindberg et al., 2002
	grassland	wetting	6 months	no change in abundance or species diversity of bacterivorous nematodes	Papatheodorou et al., 2004
	warm desert	wetting	1 year	no change in abundance of nematodes in detritus, increase in abundance in litter	Freckman et al., 1987
	grassland	wetting	4 years	no change or decrease in abundance of microbivorous nematodes	Todd et al., 1999
	polar tundra	drying and wetting *transplantation*	1 year	no change in abundance of nematodes	Sohlenius and Boström, 1999
	boreal forest	wetting	8 years	no change in abundance or species dominance of nematodes, increase in vertical distribution of nematodes upwards into litter	Sohlenius and Wasilewska, 1984
	grassland	drying and wetting	4 months	decrease in abundance of nematodes with both drying and wetting	Bakonyi and Nagy, 2000
	coniferous forest	wetting *constant vs. fluctuating*	5 months	decrease in abundance of nematodes with fluctuating irrigation	Taylor et al., 2004
	polar tundra	wetting	4 years	increase in abundance of mites and springtails	Convey et al., 2002
	deciduous forest	wetting	6 months	no change in abundance of springtails	Ferguson and Joly, 2002
	boreal forest	wetting	13 years	no change in abundance or species composition of mites and springtails	Lindberg and Persson, 2004
	coniferous forest	wetting *constant vs. fluctuating*	5 months	no change in abundance of mites or springtails	Taylor et al., 2004
	polar tundra	wetting	1 year	decrease in abundance of mites and springtails with thicker winter ice layer	Coulson et al., 2000
	coniferous forest	drying	6 years	springtails recovered faster to drought than mites	Lindberg and Bengtsson, 2005
	coniferous forest	drying and wetting	4 months	decrease in species richness of mites and springtails with drying increase in species richness of mites and springtails with wetting decrease in species richness of mites and springtails with decreasing wetting frequency	Tsiafouli et al., 2005
	boreal forest	drying and wetting	8 and 10 years	decrease in abundance of mites and springtails with drying decrease in diversity of mites with drying, increase in diversity of mites with wetting	Lindberg et al., 2002
IV. Carnivores					
	deciduous forest	wetting	6 months	no change in growth rate of predacious mites	Ferguson and Joly, 2002
	boreal forest	wetting	13 years	no change in abundance of predatory macroarthropods	Lindberg and Persson, 2004
	boreal forest	drying and wetting	8 and 10 years	decrease in abundance of macroarthropods with drying	Lindberg et al., 2002

decreases abundances of enchytraeids (Lindberg et al. 2002), nematodes (Bakonyi and Nagy 2000), and microarthropods (Lindberg et al. 2002). In the case of enchytraeids, this response is consistent with the fact that they have no special protection mechanisms against evaporation (Didden 1993). There is also evidence that microarthropod diversity decreases under reduced precipitation (Lindberg et al. 2002; Tsiafouli et al. 2005), perhaps because some species are better adapted to drier conditions. Like the second trophic level, the third trophic level appears to be negatively affected by reduced precipitation, suggesting that drier conditions may cause bottom–up resource limitation for higher trophic levels. More importantly, drier conditions can directly affect soil habitat quality (e.g., water potential and water film thickness), independently reducing biological metabolism and mobility at each trophic level.

Responses of fourth trophic level participants to climatic change are rarely studied. After more than a decade of treatments in a boreal forest, predatory macroarthropods responded negatively to drier conditions and showed no response to wetter conditions (Lindberg et al. 2002; Lindberg and Persson 2004). This pattern of response is similar to the third trophic level. Over the range of experiments conducted to date, soil organisms at all trophic levels appear to be more sensitive to reduced precipitation than elevated precipitation. Only the hardiest species survive, as demonstrated by the reduced abundance and diversity of soil organisms in the world's deserts (Wall and Virginia 1999). It should also be noted that most field studies simulate reductions in precipitation of 25 to 100%, which are more extreme than the 15 to 20% reductions predicted in some areas during the next century (Intergovernmental Panel on Climate Change [IPCC] 2001). Therefore, treatments tend to overestimate the effects of reduced precipitation. Rather, some experiments may be more useful in understanding responses to extreme drought events, which are expected to become more common in the future.

5.3.3 EFFECTS OF ELEVATED TEMPERATURE

Predicted magnitudes of global warming are unlikely to have large direct effects on the biomass or community structure of broad taxonomic groups of soil organisms (Wardle et al. 1998), except at the boundaries of a specie's tolerance range. *Tolerance range* (which may also be applied to soil moisture) describes the extent to which a species can grow and reproduce in different climates and survive through extreme weather events. If temperatures are above the upper limit of an organism's tolerance range, enzymes may be denatured by extreme temperatures. Alternatively, warming in colder ecosystems may release organisms from temperature limitation. Environmental tolerance ranges of soil organisms are rare in the literature, and knowledge of this physiological trait may help soil ecologists explain changes in trophic structure by climatic change, and help agriculturists manage these changes in temperature- or water-stressed croplands.

While larger manipulations of temperature in the laboratory tend to decrease microbial biomass (Joergensen et al. 1990; Grisi et al. 1998; Cole et al. 2002; Jonasson et al. 2004) (Table 5.3), the few longer-term field manipulations of warming either show an increase (Ruess et al. 1999) or no effect (Bardgett et al. 1999). Fungi

TABLE 5.3
Effects of Elevated Temperature on Belowground Trophic Structure

Trophic Level	Ecosystem Type	Type of Manipulation	Duration of Manipulation	Results	Reference
II. Detritivores/Herbivores					
	grassland, tropical forest	laboratory incubation (15 and 35°C)	5 months	decrease in microbial biomass with increase in temperature	Grisi et al., 1998
	subarctic heathland	laboratory incubation (10 and 12°C)	6 months	decrease in microbial biomass-C and -N with increase in temperature	Jonasson et al., 2004
	grassland	laboratory incubation (15-35°C)	8 months	decrease in microbial biomass with increase in temperature, decrease in turnover time of microbial biomass with increase in temperature	Joergensen et al., 1990
	peatland	laboratory incubation (12-18°C)	2 months	decrease in microbial biomass with increase in temperature	Cole et al., 2002
	grassland	laboratory incubation (0-25°C)	23 days	no change in microbial biomass	Contin et al., 2000
	mesocosm (weedy field)	+ 2°C	9 months	no change in microbial biomass	Bardgett et al., 1999
	coniferous forest	laboratory incubation (0-45°C)	2 days	increase in ratio of bacterial to fungal growth rates at higher temperatures	Pietikäinen et al., 2005
	subarctic heathland	+ 0.4-2°C	5 years	no change in microbial immobilization of C, N, or P	Jonasson et al., 1999
	subarctic heathland	+ 0.9-2°C	6 years	increase in microbial biomass-C, increase in active fungal biomass	Ruess et al., 1999
	deciduous forest	laboratory incubation (5-25°C)	4 months	altered microbial community composition using PLFA and LPS-OHFA	Zogg et al., 1997
	grassland	+ 1°C	1 year	increase in arbuscular mycorrhizal fungi hyphal length and colonization	Rillig et al., 2002
	grassland	+ 3°C	1 year	no change in arbuscular mycorrhizal fungi colonization	Heinemeyer et al., 2003
	wetland	laboratory incubation (15-28°C)	1 year	increase in survivorship of earthworms at intermediate temperatures, increase in fecundity of earthworms at intermediate temperatures	Presley et al., 1996
	N/A	laboratory incubation (12-35°C)	3 months	increase in growth rate of earthworms with increase in temperature	Viljoen and Reinecke, 1992
	deciduous forest	constant vs. fluctuating T	1 year	decrease in earthworm cocoon production under fluctuating temperature	Kostecka and Butt, 2001
	deciduous forest	constant vs. fluctuating T	4 months	decrease in earthworm mortality and cocoon production under fluctuating temperature	Uvarov, 1993
	grassland	+ 3°C	2 years	increase in abundance and activity of snails and slugs, no change in species composition of snails and slugs	Sternberg, 2000
	deciduous forest	laboratory incubation (5-25°C)	1 day	increase in soil respiration with increase in temperature	Kang et al., 2003
	grassland	laboratory incubation (10-35°C)	16 days	increase in soil respiration in temperature	Fierer et al., 2003
	deciduous forest	laboratory incubation (5-25°C)	4 months	increase in soil respiration with increase in temperature	Zogg et al., 1997
	polar tundra	laboratory incubation (4.3 and 9.6°C)	5 months	increase in soil respiration, increase in litter decomposition rate (mass loss)	Hobbie, 1996
	deciduous forest	laboratory incubation (5-25°C)	4 months	no change in soil respiration	Zak et al., 1999
	polar tundra, taiga	laboratory incubation (freeze-thaw cycles)	5 days	increase in soil respiration	Schimel and Clein, 1996
	boreal forest	laboratory incubation (freeze-thaw cycles)	1 month	no change in soil respiration	Sulkava and Huhta, 2003

TABLE 5.3 (CONTINUED)
Effects of Elevated Temperature on Belowground Trophic Structure

Trophic Level	Ecosystem Type	Type of Manipulation	Duration of Manipulation	Results	Reference
III. Bacterivores/Fungivores	peatland	laboratory incubation (12-18°C)	2 months	decrease in abundance of enchytraeids with increase in temperature	Cole et al., 2002
	polar tundra	+ 2°C	3 months	increase in abundance of enchytraeids	Briones et al., 1997
	boreal forest	laboratory incubation (freeze-thaw cycles)	1 month	increase in abundance of enchytraeids	Sulkava and Huhta, 2003
	coniferous forest	+ 2.4°C	6 years	no change in abundance of enchytraeids	Haimi et al., 2005
	polar tundra	+ 2.2°C	1 year	increase in abundance of nematodes, especially microbivorous nematodes	Convey and Wynn-Williams, 2002
	alpine tundra, subalpine forest	slope aspect (+ 2.2°C)	1 year	increase in abundance of nematodes on south-facing slopes (N. Hemisphere)	Hoschitz and Kaufmann, 2004
	subarctic heathland	+ 0.9-2°C	6 years	increase in abundance of nematodes, decrease in diversity of nematodes	Ruess et al., 1999
	N/A	laboratory incubation (15 and 23°C)	2 months	increase in growth and fertility rate of a nematode species with increase in temperature	Popovici, 1973
	grassland	+ 6-9°C	4 months	no change in abundance of nematodes, decrease in diversity of nematodes	Bakonyi and Nagy, 2000
	N/A	laboratory incubation (5-30°C)	6 days	no change in nematode respiration, increase in movement of nematodes with increase in temperature	Dusenbery et al., 1978
	polar tundra	transplantation mean annual T between -0.7 and +7.9°C	1 year	decrease in abundance of nematodes with increase in mean annual temperature	Sohlenius and Boström, 1999
	grassland	laboratory incubation (5-40°C)	1 month	decrease in abundance of bacterivorous nematodes at extreme low and high temperatures, optimal growth temperature different for different species	Anderson and Coleman, 1982
	N/A	laboratory incubation (10-35°C)	1.5 months	decrease in generation time of fungivorous nematodes with increase in temperature	Pillai and Taylor, 1967
	boreal forest	laboratory incubation (freeze-thaw cycles)	1 month	increase in abundance of microarthropods	Sulkava and Huhta, 2003
	polar tundra	+ 2.2°C	8 years	increase in abundance of mites and springtails, increase in proportion of small individuals	Kennedy, 1994
	polar semi-desert, heathland	10% increase in summer heat budget	3 years	no change in abundance of mites or springtails in tundra heath, increase in abundance of springtails in polar semi-desert	Coulson et al., 1996
	polar tundra	laboratory incubations (freeze-thaw cycles)	3 months	no change in abundance of mites or springtails	Sjursen et al., 2005
	polar tundra	30% increase in time above 0°C	2.25 years	no change in abundance of mites or springtails	Sinclair, 2002
	coniferous forest	+ 2-4°C	6 years	no change in abundance of orbatid mites, decrease in abundance of acaridid mites	Haimi et al., 2005
	polar tundra	laboratory incubation (2-12°C)	10 months	increase in mortality of mites with increase in temperature, increase in growth rate and molting rate of mites with increase in temperature	Convey, 1994
	polar tundra	+ 0.4°C	4 years	decrease in abundance of microarthropods (especially springtails)	Convey et al., 2002
	subalpine meadow	+ 1°C	2.5 years	increase in mesofaunal biomass and diversity during cool wet summer, decrease in mesofaunal biomass and diversity during warmer drier summer	Harte et al., 1996
IV. Carnivores	N/A	N/A	N/A	N/A	N/A

tend to show a positive response to warming (Ruess et al. 1999; Rillig et al. 2002), perhaps due to their close association with increased root production. Increased turnover rates of microbial biomass (Joergensen et al. 1990) and altered microbial community composition (Zogg et al. 1997) have also been found at elevated temperatures. Warming tends to increase soil respiration (Hobbie 1996; Zogg et al. 1997; Fierer et al. 2003; Kang et al. 2003), which indicates that soil organisms metabolize carbon at faster rates. This is consistent with increased microbial turnover rates and decreased microbial biomass. Warming also tends to increase growth rates of earthworms (Viljoen and Reinecke 1992), and increase the abundance and activity of slugs and snails (Sternberg 2000). While laboratory experiments tend to show a negative effect of warming on the second trophic level, field experiments show no effect or a positive effect. This contradiction represents an area for future research. Is 2 to 3°C warming during the next century enough to cause long-term changes in bacterial and fungal metabolic rates, or are moisture, nutrient availability, and litter quality more important controls? Responses may depend on ecosystem-specific changes in soil moisture by warming. For the time being, it is appropriate to accept the hypothesis that predicted magnitudes of warming will only have small effects on production at the second trophic level.

The third trophic level tends to be positively affected by warming. Enchytraeids, which are most important in high latitude ecosystems, can respond positively to warming in the field (Briones et al. 1997) and to freeze–thaw cycles in the laboratory (Sulkava and Huhta 2003), but effects can also be absent (Haimi et al. 2005) or negative (Cole et al. 2002). Temperature is expected to influence the species composition and population density of enchytraeids, but probably not their presence or absence, as there seem to be species adapted to most climates (Didden 1993). Warming increases nematode abundances, both in field manipulations (Ruess et al. 1999) and along environmental gradients (e.g., north- to south-facing slopes) (Hoschitz and Kaufmann 2004). Microbivorous nematodes (i.e., nematodes that eat bacteria and fungi) seem especially favored under warmer conditions (Convey and Wynn-Williams 2002). This result is consistent with the higher microbial turnover rates and lower microbial biomass. Warming also tends to decrease nematode diversity (Ruess et al. 1999; Bakonyi and Nagy 2000), increase nematode growth and fertility rates (Popovici 1973; Pillai and Taylor 1967) and increase nematode mobility (Dusenbery et al. 1978). If nematode diversity decreases, bacterivorous and fungivorous nematodes are probably the ones taking advantage of warmer conditions, and represent another example of how climatic change can modify belowground trophic structure. Responses to warming are less consistent among microarthropods. In some cases, warming increases microarthropod abundances (Kennedy 1994; Coulson et al. 1996); in other cases, warming decreases microarthropod abundances (Convey et al. 2002; Haimi et al. 2005) or has no effect (Coulson et al. 1996; Sinclair 2002; Haimi et al. 2005; Sjursen et al. 2005). This variability may be a result of interannual variation. In a subalpine meadow, experimental warming increased mesofaunal biomass during a cool wet summer, but decreased mesofaunal biomass during a warm dry summer (Harte et al. 1996). This suggests that precipitation may be more important than temperature in controlling abundances of soil fauna, and is consistent with the strong negative effects of drought described earlier.

Soil predators in the fourth trophic level are rarely investigated in warming experiments, and represent a wide open area for future research. However, it should also be noted that some of the experiments presented in the third trophic level can also apply to the fourth trophic level. Some nematodes and microarthropods also eat other soil animals. Because most studies do not differentiate between bacterivores, fungivores, and predators, it is difficult to identify exactly what changes in abundances mean in terms of trophic structure. In general, changes in abundances of fourth trophic level participants are expected to positively track changes in the third trophic level, but it is not yet known whether increases in third trophic level production by warming increases production at higher trophic levels.

5.3.4 EFFECTS OF ELEVATED ATMOSPHERIC CARBON DIOXIDE

Carbon dioxide (CO_2) enrichment experiments are numerous compared to precipitation and warming experiments, and range in scale from laboratory mesocosms to field treatments surrounding forests (Table 5.4). Fortunately, soil organisms are often studied in these experiments, with a similar pattern of focus as those found in precipitation and warming experiments: Most studies pay attention to the second trophic level, fewer studies pay attention to the third trophic level, and almost nobody pays attention to the fourth trophic level.

In most studies, second trophic level participants do not show a response to elevated CO_2. For example, no effects of elevated CO_2 were found on bacterial abundances (O'Neill et al. 1987; Klironomos et al. 1996; Schortemeyer et al. 1996; Rillig et al. 1997; Treonis and Lussenhop 1997; Rillig et al. 1999b; Schortemeyer et al. 2000) or microbial biomass (Hungate et al. 1997; Jones et al. 1998; Lussenhop et al. 1998; Schortemeyer et al. 2000; Zak et al. 2000; Niklaus et al. 2003; Sonnemann and Wolters, 2005). When effects are found, the response tends to be positive (Zak et al. 1993; Marilley et al. 1999; Insam et al. 1999; Sonnemann and Wolters 2005), especially in the rhizosphere where C exudation occurs (Cotrufo and Gorissen 1997; Jones et al. 1998). Microbial immobilization of C and N tends to increase under elevated CO_2 (Berntson and Bazzaz 1997; Hungate et al. 1997; Niklaus 1998; Hungate et al. 1999; Williams et al. 2000), but net mineralization has also been observed (e.g., Zak et al. 1993). Mycorrhizal fungi, plant symbionts that can be ambiguously placed in the second trophic level, seem to take the greatest advantage of increased root C exudation under elevated CO_2, as indicated by increased root colonization (Godbold et al. 1997; Kasurinen et al. 2005) and hyphal length (Klironomos et al. 1996; Klironomos et al. 1997; Runion et al. 1997; Sanders et al. 1998; Rillig et al. 1999b). Nonmycorrhizal fungi tend to be less affected (Klironomos et al. 1996) or negatively affected (Klironomos et al. 1997), probably reflecting their greater distance from the root source of C exudation. Organisms that are closely associated with roots and root exudates (e.g., mycorrhizal fungi and rhizospheric bacteria) seem to be especially responsive to the carbon pulse from plants at the onset of CO_2 exposure, but effects can disappear over time. Sometimes microbial diversity (Mayr et al. 1999; Phillips et al. 2002) and activity (Rillig et al. 1997; Mayr et al. 1999; Schortemeyer et al. 2000) change under elevated CO_2, while other times structure (Jones et al. 1998; Insam et al. 1999; Bruce et al. 2000; Niklaus et

TABLE 5.4
Effects of Elevated Atmospheric CO_2 on Belowground Trophic Structure

Trophic Level	Ecosystem Type	Magnitude of Manipulation (as compared to ambient reference of 370 ppm)	Duration of Manipulation	Results	Reference
II. Detritivores/Herbivores					
	mesocosm (deciduous)	+35 Pa (partial pressure)	5 months	no change in microbial biomass in rhizosphere	Lussenhop et al., 1998
	mesocosm (weedy field)	+200 ppm	9 months	no change in microbial biomass	Jones et al., 1998
	grassland	+360 ppm	1 year	no change in microbial biomass	Hungate et al., 1997
	scrub oak woodland	+390 ppm	2 years	no change in microbial biomass in bulk soil	Schortemeyer et al., 2000
	mesocosm (deciduous)	+35 Pa	2.5 years	no change in microbial biomass	Zak et al., 2000
	grassland	+230 ppm	6 years	no change in microbial biomass	Niklaus et al., 2003
	mesocosm (grassland)	+330 ppm	2 months	increase in microbial biomass in rhizosphere	Cotrufo and Gorissen, 1997
	mesocosm (deciduous)	+322 ppm	5 months	increase in microbial biomass C in rhizosphere and bulk soil	Zak et al., 1993
	mesocosm (herbaceous)	+330 ppm	1 month	no change in abundance of bacteria	Treonis and Lussenhop, 1997
	mesocosm (sagebrush)	+330 ppm	3 months	no change in abundance of bacteria	Klironomos et al., 1996
	mesocosm (woody shrub)	+380 ppm	4 months	no change in abundance of bacteria in rhizosphere	Rillig et al., 1997
	mesocosm (deciduous)	+320 ppm	6 months	no change in abundance of bacteria in rhizosphere	O'Neill et al., 1987
	mesocosm (grassland)	+230 ppm	1.5 years	no change in abundance or biomass of bacteria in rhizosphere	Schortemeyer et al., 1996
	scrub oak woodland	+390 ppm	2 years	no change in abundance of bacteria in rhizosphere	Schortemeyer et al., 2000
	grassland	+74 ppm	3 years	increase in bacterial biomass	Sonnemann and Wolters, 2005
	grassland	+330 ppm	6 years	no change in abundance of bacteria	Rillig et al., 1999b
	mesocosm (grassland)	+35 Pa	3 months	increase in abundance of bacteria in rhizosphere	Marilley et al., 1999
	mesocosm (tropical)	+240 ppm	1.5 years	increase in abundance of bacteria	Insam et al., 1999
	mesocosm (weedy field)	+200 ppm	9 months	increase in abundance of saprophagous fungi	Jones et al., 1998
	mesocosm (deciduous/coniferous)	+330 ppm	8 months	increase in ectomycorrhizal colonization	Godbold et al., 1997
	mesocosm (grassland)	+230 ppm	5 months	increase in length of arbuscular mycorrhizal fungi	Sanders et al., 1998
	mesocosm (deciduous)	+350 ppm	3 years	increase in mycorrhizal fungi infection rate; no change in ratio of fungi to bacteria	Kasurinen et al., 2005
	grassland	+74 ppm	3 years	no change in fungal biomass	Sonnemann and Wolters, 2005
	grassland	+330 ppm	6 years	increase in total fungal hyphal length	Rillig et al., 1999b
	mesocosm (coniferous)	+350 ppm	1.6 years	increase in density of ectomycorrhizal fungi	Runion et al., 1997
	mesocosm (deciduous)	+330 ppm	1.2 years	increase in hyphal length of arbuscular mycorrhizal fungi, decrease in hyphal length of non-mycorrhizal fungi	Klironomos et al., 1997
	grassland	+190 ppm	3 years	no change in arbuscular mycorrhizal fungal spore production	Wolf et al., 2003
	mesocosm (sagebrush)	+330 ppm	3 months	no change in abundance of non-mycorrhizal fungi; increase in abundance of arbuscular mycorrhizal fungi	Klironomos et al., 1996
	mesocosm (grassland)	+350 ppm	4 months	increase, no change, or decrease in arbuscular mycorrhizal fungi depending on plant species	Rillig et al., 1998
	mesocosm (weedy field)	+200 ppm	9 months	no change in bacterial community composition using DNA profiles	Jones et al., 1998
	mesocosm (annuals)	+200 ppm	10 months	no change in bacterial community structure using DGGE	Bruce et al., 2000
	mesocosm (tropical)	+240 ppm	1.5 years	no change in microbial community composition using CLPPs and PLFA	Insam et al., 1999
	grassland	+230 ppm	6 years	no change in microbial community composition using PLFA	Niklaus et al., 2003
	deciduous forest	+190 ppm	3 years	altered microbial community composition using PLFA	Phillips et al., 2002
	alpine meadow	+310 ppm	4 years	altered microbial community composition using CLPPs	Mayr et al., 1999
	grassland	+360 ppm	1 year	increase in microbial N immobilization	Hungate et al., 1997
	mesocosm (deciduous)	+330 ppm	1 year	increase in microbial N immobilization	Berntson and Bazzaz, 1997

TABLE 5.4 (CONTINUED)
Effects of Elevated Atmospheric CO_2 on Belowground Trophic Structure

Trophic Level	Ecosystem Type	Magnitude of Manipulation (as compared to ambient reference of 370 ppm)	Duration of Manipulation	Results	Reference
	scrub oak woodland	+350 ppm	1.2 years	increase in microbial immobilization rate	Hungate et al., 1999
	mesocosm (deciduous)	+35 Pa	2 years	increase in microbial C immobilization	Mikan et al., 2000
	grassland	+230 ppm	3 years	increase in microbial N immobilization	Niklaus, 1998
	grassland	+370 ppm	8 years	increase in microbial C and N immobilization	Williams et al., 2000
	deciduous forest coniferous forest	+195 ppm	3 years	no change in microbial N immobilization	Zak et al., 2003
	mesocosm (deciduous)	+322 ppm	5 months	increase in N mineralization	Zak et al., 1993
	mesocosm (grassland)	+330 ppm	8 months	no change in microbial substrate utilization in rhizosphere	Van Ginkel et al., 2000
	deciduous forest	+195 ppm	2 years	no change in soil extracellular enzyme activities, no change in microbial substrate utilization	Sinsabaugh et al., 2003
	scrub oak woodland	+390 ppm	2 years	decrease in microbial activity using FDA hydrolysis	Schortemeyer et al., 2000
	mesocosm (woody shrub)	+380 ppm	4 months	altered microbial substrate utilization using CLPPs	Rillig et al., 1997
	alpine meadow	+310 ppm	4 years	altered microbial enzyme activity	Mayr et al., 1999
	grassland	+74 ppm	3 years	no change in abundance of herbivorous nematodes	Sonnemann and Wolters, 2005
	mesocosm (deciduous)	+330 ppm	5 months	decrease in abundance of earthworms	Lussenhop et al., 1998
	mesocosm (grassland)	+330 ppm	1.3 years	increase in abundance of earthworms	Yeates et al., 1997
	deciduous forest	+190 ppm	3 years	increase in soil respiration	Phillips et al., 2002
	grassland	+370 ppm	8 years	increase in soil respiration	Williams et al., 2000
III. Bacterivores/Fungivores	mesocosm (deciduous)	+35 Pa	5 months	increase in abundance of protozoa	Lussenhop et al., 1998
	mesocosm (herbaceous)	+330 ppm	1 month	increase in abundance of flagellates, decrease in abundance of amoebae	Treonis and Lussenhop, 1997
	grassland	+330 ppm	6 years	no change in abundance of protozoa	Rillig et al., 1999b
	mesocosm (grassland)	+330 ppm	1.3 years	increase in abundance of enchytraeids	Yeates et al., 1997
	coniferous forest	+350 ppm	6 years	no change in abundance of enchytraeids	Haimi et al., 2005
	grassland	natural CO_2 vent (370 to 3900 ppm)	N/A	decrease in abundance and diversity of nematodes, increase in dominance of bacterivorous nematodes	Yeates et al., 1999
	mesocosm (grassland)	+330 ppm	1.3 years	increase in abundance of bacterivorous nematodes	Yeates et al., 1997
	grassland	+330 ppm	4 years	no change in abundance of bacterivorous or fungivorous nematodes, decrease in nematode diversity	Hungate et al., 2000
	mesocosm (sagebrush)	+330 ppm	3 months	increase in abundance of nematodes	Klironomos et al., 1996
	mesocosm (weedy field)	+200 ppm	9 months	increase in abundance of springtails, altered species composition of springtails	Jones et al., 1998
	coniferous forest	+350 ppm	6 years	no change in abundance of of orbatid mites, decrease in abundance of acaridid mites	Haimi et al., 2005
	grassland	+230 ppm	6 years	no change in abundance of mites or springtails	Niklaus et al., 2003
	mesocosm (deciduous)	+330 ppm	1.2 years	no change in abundance of mites, decrease in abundance of springtails	Klironomos et al., 1997
	grassland	+330 ppm	6 years	increase in abundance of mites and springtails	Rillig et al., 1999b
	mesocosm (grassland)	+180 ppm	N/A	increase in attractivity of invertebrate (sciarid fly) to fungi in litter	Frouz et al., 2002
	mesocosm (sagebrush)	+330 ppm	3 months	no change in abundance of microbivorous arthropods	Klironomos et al., 1996
IV. Carnivores	mesocosm (grassland)	+330 ppm	1.3 years	increase in abundance of predatory and omnivorous nematodes	Yeates et al., 1997
	mesocosm (sagebrush)	+330 ppm	3 months	no change in abundance of predatory arthropods	Klironomos et al., 1996
	grassland	+74 ppm	3 years	decrease in abundance of predatory nematodes	Sonnemann and Wolters, 2005
	grassland	+230 ppm	6 years	decrease in abundance of predatory and omnivorous nematodes	Niklaus et al., 2003

al. 2003) and activity (van Ginkel et al. 2000; Sinsabaugh et al. 2003) remain the same. Sometimes earthworm abundances decrease (Lussenhop et al. 1998), while other times abundances increase (Yeates et al. 1997).

These varied results, and results from precipitation and warming experiments, stress that climatic change effects on belowground trophic structure are strongly ecosystem dependent. The belowground food web operates with the context of a particular set of climatic variables, plant community properties, and physical soil properties. And even though all dirt kind of looks the same, the communities living within different soils do not have the same responses to climatic change factors. There are quite a few examples of the same trophic level responding differently to the same climatic change factor in the same type of ecosystem. This suggests that responses of belowground structure to climatic change may be controlled by different bottom–up and top–down forces. And these different trophic forces can vary within the same soil over time, probably due to seasonal and annual variation in microclimate.

Now, let us return to elevated CO_2 effects on the third trophic level, where there are again few consistent responses. The longer-term field studies, however, tend to find a lack of response among third trophic level participants. In some instances, abundances of protozoa (flagellates) increase (Treonis and Lussenhop 1997; Lussenhop et al. 1998), in other instances abundances (amoebae) decrease (Treonis and Lussenhop 1997). Enchytraeid abundances have been shown to increase (Yeates et al. 1997) or not change (Haimi et al. 2005). Nematode abundances may increase (Klironomos et al. 1996), decrease (Yeates et al. 1999), or not change (Hungate et al. 2000). Nematode diversity tends to decrease (Yeates et al. 1999; Hungate et al. 2000). Decreased diversity could reflect a greater dominance of a few species of bacterivorous nematodes (Yeates et al. 1997; Yeates et al. 1999), and illustrates how changes in trophic diversity may be hidden beneath undetectable changes in abundances or biomass. Elevated CO_2 can increase (Rillig et al. 1999b), have no effect (Klironomos et al. 1997; Niklaus et al. 2003; Haimi et al. 2005), or decrease (Haimi et al. 2005) abundances of mites. Springtails respond positively (Jones et al. 1998; Rillig et al. 1999b), negatively (Klironomos et al. 1997), or not at all (Niklaus et al. 2003). And microbivorous arthropods show no response (Klironomos et al. 1996). There is interesting evidence that elevated CO_2 can affect the third trophic level through changes in the second trophic level (Jones et al. 1998). In a model system, elevated CO_2 increased abundances of saprophagous fungi, probably due to increased soil carbon availability. This change in fungal composition affected the next highest trophic level, as indicated by increased springtail abundances, and a change in springtail diversity to a community probably dominated by species well suited at eating newly dominant saprophagous fungal species. This is a rare example of how climatic change may affect belowground trophic structure in a single (model) ecosystem.

Once again, climatic change effects on the fourth trophic level are understudied in comparison to lower trophic levels, and no consistent responses to elevated CO_2 are apparent yet. Predatory and omnivorous nematodes have been found to increase (Yeates et al. 1997) or decrease (Niklaus et al. 2003; Sonnemann and Wolters 2005) in abundance. These different responses may be caused by differences in length of experimentation, with the increase reflecting about a year of treatments and the decrease reflecting 6 years of treatments. Increased microbial turnover may be a relatively short-lived source of energy for bacterivores, fungivores, and predators, and trophic structure may eventually return to more of an upright biomass pyramid as omnivory compensates for increased abundance of predators. On a shorter time scale, predatory arthropods showed no response after three months at elevated CO_2 (Klironomos et al. 1996).

5.4 CONCLUSIONS

The belowground food web is biologically complex, functionally important, and apparently responsive to climatic change. However, knowledge of belowground food webs, especially in unmanaged ecosystems, lags behind that of aboveground and aquatic food webs. While soil organisms sometimes respond to climatic change through adjustments in population size, adjustments may also occur in activity (e.g., trophic efficiencies), diversity, and spatial distribution—properties of belowground food webs that are rarely examined.

Figure 5.3 attempts to summarize results from over 100 studies. Organisms were sampled in different ecosystems, with different soil types, and at a particular micro-climate in time. They had also experienced a wide variety of experimental designs. The responses of the second, third, and fourth trophic levels are based on changes in abundance or biomass, and the consistency and availability of data. The first conclusion is that there does not seem to be a common global response of trophic structure to climatic change. There are some cases (e.g., reduced precipitation) in which most studies agree, but there is usually an exception. If there are already exceptions, given the low availability of studies, we can expect to find more excep-tions in the future. Because of context-dependent responses, the generalized responses in Figure 5.3 are expected results, but may not be the case in every ecosystem on every sampling date. The benefit of the doubt is given to a "no response," if there is a large proportion of studies that found a lack of response or if there is not strong enough evidence to assign a common directional response. However, arrows are displayed if multiple studies agree on a directional response.

5.4.1 SUMMARY OF CLIMATIC CHANGE EFFECTS

It is expected that belowground food web responses to climatic change are dominated by direct effects in the cases of altered precipitation and elevated temperature (e.g., changes in metabolism, mobility, and vertical distribution), and by indirect effects in the case of elevated atmospheric CO_2 (e.g., changes in root production, soil structure, and soil moisture).

Despite greater abundances of second trophic level participants under elevated precipitation, population sizes in the third and fourth trophic levels tend to remain the same. What is the fate of this increased production at the bottom of the food web? Higher trophic levels probably use this extra production to metabolize faster or move around more, which may result in a change in ecosystem function (e.g., faster decomposition). Because most of the data on belowground food web participants are based on snapshot views of abundances, we need to be careful in

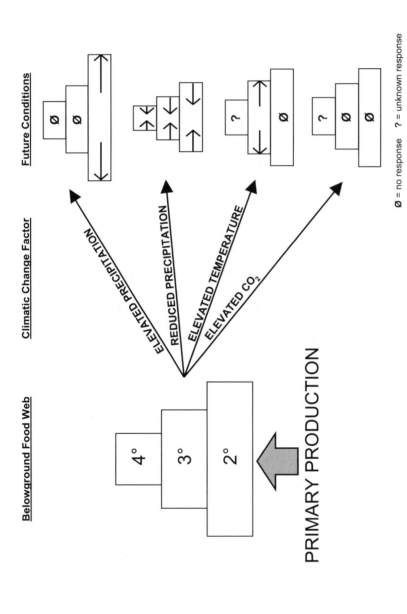

FIGURE 5.3 Climatic change effects on belowground trophic structure showing common or unknown responses of second, third, and fourth, trophic levels. It is important to note that these responses are generalized and that different ecosystems can display different responses.

interpreting a lack of response. There could be an associated change in activity, which is the main concern of any land manager.

All trophic levels tend to be negatively affected by reduced precipitation, but it is unknown whether this is caused by individual responses of organisms to water stress and reduced mobility, or a bottom–up effect associated with a reduction in secondary production. Warming tends to have little effect on the second trophic level and a positive effect on the third trophic level. The third trophic level may respond positively to warming because of physiological preferences or accelerated microbial turnover rates. Under elevated atmospheric CO_2, no consistent changes are observed in abundances of second trophic level participants, but there are instances when increased carbon input by plants under elevated CO_2 cascade up to higher trophic levels. When no change is found, it is possible that third trophic level grazing and fourth trophic level predation hide any increases in second trophic level abundances. By understanding mechanisms for climatic change effects on belowground trophic structure, we can increase our understanding of more general mechanisms for climatic change effects on ecosystem functions.

5.4.2 RESEARCH NEEDS

Sampling belowground structure requires a significant amount of time and effort for isolating and identifying the high diversity of dirty microscopic organisms. Results from different ecosystems show that it is more likely to be worth the time and effort if multiple trophic levels are studied in a single ecosystem. Other issues include biological dormancy, which allows organisms to survive through extremely hot, cold, wet, and dry conditions. Lateral gene transfer can occur between microorganisms, allowing a species to acquire a novel gene, and perhaps participate in a different trophic level or tolerate a new climate. Organisms can be omnivorous or picky in their food selection. And bottom–up and top–down trophic interactions appear to have different effects in different ecosystems. While abundances have laid the foundation for our understanding of belowground food webs, methods accounting for net energy flow and greater soil biodiversity will improve our ability to predict responses to climatic change. We need to examine all trophic groups at the same time, in the same ecosystem, and under different microclimates. Climatic change does not always change belowground trophic structure, but it can in certain contexts. And climatic change factors may interact to produce unique responses. Through long-term ecosystem-specific studies, we can figure out where and when changes in belowground trophic structure will be most important in the future.

REFERENCES

Adams, G.A. and Wall, D.H. (2000) Biodiversity above and below the surface of soils and sediments: linkages and implications for global change. *Bioscience*, 50, 1043–1048.

Anderson, R.V. and Coleman, D.C. (1982) Nematode temperature responses: a niche dimension in populations of bacterial-feeding nematodes. *Journal of Nematology*, 14, 69–76.

André, H.M., Ducarme, X., and Lebrun, P. (2002) Soil biodiversity: myth, reality or conning? *Oikos*, 96, 3–24.

Arnone, J.A., III, Zaller, J.G., Spehn, E.M., Niklaus, P.A., Wells, C.E., and Körner, C. (2000) Dynamics of root systems in native grasslands: effects of elevated atmospheric CO_2. *New Phytologist*, 147, 73–85.

Badeck, F.W., Bondeau, A., Böttcher, K., Doktor, D., Lucht, W., Schaber, J., and Sitch, S. (2004) Responses of spring phenology to climate change. *New Phytologist*, 162, 295–309.

Bakonyi, G. and Nagy, P. (2000) Temperature- and moisture-induced changes in the structure of the nematode fauna of a semiarid grassland—patterns and mechanisms. *Global Change Biology*, 6, 697–707.

Bardgett, R.D., Kandeler, E., Tscherko, D., Hobbs, P.J., Bezemer, T.M., Jones, T.H., and Thompson, L.J. (1999) Below-ground microbial community development in a high temperature world. *Oikos*, 85, 193–203.

Bengtsson, J., Jones, H., and Setälä, H. (1997) The value of biodiversity. *TREE*, 12, 334–336.

Berntson, G.M. and Bazzaz, F.A. (1997) Nitrogen cycling in microcosms of yellow birch exposed to elevated CO_2: simultaneous positive and negative below-ground feedbacks. *Global Change Biology*, 3, 247–258.

Bouwman, L.A., Bloem, J., van den Boogert, P.H.J.F., Bremer, F., Hoenderboom, G.H.J., and de Ruiter, P.C. (1994) Short-term and long-term effects of bacterivorous nematodes and nematophagous fungi on carbon and nitrogen mineralization in microcosms. *Biology and Fertility of Soils*, 17, 249–256.

Briones, M.J.I., Ineson, P., and Piearce, T.G. (1997) Effects of climate change on soil fauna: responses of enchytraeids, Diptera larvae and tardigrades in a transplant experiment. *Applied Soil Ecology*, 6, 117–134.

Bruce, K.D., Jones, T.H., Bezemer, T.M., Thompson, L.J., and Rotchie, D.A. (2000) The effect of elevated atmospheric carbon dioxide levels on soil bacterial communities. *Global Change Biology*, 6, 427–434.

Chapin, F.S., III, Shaver, G.R., Giblin, A.E., Nadelhoffer, K.J., and Laundre, J.A. (1995) responses of arctic tundra to experimental and observed changes in climate. *Ecology*, 76, 694–711.

Chapin, F.S., III, Walker, B.H., Hobbs, R.J., Hooper, D.U., Lawton, J.H., Sala, O.E., and Tilman, D. (1997) Biotic control over the functioning of ecosystems. *Science*, 277, 500–503.

Clarholm, M. (1981) Protozoan grazing of bacteria in soil—impact and importance. *Microbial Ecology*, 7, 343–350.

Cole, L., Bardgett, R.D., Ineson, P., and Hobbs, P.J. (2002) Enchytraeid worm (Ologochaeta) influences on microbial community structure, nutrient dynamics and plant growth in blanket peat subjected to warming. *Soil Biology and Biochemistry*, 34, 83–92.

Colinvaux, P.A. and Barnett, B.D. (1979) Lindeman and the ecological efficiency of wolves. *American Naturalist*, 114, 707–718.

Contin, M., Corcimaru, S., De Nobili, M., and Brookes, P.C. (2000) Temperature changes and the ATP concentration of the soil microbial biomass. *Journal Soil Biology and Biochemistry*, 32, 1219–1225.

Convey, P. (1994) The influence of temperature on individual growth rates of the Antarctic mite *Alaskozetes antarcticus*. *Acta Oecologica*, 15, 43–53.

Convey, P., Pugh, A., Jackson, C., Murray, A.W., Ruhland, C.T., Xiong, F.S., and Day, T,A, (2002) Response of Antarctic terrestrial microarthropods to long-term climate manipulations. *Ecology*, 83, 3130–3140.

Convey, P. and Wynn-Williams, D.D. (2002) Antarctic soil nematode response to artificial climate amelioration. *European Journal of Soil Biology*, 38, 255–259.

Cotrufo, M.F. and Gorissen, A. (1997) Elevated CO_2 enhances below-ground C allocation in three perennial grass species at different levels of N availability. *New Phytologist*, 137, 421–431.

Coulson, S.J., Hodkinson, I.D., Webb, N.R., Block, W., Bale, J.S., Strathdee, A.T., Worland, M.R., and Wooley, C. (1996) Effects of experimental temperature elevation on high-arctic soil microarthropod populations. *Polar Biology*, 16, 147–153.

Coulson, S.J., Leinaas, H.P., Ims, R.A., and Søvik, G. (2000) Experimental manipulation of the winter surface ice layer: The effects on a high Arctic soil microarthropod community. *Ecography*, 23, 299–306.

Darnell, R.M. (1961) Trophic spectrum of an estuarine community, based upon studies of Lake Ponchartrain, LA. *Ecology*, 42, 553–568.

de Ruiter, P.C., Neutel, A,M,, and Moore, J.C. (1998) Biodiversity in soil ecosystems: the role of energy flow and community stability. *Applied Soil Ecology*, 10, 217–228.

Didden, W.A.M. (1993) Ecology of terrestrial Enchytraeidae. *Pedobiologia*, 37, 2–29.

Dijkstra, P., Hymus, G., Colavito, D., Vieglais, D.A., Cundari, C.M., Johnson, D.P., Hungate, B.A., Hinkle, C.R., and Drake, B.G. (2002) Elevated atmospheric CO_2 stimulates aboveground biomass in a fire-regenerated scrub-oak ecosystem. *Global Change Biology*, 8, 90–103.

Dusenbery, D.B., Anderson, G.L., and Anderson, E.A. (1978) Thermal acclimation more extensive for behavioral parameters than for oxygen consumption in the nematode *Caenorhabditis elegans*. *Journal of Experimental Zoology*, 206, 191–198.

Ekschmitt, K. and Griffiths, B.S. (1998) Soil biodiversity and its implications for ecosystem functioning in a heterogeneous and variable environment. *Applied Soil Ecology*, 10, 201–215.

Farnsworth, E.J., Núñez-Farfán, J., Careaga, S.A., and Bazzaz, F.A. (1995) Phenology and growth of three temperate forest life forms in response to artificial soil warming. *Journal of Ecology*, 83, 967–977.

Ferguson, S.H. and Joly, D.O. (2002) Dynamics of springtail and mite populations: the role of density dependence, predation, and weather. *Ecological Entomology*, 27, 565–573.

Ferris, R., Sabatti, M., Miglietta, F., Mills, R.F., and Taylor, G. (2001) Leaf area is stimulated in *Populus* by free air CO_2 enrichment (POPFACE), through increased cell expansion and production. *Plant, Cell and Environment*, 24, 305–315.

Field, C.B., Lund, C.P., Chiariello, N.R., and Mortimer, B.E. (1997) CO_2 effects on the water budget of grassland microcosm communities. *Global Change Biology*, 3, 197–206.

Fierer, N. and Schimel, J.P. (2002) Effects of drying-rewetting frequency on soil carbon and nitrogen transformations. *Journal Soil Biology and Biochemistry*, 34, 777–787.

Fierer, N., Allen, A.S., Schimel, J.P., and Holden, P.A. (2003) Controls on microbial CO_2 production: a comparison of surface and subsurface soil horizons. *Global Change Biology*, 9, 1322–1332.

Freckman, D.W., Whitford, W.G., and Steinberger, Y. (1987) Effect of irrigation on nematode population dynamics and activity in desert soils. *Biology and Fertility of Soils*, 3, 3–10.

Frouz, J., Nováková, A., and Jones, T.H. (2002) The potential effect of high atmospheric CO_2 on soil fungi-invertebrate interactions. *Global Change Biology*, 8, 339–344.

Gallardo, A. and Schlesinger, W.H. (1995) Factors determining soil microbial biomass and nutrient immobilization in desert soils. *Biogeochemistry*, 28, 55–68.

Godbold, D.L., Berntson, G.M., and Bazzaz, F.A. (1997) Growth and mycorrhizal colonization of three North American tree species under elevated atmospheric CO_2. *New Phytologist*, 137, 433–440.

Goyal, R.K. (2004) Sensitivity of evapotranspiration to global warming: a case study of arid zone of Rajasthan (India). *Agricultural Water Management*, 69, 1–11.

Grisi, B., Grace, C., Brookes, P.C., Benedetti, A., and Dell'Abate, M.T. (1998) Temperature effects of organic matter and microbial biomass dynamics in temperate and tropical soils. *Journal Soil Biology and Biochemistry*, 30, 1309–1315.

Gulledge, J. and Schimel, J.P. (1998) Moisture control over atmospheric CH_4 consumption and CO_2 production in diverse Alaskan soils. *Journal Soil Biology and Biochemistry*, 30, 1127–1132.

Haimi, J., Laamanen, J., Penttinen, R., Räty, M., Koponen, S., Kellomäki, S., and Niemelä, P. (2005) Impacts of elevated CO_2 and temperature on the soil fauna of boreal forests. *Applied Soil Ecology*, 30, 104–112.

Hairston, N., Smith, F., and Slobodkin, L. (1960) Community structure, population control, and competition. *American Naturalist*, 94, 421–425.

Harte, J., Rawa, A., and Price, V. (1996) Effects of manipulated soil microclimate on meso-faunal biomass and diversity. *Journal Soil Biology and Biochemistry*, 28, 313–322.

Heinemeyer, A., Ridgway, K.P., Edwards, E.J., Benham, D.G., Young, J.P.W., and Fitter, A.H. (2003) Impact of soil warming and shading on colonization and community structure of arbuscular mycorrhizal fungi in roots of a native grassland community. *Global Change Biology*, 10, 52–64.

Hobbie, S.E. (1996) Temperature and plant species control over litter decomposition in Alaskan tundra. *Ecological Monographs*, 66, 503–522.

Hoschitz, M. and Kaufmann, R. (2004) Soil nematode communities of Alpine summits—site differentiation and microclimatic influences. *Pedobiologia*, 48, 313–320.

Hungate, B.A., Lund, C.P., Pearson, H.L., and Chapin, F.S., III (1997) Elevated CO_2 and nutrient addition alter soil N cycling and N trace gas fluxes with early season wet-up in a California annual grassland. *Biogeochemistry*, 37, 89–109.

Hungate, B.A., Dijkstra, P., Johnson, D.W., Hinkle, C.R., and Drake, B.G. (1999) Elevated CO_2 increases nitrogen fixation and decreases soil nitrogen mineralization in Florida scrub oak. *Global Change Biology*, 5, 781–789.

Hungate, B.A., Jaeger, C.H., III, Gamara, G., Chapin, F.S., III, and Field, C.B. (2000) Soil Microbiota in two grasslands: responses to elevated atmospheric CO_2. *Oecologia*, 124, 589–598.

Hungate, B.A., Reichstein, M., Dijkstra, P., Johnson, D., Hymus, G., Tenhunen, J.D., Hinkle, C.R., and Drake, B.G. (2002) Evapotranspiration and soil water content in a scrub-oak woodland under carbon dioxide enrichment. *Global Change Biology*, 8, 289–298.

Ingram, J. and Freckman, D.W. (1998) Soil biota and global change—preface. *Global Change Biology*, 4, 699–701.

Insam, H., Bååth, E., Berreck, M., Frostgård, A., Gerzabeck, M.H., Kraft, A., Schinner, F., Schweiger, P., and Tschuggnall, G. (1999) Responses of soil microbiota to elevated CO_2 in an artificial tropical ecosystem. *Journal of Microbiological Methods*, 36, 45–54.

Intergovernmental Panel on Climate Change (IPCC) (2001) Contribution of Working Group I to the Third Assessment Report of the Intergovernmental Panel on Climate Change. In *Climate Change 2001: The Scientific Basis*, Houghton, J.T., Ding, Y., Griggs, D.J., Noguer, M., van der Linden, P.J., Dai, X., Maskell, K., and Johnson, C.A., Eds., Cambridge University Press, Cambridge, U.K.

Joergensen, R.G., Brookes, P.C., and Jenkinson, D.S. (1990) Survival of the soil microbial biomass at elevated temperatures, *Journal Soil Biology and Biochemistry*, 22, 1129–1136.

Jonasson, S., Michelsen, A., Schmidt, I.K., and Nielsen, E.V. (1999) Responses in microbes and plants to changes temperature, nutrient, and light regimes in the Arctic. *Ecology*, 80, 1828–1843.

Jonasson, S., Castro, J., and Michelsen, A. (2004) Litter, warming and plants affect respiration and allocation of soil microbial and plant C, N and P in arctic mesocosms. *Journal Soil Biology and Biochemistry*, 36, 1129–1139.

Jones, T.H., Thompson, L.J., Lawton, J.H., Bezemer, T.M., Bardgett, R.D., Blackburn, T.M., Bruce, K.D., Cannon, P.F., Hall, G.S., Hartley, S.E., Howson, G., Jones, C.G., Kampichler, C., Kandeler, E., and Ritchie, D.A. (1998) Impacts of rising atmospheric carbon dioxide on model terrestrial ecosystems. *Science*, 280, 441–443.

Kandeler, E., Tscherko, D., Bardgett, R.D., Hobbs, P.J., Kampichler, C., and Jones, T.H. (1998) The response of soil microorganisms and roots to elevated CO_2 and temperature in a terrestrial model ecosystem. *Plant and Soil*. 202, 251–262.

Kang, S., Doh, S., Lee, D., Lee, D., Jin, V.L., and Kimball, J.S. (2003) Topographic and climate controls on soil respiration in six temperate and mixed hardwood forest slops, Korea. *Global Change Biology*, 9, 1427–1437.

Kasurinen, A., Keinänen, M.M., Kaipainen, S., Nilsson, L.O., Vapaavuori, E., Kontro, M.H., and Holopainen, T. (2005) Below-ground responses of silver birch trees exposed to elevated CO_2 and O_3 levels during three growing seasons. *Global Change Biology*, 11, 1167–1179.

Kennedy, A.D. (1994) Simulated climate change: a field manipulation study of polar microarthropod community response to global warming. *Ecography*, 17, 131–140.

Kieft, T.L., Soroker, E., and Firestone, M.K. (1987) Microbial biomass response to a rapid increase in water potential when dry soil is wetted. *Journal Soil Biology and Biochemistry*, 19, 119–126.

King, J.S., Thomas, R.B., and Strain, B.R. (1996) Growth and carbon accumulation in root systems of *Pinus taeda* and *Pinus ponderosa* seedlings as affected by varying CO_2, temperature, and nitrogen. *Tree Physiology*, 16, 635–642.

Klironomos, J.N., Rillig, M.C., and Allen, M.F. (1996) Below-ground microbial and microfaunal responses to *Artemisia tridentate* grown under elevated atmospheric CO_2. *Functional Ecology*, 10, 527–534.

Klironomos, J.N., Rillig, M.C., Allen, M.F., Zak, D.R., Kubiske, M., and Pregitzer, K.S. (1997) Soil fungal-arthropod responses to *Populus temuloides* grown under enriched atmospheric CO_2 under field conditions. *Global Change Biology*, 3, 473–478.

Kostecka, J. and Butt, K.R. (2001) Ecology of the earthworm *Allolobophora carpathica* in field and laboratory studies. *European Journal of Soil Biology*, 37, 255–258.

Laakso, J. and Setälä, H. (1999) Sensitivity of primary production to changes in the architecture of belowground food webs. *Oikos*, 87, 57–64.

Lavelle, P., Bignell, D., Lepage, M., Wolters, V., Roger, P., Ineson, P., Heal, O.W., and Dhillion, S. (1997) Soil function in a changing world: The role of invertebrate ecosystem engineers. *European Journal of Soil Biology*, 33, 159–193.

Lindberg, N., Engtsson, J.B., and Persson, T. (2002) Effects of experimental irrigation and drought on the composition and diversity of soil fauna in a coniferous stand. *Journal of Applied Ecology*, 39, 924–936.

Lindberg, N. and Persson, T. (2004) Effects of long-term nutrient fertilization and irrigation on the microarthropod community in a boreal Norway spruce stand. *Forest Ecology and Management*, 188, 125–135.

Lindberg, N. and Bengtsson, J. (2005) Population responses of oribatid mites and collembolans after drought. *Applied Soil Ecology*, 28, 163–174.

Lloret, F., Peñuelas, J., and Estiarte, M. (2004) Experimental evidence of reduced diversity of seedlings due to climate modification in a Mediterranean-type community. *Global Change Biology*, 10, 248–258.

Loreau, M. (2004) Does functional redundancy exist? *Oikos*, 104, 606–611.

Lundgren, B. and Söderström, B. (1983) Bacterial numbers in a pine forest soil in relation to environmental factors. *Journal Soil Biology and Biochemistry*, 15, 625–630.

Lussenhop, J., Treonis, A., Curtis, P.S., Teeri, J.A., and Vogel, C.S. (1998) Response of soil biota to elevated atmospheric CO_2 in poplar model systems. *Oecologia*, 113, 247–251.

Marilley, L., Hartwig, U.A., and Aragno, M. (1999) Influence of an elevated atmospheric CO_2 content on soil and rhizosphere bacterial communities beneath *Lolium perenne* and *Trifolium repens* under field conditions. *Microbial Ecology*, 38, 39–49.

Mayr, C., Miller, M., and Insam, H. (1999) Elevated CO_2 alters community-level physiological profiles and enzyme activities in alpine grassland. *Journal of Microbiological Methods*, 36, 35–43.

McLaren, B.E. and Peterson, R.O. (1994) Wolves, moose, and tree rings on Isle Royale. *Science*, 266, 1555–1558.

Mikan, C.J., Zak, D.R., Kubiske, M.E., and Pregitzer, K.S. (2000) Combined effects of atmospheric CO_2 and N availability on the belowground carbon and nitrogen dynamics in aspen mesocosms. *Oecologia*, 124, 432–445.

Mikola, J. and Setälä, H. (1998) No evidence of trophic cascades in an experimental microbial-based soil food web. *Ecology*, 79, 153–164.

Nelson, J.A., Morgan, J.A., LeCain, D.R., Mosier, A.R., Milchunas, D.G., and Parton, B.A. (2004) Elevated CO_2 increases soil moisture and enhances plant water relations in a long-term field study in a semi-arid shortgrass steppe of Colorado. *Plant and Soil*, 259, 169–179.

Niklaus, P.A., Spinnler, D., and Körner, C. (1998) Soil moisture dynamics of calcareous grassland under elevated CO_2. *Oecologia*, 117, 201–208.

Niklaus, P.A. (1998) Effects of elevated atmospheric CO_2 on soil microbiota in calcareous grassland. *Global Change Biology*, 4, 451–458.

Niklaus, P.A., Alphei, J., Ebersberger, D., Kampichler, C., Kandeler, E., and Tscherko, D. (2003) Six years of *in situ* CO_2 enrichment evoke changes in soil structure and soil biota of nutrient-poor grassland. *Global Change Biology*, 9, 585–600.

Niklaus, P.A. (2006) Climate change effects on biogeochemical cycles, nutrients, and water supply. In *Agroecosystems in a Changing Environment*, Newton. P.C.D., Carran, R.A., Edwards, G.R., and Niklaus, P.A., Eds., CRC Press, Boca Raton, FL, 11–54.

Norby, R.J. and Jackson, R.B. (2000) Root dynamics and global change: seeking an ecosystem perspective. *New Phytologist*, 147, 3–12.

Norby, R.J., Hartz-Rubin, J.S., and Verbrugge, M.J. (2003) Phenological responses in maple to experimental atmospheric warming and CO_2 enrichment. *Global Change Biology*, 9, 1792–1801.

O'Neill, E.G., Luxmoore, R.J., and Norby, R.J. (1987) Elevated atmospheric CO_2 effects on seedling growth, nutrient uptake, and rhizosphere bacterial populations of *Liriodendron tulipifera* L. *Plant and Soil*, 104, 3–11.

Owensby, C.E., Ham, J.M., Knapp, A.K., Auen, L.M. (1999) Biomass production and species composition change in a tallgrass prairie ecosystem after long-term exposure to elevated atmospheric CO_2. *Global Change Biology*, 5, 497–506.

Oztas, T. and Fayetorbay, F. (2003) Effect of freezing and thawing processes on soil aggregate stability. *Catena*, 52, 1–8.

Pace, M.L., Cole, J.J., Carpenter, S.R., and Kitchell, J.F. (1999) Trophic cascades revealed in diverse ecosystems. *TREE*, 14, 483–488.

Papatheodorou, E.M., Argyropoulou, M.D., and Stamou, G.P. (2004) The effects of large- and small-scale differences in soil temperature and moisture on bacterial functional diversity and the community of bacterivorous nematodes. *Applied Soil Ecology*, 25, 37–49.

Paterson, E., Hall, J.M., Rattray, E.A.S., Griffiths, B.S., Ritz, K., and Killham. K. (1997) Effect of elevated CO_2 on rhizosphere carbon flow and soil microbial processes. *Global Change Biology*, 3, 363–377.

Paustian, K. (1994) Modeling soil biology and biochemical processes for sustainable agricultural research. In *Soil Biota Management in Sustainable Farming Systems*, Pankhurst, C.E., Doube, B.M., Gupta, V.V.S.R., and Grace, P.R., Eds., CSIRO Information Services, Melbourne, Australia, 182.

Phillips, R.L., Zak, D.R., Holmes, W.E., and White, D.C. (2002) Microbial community composition and function beneath temperate trees exposed to elevated atmospheric carbon dioxide and ozone. *Oecologia*, 131, 236–244.

Pietikäinen, J., Pettersson, M., and Bååth, E. (2005) Comparison of temperature effects on soil respiration and bacterial and fungal growth rates. *FEMS Microbiology Ecology*, 52, 49–58.

Pillai, J.K. and Taylor, D.P. (1967) Effect of temperature on the time required for hatching and duration of life cycle of five mycophagous nematodes. *Nematologica*, 13, 512–516.

Polis, G.A. (1984) Age structure component niche width and intraspecific resource partitioning: can age groups function as ecological species? *American Naturalist*, 123, 541–546.

Polis, G.A. and Strong, D.R. (1996) Food web complexity and community dynamics. *American Naturalist*, 147, 813–846.

Popovici, I. (1973) The influence of temperature and of nutrient medium on populations of *Cephalobus nanus* (Nematoda, Cephalobidae). *Pedobiologia*, 13, 401–409.

Pregitzer, K.S., King, J.S., Burton, A.J., and Brown, S.E. (2000a) Responses of tree fine roots to temperature. *New Phytologist*, 147, 105–115.

Pregitzer, K.S., Zak, D.R., Maziasz, J., DeForest, J., Curtis, P.S., and Lussenhop, J. (2000b) Interactive effects of atmospheric CO_2 and soil-N availability on fine roots of *Populus tremuloides*. *Ecological Applications*, 10, 18–33.

Presley, M.L., McElroy, T.C., and Diehl, W.J. (1996) Soil moisture and temperature interact to affect growth, survivorship, fecundity, and fitness in the earthworm *Eisenia fetida*. *Comparative Biochemical Physiology*, 114A, 319–326.

Price, M.V. and Waser, N.M. (1998) Effects of experimental warming on plant reproductive phenology in a subalpine meadow. *Ecology*, 79, 1261–1271.

Rillig, M.C., Scow, K.M., Klironomos, J.N., and Allen, M.F. (1997) Microbial carbon-substrate utilization in the rhizosphere of *Gutierrezia sarothrae* grown in elevated atmospheric carbon dioxide. *Journal Soil Biology and Biochemistry*, 29, 1387–1394.

Rillig, M.C., Allen, M.F., Klironomos, J.N., Chiariello, N.R., and Field, C.B. (1998) Plant species-specific changes in root-inhabiting fungi in a California annual grassland: responses to elevated CO_2 and nutrients. *Oecologia*, 113, 252–259.

Rillig, M.C., Wright, S.F., Allen, M.F., and Field, C.B. (1999a) Rise in carbon dioxide changes soil structure. *Nature*, 400, 628.

Rillig, M.C., Field, C.B., and Allen, M.F. (1999b) Soil biota responses to long-term atmospheric CO_2 enrichment in two California annual grasslands. *Oecologia*, 119, 572–577.

Rillig, M.C., Wright, S.F., Shaw, M.R., and Field, C.B. (2002) Artificial climate warming positively affects arbuscular mycorrhizae but decreases soil aggregate stability in an annual grassland. *Oikos*, 97, 52–58.

Ruess, L., Michelsen, A., Schmidt, I.K., and Jonasson, S. (1999) Simulates climate change affecting microorganisms, nematode density and biodiversity in subarctic soils. *Plant and Soil*, 212, 63–73.

Runion, G.B., Mitchell, R.J., Rogers, H.H., Prior, S.A., and Counts, T.K. (1997) Effects of nitrogen and water limitation and elevated atmospheric CO_2 on ectomycorrhiza of longleaf pine. *New Phytologist*, 137, 681–689.

Rutherford, P.M. and Juma, N.G. (1992) Influence of texture on habitable pore space and bacterial-protozoan populations in soil. *Biology and Fertility of Soils*, 12, 221–227.

Rypstra, A.L. and Carter, P.E. (1995) Top-down effects in soybean agroecosystems: spider density affects herbivore damage. *Oikos*, 72, 433–439.

Salamanca, E.F., Kaneko, N., and Katagiri, S. (2003) Rainfall manipulation effects on litter decomposition and the microbial biomass of the forest floor. *Applied Soil Ecology*, 22, 271–281.

Sanders, I.R., Streitwolf-Engel, R., van der Heijden, M.G.A., Boller, T., and Wiemken, A. (1998) Increased allocation to external hyphae of arbuscular mycorrhizal fungi under CO_2 enrichment. *Oecologia*, 117, 496–503.

Santos, P.F., Phillips, J., and Whitford, W.G. (1981) The role of mites and nematodes in early stages of buried litter decomposition in a desert. *Ecology*, 62, 664–669.

Scheu, S. (2002) The soil food web: structure and perspectives. *European Journal of Soil Biology*, 38, 11–20.

Schimel, J.P. and Clein, J.S. (1996) Microbial response to freeze-thaw cycles in tundra and taiga soils. *Journal Soil Biology and Biochemistry*. 28, 1061–1066.

Schortemeyer, M., Hartwig, U.A., Hendrey, G.R., and Sadowsky, M.J. (1996) Microbial community changes in the rhizospheres of white clover and perennial ryegrass exposed to free air carbon dioxide enrichment (FACE). *Journal Soil Biology and Biochemistry*, 28, 1717–1724.

Schortemeyer, M., Dijkstra, P., Johnson, D.W., and Drake, B.G. (2000) Effects of elevated atmospheric CO_2 concentration on C and N pools and rhizosphere processes in a Florida scrub oak community, *Global Change Biology*, 6, 383–391.

Setälä, H. (2002) Sensitivity of ecosystem functioning to changes in trophic structure, functional group composition and species diversity in belowground food webs. *Ecological Research*, 17, 207–215.

Sinclair, B.J. (2002) Effect of increased temperatures simulating climate change on terrestrial invertebrates on Ross Island, Antarctica. *Pedobiologia*, 46, 150–160.

Sinsabaugh, R.L., Saiya-Cork, K., Long, T., Osgood, M.P., Neher, D.A., Zak, D.R., and Norby, R.J. (2003) Soil microbial activity in a *Liquidambar* plantation unresponsive to CO_2-driven increases in primary production. *Applied Soil Ecology*, 24, 263–271.

Sjursen, H., Michelsen, A., and Holmstrup, M. (2005) Effects of freeze-thaw cycles on microarthropods and nutrient availability in a sub-Arctic soil. *Applied Soil Ecology*, 28, 79–93.

Sohlenius, B. and Wasilewska, L. (1984) Influence of irrigation and fertilization on the nematode community in a Swedish pine forest soil. *Journal of Applied Ecology*, 21, 327–342.

Sohlenius, B. and Boström, S. (1999) Effects of climate change on soil factors and metazoan microfauna (nematodes, tardigrades and rotifers) in a Swedish tundra soil—a soil transplantation experiment. *Applied Soil Ecology*, 12, 113–128.

Sonnemann, I. and Wolters, V. (2005) The microfood web of grassland soils responds to a moderate increase in atmospheric CO_2. *Global Change Biology*, 11, 1148–1155.

Soulides, D.A. and Allison, F.E. (1961) Effects of drying and freezing soils on carbon dioxide production, available mineral nutrients, aggregation, and bacterial population. *Soil Science*, 91, 291–298.

Sternberg, M. (2000) Terrestrial gastropods and experimental climate change: a field study in a calcareous grassland. *Ecological Research*, 15, 73–81.

Sulkava, P. and Huhta, V. (2003) Effect of hard frost and free-thaw cycles on decomposer communities and N mineralization in boreal forest soil. *Applied Soil Ecology*, 22, 225–239.

Taylor, A.R., Schröter, D., Pflug, A., and Wolters, V. (2004) Response of different decomposer communities to the manipulation of moisture availability: potential effects of changing precipitation patterns. *Global Change Biology*, 10, 1313–1324.

Todd, T.C., Blair, J.M., and Milliken, G.A. (1999) Effects of altered soil-water availability on a tallgrass prairie nematode community. *Applied Soil Ecology*, 13, 45–55.

Treonis, A.M. and Lussenhop, J.F. (1997) Rapid response of soil protozoa to elevated CO_2. *Biology and Fertility of Soils*, 25, 60–62.

Tsiafouli, M.A., Kallimanis, A.S., Katana, E., Stamou, G.P., and Sgardelis, S.P. (2005) Responses of soil microarthropods to experimental short-term manipulations of soil moisture. *Applied Soil Ecology*, 29, 17–26.

Uvarov, A.V. (1993) A microcosmic approach to compare effects of constant and varying temperature conditions on soil structure/soil biota interrelationships. *Geoderma*, 56, 609–615.

Van Ginkel, J.H., Gorissen, A., and Polci, D. (2000) Elevated atmospheric carbon dioxide concentration: effects of increased carbon input in a *Lolium perenne* soil on microorganisms and decomposition. *Journal Soil Biology and Biochemistry*, 32, 449–456.

Viljoen, S.A. and Reinecke, A.J. (1992) The temperature requirements of the epigeic earthworm species *Eudrilus eugeniae* (Oligochaeta)—a laboratory study. *Soil Biology and Biochemistry*, 24, 1345–1350.

Wall, D.H. and Virginia, R.A. (1999) Controls on soil biodiversity: insights from extreme environments. *Applied Soil Ecology*, 13, 137–150.

Wardle, D.A., Verhoef, H.A., and Clarholm, M. (1998) Trophic relationships in the soil microfood-web: predicting the responses to a changing global environment. *Global Change Biology*, 4, 713–727.

Wilkinson, S.C., Anderson, J.M., Scardelis, S.P., Tisiafouli, M., Taylor, A., and Wolters, V. (2002) PLFA profiles of microbial communities in decomposing conifer litters subject to moisture stress. *Journal Soil Biology and Biochemistry*, 34, 189–200.

Williams, M.A., Rice, C.W., and Owensby, C.E. (2000) Carbon dynamics and microbial activity in tallgrass prairie exposed to elevated CO_2 for 8 years. *Plant and Soil*, 227, 127–137.

Wilson, E.O. (1994) *Naturalist*. Island Press, Washington, D.C., 364.

Wolf, J., Johnson, N.C., Rowland, D.L., and Reich, P.B. (2003) Elevated CO_2 and plant species richness impact arbuscular mycorrhizal fungal spore communities. *New Phytologist*, 157, 579–588.

Yeates, G.W., Tate, K.R., and Newton, P.C.D. (1997) Response of the fauna of a grassland soil to doubling of atmospheric carbon dioxide concentration. *Biology and Fertility of Soils*, 25, 307–315.

Yeates, G.W., Newton, P.C.D., and Ross, D.J. (1999) Response of soil nematode fauna to naturally elevated CO_2 levels influenced by soil pattern. *Nematology*, 1, 285–293.

Young, I.M., Blanchart, E., Chenu, C., Dangerfield, M., Fragoso, C., Grimaldi, M., Ingram, J., and Monrozier, L.J. (1998) The interaction of soil biota and soil structure under global change. *Global Change Biology*, 4, 703–712.

Zak, D.R., Pregitzer, K.S., Curtis, P.S., Teeri, J.A., Fogel, R., and Randlett, D.L. (1993) Elevated atmospheric CO_2 and feedback between carbon and nitrogen cycles. *Plant and Soil*, 151, 105–117.

Zak, D.R., Holmes, W.E., MacDonald, N.W., and Pregitzer, K.S. (1999) Soil temperature, matric potential, and the kinetics of microbial respiration and nitrogen mineralization. *Soil Science Society of America Journal*, 63, 575–584.

Zak, D.R., Pregitzer, K.S., Curtis, P.S., Holmes, W.E. (2000) Atmospheric CO_2 and the composition and function of soil microbial communities. *Ecological Applications*, 10, 47–59.

Zak, D.R., Holmes, W.E., Finzi, A.C., Norby, R.J., and Schlesinger, W.H. (2003) Soil nitrogen cycling under elevated CO_2: a synthesis of forest FACE experiments. *Ecological Applications*, 13, 1508–1514.

Zavaleta, E.S., Thomas, B.D., Chiariello, N.R., Asner, G.P., Shaw, M.R., and Field, C.B. (2003a) Plants reverse warming effects on ecosystem water balance. *PNAS*, 100, 9892–9893.

Zavaleta, E.S., Shaw, M.R., Chiariello, N.R., Mooney, H.A., and Field, C.B. (2003b) Additive effects of simulated climate changes, elevated CO_2, and nitrogen deposition on grassland diversity. *PNAS*, 100, 7650–7654.

Zogg, G.P., Zak, D.R., Ringelberg, D.B., MacDonald, N.W., Pregitzer, K.S., and White, D.C. (1997) Compositional and functional shifts in microbial communities due to soil warming. *Soil Science Society of America Journal*, 61, 475–481.

6 Herbivory and Nutrient Cycling

R. Andrew Carran and Vincent Allard

CONTENTS

6.1 INTRODUCTION

From early in our agricultural history, animals have been used to convert coarse and low-quality plant materials into high-protein foods and fibres and hides. In the modern agricultural world, animal- and grass-based agroecosystems use about 20% of the terrestrial area at various levels of intensity (Hadley 1993). Such systems have important features that distinguish them from cropping systems, from managed forests, and from natural or undisturbed ecosystems. These distinguishing features have at their core biologically mediated processes that are, or may be, sensitive to environmental change. As a consequence, the effects, interactions, and consequences of change in environmental conditions may be more fully expressed in these systems than in other agroecosystems. In this chapter, we focus attention on the unique processes of grazed agroecosystems, particularly nutrient cycling, and explore the sensitivity to climatic change associated with them.

Livestock production from agroecosystems may involve herbivores, mostly ruminants, either foraging in pasture and rangeland *in situ* or eating a supplied diet in barns or feedlots. Frequently, some combination of the two will be used. This chapter is concerned only with *in situ* foraging, the associated nutrient returns to the system, and their impact on the response of such systems to changing environmental conditions.

There is the potential for direct effects of climate change on grazed systems; for instance, summer temperatures can determine the choice and economic success of some animal species or types (Kadzere et al. 2002). Shifting margins of forest, rangeland, and desert in response to climatic changes can also impact directly on the viability of some extensive systems. Responses to these effects by farmers are possible in short time scales, and separation of systematic changes and stochastic events may not be possible anyway. Of more interest to us in this chapter are the complex indirect responses that grazed systems can show to a changing environment, in particular, changes involving nutrient cycling.

6.2 IMPORTANT DISTINCTIONS BETWEEN CROP AND GRAZED SYSTEMS

Crop production systems are characterized by simple input–output dynamics for nutrients and organic matter; nutrients are added to the soil (or exploited from the resource in the soil), and removed in the harvested components of the crop while accumulation, usually modest, in the soil occurs in roots and residues together with organic matter. Eventually, exported nutrients move through the food chain and are ultimately excreted, often well beyond the boundaries of the production system. Grazing systems, however, are characterized by small inputs and productive outputs at the system boundary, and large fluxes of nutrients and organic matter through excreta and decomposition inside the system boundaries. High rates of nonproductive output to the environment can be associated with these large fluxes. This exaggeration of biologically mediated pathways distinguishes grazing systems from crop and natural systems.

Pasture and rangeland can be loosely grouped according to function. In temperate regions, pastures may be intensively used (that is, a large proportion of net primary production is consumed or removed) and fertilized to remove nutrient limitations. They are often sown with one or a few perennial C3 species or cultivars, and may be harvested for conservation and produce forage and plant residues that are of high quality. By contrast, rangeland and native grassland systems are widespread in subhumid and semiarid regions of all the continents and have limiting water and nutrient supplies, may be species rich or poor depending on management and the return of nutrients in litter, above- and belowground, and may be large compared to nutrient flows through the excreta of grazing animals. A third group falls between the two extremes. It may be intensively used within constraints of water and nutrient availability, but symbiotic nitrogen fixation may be a significant process and nutrient cycling through excreta is an important determinant of productivity. These pastures may be species rich and spatially and temporally heterogeneous in nutrient supply, and because of these, they are ideal study units for expression of response to environmental change.

The key elements of these groups, or this sequence of grazed systems, is shown in Table 6.1 in comparison to a cropping system, and the sensitivity of various processes to environmental change are identified. The emphasis on the cycling of nutrients as excreta rather than litter return perhaps needs to be amplified. Animal digestion and metabolism separates the nitrogen (N) and carbon (C) brought together during net primary production (NPP) (Ball and Ryden 1984); C not retained in body mass or product is metabolized to CO_2 or excreted in faeces, while nonretained N is increasingly excreted as herbage N content increases (Haynes and Williams 1993). Figure 6.1 illustrates both the relative fluxes of C and N in a grazed, temperate pasture and their separation by respiration and excretion from herbivores. Retention of N within the soil is limited by the spatial separation from C and losses to air and water are promoted. The contrast with combined C and N cycling through litter, therefore, is stark. The contrast with natural systems is also pronounced where significant nutrient cycling occurs through small animal or arthropod herbivory. In these situations, the scale of excreta deposition is relatively fine, being more or less homogeneous from a plant perspective, whereas excreta deposition by large ruminants can be considered as a heterogeneous process at the plant scale.

6.3 CREATION OF HETEROGENEITY IN AVAILABLE SOIL N DISTRIBUTION

In the grazed pasture almost 50% of the N cycling through the system does so through animal urine (Figure 6.1). This element of nutrient cycling is one of the major differences between grazed and nongrazed systems; it is a major determinant of pasture productivity (Figure 6.2) and provides a sensitive phase for global change drivers to operate. In the first instance, there are a number of environmental factors that can alter the volume of urine and the concentration of N in the urine, including moisture content of the herbage (Smith and Frost 2000), chemical composition of the diet (Haynes and Williams 1993), and the herbage plant species identity in the diet (Hutton et al. 1967; Orr et al. 1995; Allard et al. 2003). Temperature, moisture, and elevated CO_2 can all modify these factors, and temperature and water availability can directly affect animal metabolism and N excretion (Kadzere et al. 2002).

The return of N in urine results in a mosaic of urine patches of different ages and of different N status. The components of heterogeneity are summarised in Table 6.2 by considering the patches as falling into one of three categories:

- Patch 1: Recent urine patches less than 60 days from deposition characterized by excessive amounts of mineral N. Plant growth rates and immobilization of N can occur rapidly.
- Patch 2: Older patches with small pools of mineral N but somewhat elevated fluxes of N from mineralization of urinary N previously immobilized. Plant growth is generally N limited.
- Patch 3: Areas of low N availability that can be considered unaffected by urine in a functional sense.

The proportions of these units varies depending on the grazing management (Steele and Brock 1978) as greater frequency or intensity of grazing leads to a higher proportion of areas in Patch 1 or Patch 2 state. The transition between patches is controlled by the environmental conditions, in particular the rainfall conditions

TABLE 6.1

A Comparison of Some C and N Fluxes in a Corn Crop and in Three Animal Production Systems and the Direct Sensitivity of the Processes to Environmental Changes or Secondary (2°) Response, i.e., Forage Quality Is Affected by CO_2 Concentration in the Atmosphere and This Influences Partitioning of N between Dung and Urine

	Corn Crop	Rangeland	Intensive N Fix N Inputs	Intensive N Fertiliser	Sensitive To
N fixation	<20	20	30–250	20–	T , SW*, CO_2
Fertiliser N	200	0	0	200+	
Deposition	10	10	10	10	Location
N Losses					
Product N	150–250	<10	15–40	20–50	2° Forage quantity, quality
NH_3N	low	low	50	80	2° Forage quantity, quality
Leached N (NH_4, NO_3, DON)	50–200	low	5–20	100	T , SW*, CO_2, and 2°
Leached C (DOC)	100	?but low	100	100	T , SW, CO_2, and 2°
N_2, N_2O	20	1	10	20	TSW 2° Forage quantity, quality
Internal Processes					
Plant residues	12t	2.5	5t	6	CO_2
Root turnover	20t	5	11	15	CO_2
Dung C and N	0	15 N, 320 C	65N, 1300 C	100N, 2000 C	2° Forage quantity, quality

Urine N	0	0–5	200–300	350
Spatial variation in N supply	no	yes	large	no
Properties of the system affecting sensitivity to change	Single C4 cultivar. Weeds pests and disease may be issues. No nutrient limitation.	Species rich C3, C4, legumes, algal crusts. Nutrient limitation. Sensitive topsoil moisture regime.	Species rich, C3, C4, Legumes high MIT* Sensitive to nutritional limitations for N fixation.	Maybe temperate or tropical. Nutrient supply and botanical properties; highly managed.
2° Forage quantity, quality				

Notes: MIT = mineralization turnover, the recycling of nutrients through mineral, soil biomass, and highly labile organic pools.
SW = soil water regime and is shorthand for rainfall, evapotranspiration and their seasonal changes, which can affect the system positively or negatively.
The corn crop assumes 12 tonnes/ha of harvested grain and that all residues are returned to the soil.
Yields are 2.5, 11, and 15 tonnes/ha in the rangeland legume and N fertilised systems, respectively.
All quantities are kg/ha of N or C except plant residues and root turnover, which are tonnes/ha.

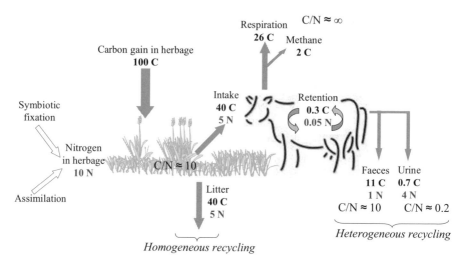

FIGURE 6.1 Relative fluxes of carbon (C) and nitrogen (N) in a grazed grassland. Total C fixed in the herbage is normalized to 100. Numbers are a proportion of C or N fixed by the vegetation in a year. This emphasises the N concentrating effect of ruminants by (1) decoupling C and N, and (2) concentrating spatially the N returns.

(through N leaching losses), but also the temperature (through mineralization rates) and, as we will see below, by the atmospheric CO_2 concentration operating through immobilisation. To describe how global change might interact with grazed pastures showing this heterogeneity, we will use the example of elevated CO_2. In Table 6.2 these pasture units are ascribed a notional but realistic proportion of area and growth rate at both ambient and elevated CO_2 concentration. Response of the pasture as a whole to elevated CO_2 is modified by the unresponsive areas limited by nitrogen, and the quite subtle effects of CO_2 on excreta N partitioning. Nitrogen yield appears to be very sensitive to these effects also (Table 6.2).

An even more complex range of outcomes could follow if increased immobilisation of urine-N occurs at elevated CO_2, and the progressive nitrogen limitation (PNL) hypothesis (Luo et al. 2004) suggests this is possible. An example of this occurring in a Patch 1 situation is shown in Figure 6.3. After the application of urine, there is a marked decline in NO_3N and total mineral N availability at elevated CO_2 compared to the urine patches at ambient CO_2. However, in this very high-N situation, plant growth is unlikely to be markedly affected as N levels are probably adequate. In a Patch 3 situation, N availability at elevated CO_2 has been shown to be significantly reduced (Newton et al. 2006), and plant growth is very likely to be restricted at elevated CO_2. This may well be a situation in which negative effects of elevated CO_2 on plant growth are expressed (Diaz et al. 1993; Newton et al. 2001; Shaw et al. 2002).

Understanding the degree of recalcitrance of N immobilized in a high-CO_2 world and the functioning of mineralization immobilisation turnover (MIT) is a key to a proper understanding of the impact of CO_2 on nutrient availability and productivity of farmed grasslands. With only a slight change in perspective, it will also unlock

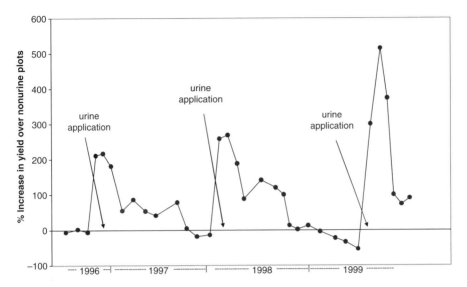

FIGURE 6.2 Relative yield response of a grass and clover pasture to the application of synthetic urine supplying 100 gN/m2N (80% as urea-N, 20% glycine-N) and 100 g/m² potassium (49% as Cl, 16% as SO_4^2, and 35% as HCO_3). Urine was reapplied at intervals that approximated the return times for urine events in a system with year-round grazing in the Manawatu region of New Zealand. Yields from the nonurine-treated areas were 10.9, 8.8, and 7.9 tonnes dry matter/ha, and from the urine-treated areas were 18.4, 16.4, and 15.1 tonnes of dry matter/ha. The complex and prolonged yield responses following the summer application is particularly interesting as it suggests immobilization and subsequent mineralization of urine is an active process.

many of the issues surrounding C sequestration in a changing climate. The link is the simple recognition that C cannot be readily sequestered without simultaneously sequestering N, S, and P, and that this process is identical to PNL. In agroecosystems, these components of organic matter are often purchased inputs, so this issue moves beyond the purely biological and into the economic dimensions of the farm and its function.

Biological nitrogen fixation (BNF) provides an important input of N into many grazed agroecosystems (see Table 6.1) and may assume greater importance where a changing climate and atmosphere system is reducing N availability or promoting higher N losses. Where N fertiliser is used in limited amounts, or not at all, BNF makes good losses in stable systems and legumes contribute another layer of heterogeneity as they exploit areas of low N availability in the sward (Schwinning and Parsons 1996). Annual inputs up to several hundred kg N are recorded (Ball and Ryden 1984) in temperate, well-watered grasslands. In less favoured sites and in extensive grassland, the contribution of fixed N to the ecosystem may be equally important, however. Seasonal inputs of fixed N may underpin the N economy and productivity of those systems, although they may be only a few tens of kg N/ha.

TABLE 6.2
The Effect of Elevated CO_2 on Some Components of a Hypothetical Grazed Pasture

	Mean Annual Growth Rate, kg/ha/day	Mean Annual N%	Proportion of Area Occupied	Above Ground Yield	Total N Yield
Ambient					
P 1 G+L	60	4–5	0.25	5500	245
P 2 G+L	45	3.5	0.20	3300	115
P3 G	30	2	0.45	4900	100
P3 L	20	3.5	0.1	700	25
Total			1.0	14400	685
Elevated					
P 1 G+L	80	4–5	0.2	5800	230
P 2 G+L	60	3.5	0.25	5600	195
P3 G	30	2	0.4	4380	88
P3 L	30	3–3.5	0.15	1600	50
Total			1.0	17380	563

Notes: The components are defined functionally as areas with grossly different N supply from urine deposition. P1 (Patch 1) refers to areas in the first phase of N dynamics (see also Figure 6.1 and Figure 6.2) characterised by excess mineral N. P2 (Patch 2) is characterised by small pools of mineral N, but a background state of low-level N availability. Within P1 and P2, legumes are assumed to contribute to yield but to be largely mineral dependent while symbiotic N fixation is assumed to supply all N in P3 (Patch 3). In the hypothetical case, yield in P1 and P2 is assumed to increase by 30% at elevated CO_2 concentration, and the grass yield in P3 is assumed to be N limited and not CO_2 responsive. Legume yield is, however, assumed to increase in P3. In the whole unit, pasture yield increases, but the response is modified by the area of P3. In our example, we also assume that the area of P3 increases because of the effect of CO_2 on N partitioning into feces rather than urine (Allard et al. 2003) and this further reduces the expression of a CO_2 effect on whole sward yield to 21%. Estimation of total N yield shows that this term is very sensitive to quite small changes in area of P1, P2, and P3 in response to N partitioning in excreta.

The responses of biological nitrogen fixation to environmental changes are discussed by Thomas et al. (Chapter 4, this volume), and here we are concerned only with any special cases associated with grazed ecosystems.

Any impact of climatic change on N fixation (Zanetti et al. 1996) and legume productivity will also translate into improved forage quality and thus the main secondary effects of the driver will be expressed as N recycles through excreta. Overall, the impact at a system level will depend on the partitioning of the fixed N in highly labile forms and less mobile forms that may, over time, elevate the base level of N mineralization.

This argument assumes that the quite specific requirements for N fixation are met; that the soil is of appropriate pH and molybdenum, (Mo) phosphate (PO_4), potassium (K), and sulfate (SO_4^2) supply is adequate. It also requires that the drivers of change

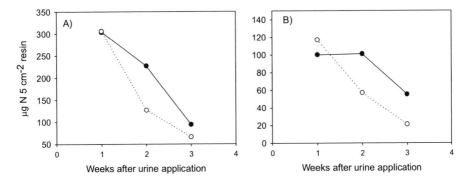

FIGURE 6.3 Changes in (A) mineral-N and (B) NO_3-N after application of 150 ml of sheep urine containing 0.8% N to soil in control (closed symols) and elevated CO_2 (FACE) (open symbols) treatments. N was measured with ion exchange membranes exposed in the soil for 7-day periods. Differences are significant ($p < 0.01$) at week 2 in both cases and week 3 for NO_3-N. Ammonia volatilisation did not differ between treatments. Values plotted are mean of 14 applications made in each of 3 rings at ambient and 3 rings at elevated (475 ppm) CO_2. Urine was poured from a spouted beaker 250 mm above the soil and allowed to infiltrate or spread unhampered.

do not impact on soil properties in ways that shift them across the boundaries between adequate and limiting supply. Examples of this are given in Hungate et al. (2004), who show down regulation of N fixation as a consequence of a decrease in Mo supply driven by elevated CO_2, and Niklaus and Körner (2004), who attribute a lack of CO_2 response in legumes to an induced P deficiency.

6.4 NUTRIENT LOSS AND STORAGE IN GRAZED SYSTEMS AND THEIR SENSITIVITY TO CHANGE

6.4.1 NUTRIENT LOSS

Loss of nutrients from grazed systems is a well researched topic (Ball and Ryden 1984) and is one of those issues that set grazed agroecosystems apart from cropping systems. Return of excreta, and especially urine, creates areas where nutrient availability greatly exceeds demand by plants and microorganisms (the Patch 1 situation), and where high rates of loss can occur. The issue that needs to be addressed here, though, is whether any of the processes controlling nutrient loss are sensitive to global change drivers or feedbacks from them. Nitrogen loss is the best exemplar here because of the multiple possibilities for loss as it is transformed among oxidation states that confer mobility as ions or gases or stability in combination with carbon.

In Figure 6.1 the major loss pathways for N are shown and the sensitivity to global change drivers shown in Table 6.1. Ammonia volatilization is a process strongly promoted by the hydrolysis of urea when urine is voided in the field or during storage of manures from feeding facilities. A straightforward chemical process, it responds positively to temperature increase and is also sensitive to evaporative conditions (Bolan et al. 2004). The concentration of urea in urine-affected soil is a key driver of loss of NH_3 to the atmosphere, and this in turn is influenced by atmospheric CO_2 concentration through the mass of herbage on offer to herbivores and its protein content and digestibility. Protein content is consistently lower at elevated CO_2 (see Figure 6.4), but legumes, if present, can flourish at elevated CO_2 producing a diet richer overall in protein (Allard et al. 2003). In addition, a more digestible diet, with less structural carbohydrate, will result in more N partitioned to urine rather than dung (Haynes and Williams 1993).

Generalizing an impact on NH_3 volatilisation in the face of such a complex trail of response and consequence is difficult, but the potential for significant loss of N through this pathway in grazed ecosystems makes a detailed analysis of soil, environment, and dietary responses essential.

Leaching, at any given level of ions in soil solution, is driven by downward water movement, and any net change in rainfall and evapotranspiration will have a readily calculated effect where the changing components of water balance and its seasonality can be predicted. In the case of NO_3–N the effects of environmental change on availability may override other considerations.

Less obvious though are the impacts of change on biologically mediated returns to the atmosphere (N_2 and N_2O) and the biologically mediated changes in the size of soil N pools vulnerable to leaching or volatilization. Atmospheric CO_2 is a key driver in this regard as increased allocation of C below ground in grazed systems exposed to elevated CO_2 (Allard et al. 2005) could result in rearrangements of trophic relationships and habitats within the organomineral fabric of the soil. Atmospheric CO_2 content also impacts on animal diets, excreta return (Allard et al. 2003), and ultimately this should impact on nutrient forms and pool sizes.

Emission of N_2O and N_2 are interesting examples of N loss processes that can be stimulated or repressed by elevated atmospheric CO_2. Biological demand for oxygen is an important promoter of the anaerobic conditions that favour N_2O production in soils and its further reduction to N_2. Where the availability of C is limited, denitrification may also be limited, and an increase in atmospheric CO_2 would increase C allocation below ground and cause a consequent increase in denitrification. Ineson et al. (1998) provide an example of this from a cut grass sward fertilised with N. On the other hand, grassland soils are often C rich and further C inputs may impact on nitrogen availability, nitrification (DeLuca and Keeney 1994), or microaggregation (Rillig et al. 1999; Niklaus et al. 2001, 2003; Six et al. 2001) reducing emissions of N_2O. Two soils, a freely draining sandy loam and a poorly draining silt loam, exposed to ambient or elevated CO_2 (475 pppm) in mesocolumns within a free air CO_2 enrichment (FACE) experiment illustrates this sort of response. The freely drained soil emitted 9.6 and 3.8 mg N_2O-N.m^2 ($p < 0.05$) at ambient and elevated CO_2, and the poorly drained soil 21.0 and 7.2 mg N_2O-N.m^2($p < 0.05$). The mechanisms are not clear, but we need to invoke more complexity than a simple reduction in N availability, as these data are from urine-treated soils and the substrates are greatly in excess of those required to support the emission of N_2O and N_2. Analysis of soil gas samplers treated or not with acetylene to inhibit N_2O-reductase provided no clear evidence for significant formation of N_2 in this experiment. An increase in carbon allocation below ground or a decrease in NO_3-N in soil solution would both predispose to a higher ratio of N_2:N_2O in denitrification products.

The sensitivity of denitrification to environmental factors, notably soil water content and temperature, do allow some predictions about sensitivity to changes other than atmospheric CO_2 concentration. Where the soil water threshold (commonly more than 70 to 80% water-filled pore space) is exceeded with warm soils, rates of denitrification will exceed those where the soil water criteria are only met from autumn to spring. An increase in soil temperature overall or a shift toward summer precipitation will predispose to losses and possibly strong forcing from increased emission of N_2O.

Does environmental change pose a threat to nutrient budgets in agroecosystems through increased rates of nutrient loss? The answer to this again cannot be generalized, but the unifying principle that carbon sequestration is demanding of nutrients must apply. Pools of nutrients in excess of plant demand will create conditions where carbon can be assimilated in soil organic matter and those pools reduced together with the potential for loss. Rapid loss of NH_3, which escapes before becoming enmeshed with the carbon cycle may be promoted in some circumstances that encourage increased deposition of urinary N. Similarly, changes to the soilwater balances may promote rapid leaching of nutrients.

While the focus here has been on N, the impacts on loss of SO_4^{2-}-S should follow the stoichiometry of these to elements in nature and that of P, although that element has fewer avenues of escape from the system.

6.4.2 Soil Organic Matter Accumulation and Carbon Sequestration

Jenny's model of soil organic matter (SOM) accumulation in soils, first published in 1941 (Jenny 1941), remains a good foundation for considering how changing soil water status and temperature will affect SOM status in a changing climate. It falls short, though, where other components of the climate atmosphere system, like CO_2 concentration, need to be considered. Indeed the concept of an equilibrium soil SOM may have little relevance in an environment of continuous change in CO_2 concentration. The concept of biota as a soil-forming factor needs careful thought before application to herbivory in managed grasslands. The balance between consumption and decomposition and, of course, yield must be considered in agroecosytems.

Grassland soils that have remained uncultivated for decades contain large quantities of C even if intensively used for animal production, and these systems attract attention as potential C sinks; however, these grasslands have little potential for further SOM accumulation in a stable environment. This position needs reviewing, however, against the background of environmental change. Three issues quickly emerge:

- The first relates to understanding the status of the soil resource and nutrient availability, and is in essence the converse of PNL; nitrogen (and P and S) is sequestered in SOM along with C, and PNL is thus a consequence of C sequestration and at some point a limitation to C sequestration.
- The second is the interaction of grazing animals with the elements of environmental change. This has been discussed in previous sections, but

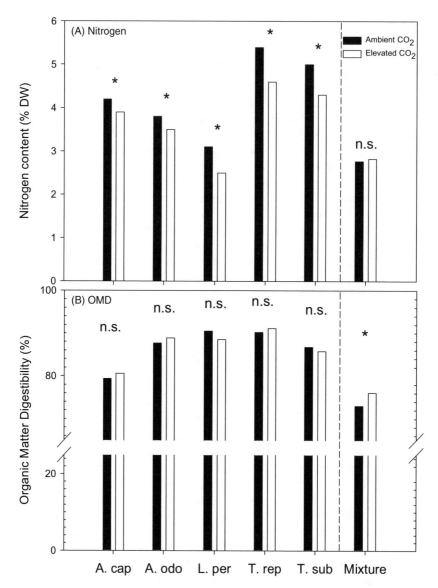

FIGURE 6.4 (A) Nitrogen content and (B) organic matter digestibility (OMD) of five dominant plant species (A. cap = *Agrostis capillaris*, A. odo = *Anthoxanthum odoratum*, L. per = *Lolium perenne*, T. rep = *Trifolium repens,* and T. sub = *Trifolium subterraneum*), and of the mixed herbage grow under ambient (full bars) and elevated (open bars) CO_2 concentrations. Data were obtained in November 2000 after 3 years of CO_2 enrichment. Note that the chemical composition of single species were obtained on young green leaves while the chemical composition of the mixture was obtained from bulk samples after mowing. This illustrates the importance of second order CO_2 effects: At the single species scale elevated CO_2 decreases the average N concentration, but a parallel increase in the proportion of N-rich species (legumes) counterbalance the first order effect.

here it is worth noting the importance of dung carbon in maintaining SOM levels in grazing systems (Carran and Theobald 2000) and the effects of elevated CO_2 on plant composition and feedbacks to dung.

- The third recognises the heterogeneous nutrient status and asks whether sequestration will be freed at least partially from the constraints of PNL where parts of the system have a surplus of N and where P, S, and trace elements are managed for productive purposes.

These are research areas poorly served with experimental work at present. They will be of agricultural importance because of the central role SOM has in soil management as well as climate change issues.

6.5 CONCLUSIONS

Within grazed pastures there is a perpetually shifting mosaic of areas of nutrient excess and poverty and all the transitional states between those extremes. The frequency of exposure to the causal agents depends on productivity and stocking rates in the system, and is greatest in productive temperate pastures. At any time, grazed pasture will contain areas responding to environmental changes in a range of ways; extreme limitation of response through nitrogen deficiency and the full-blown expression in the absence of nutrient limitations may occur at scales less than 1 m. Understanding the community response to changing climate is difficult and unifying principles for these diverse agroecosystems are needed.

One of the key distinctions between agroecosystems and natural ecosystems is that nutrient limitations are, to a greater or lesser degree, managed out of the system. Nutrients are supplied specifically to remove constraints to biological N fixation in many agricultural systems. A vigilant approach to nutrient management may well prove to be a requirement of operating in a changing world, and increases in the costs of farming a consequence of change. Movement in the margins of nutrient sufficiency and deficiency will impact on farm profitability and may shift the distribution of the agricultural enterprise as much as, say, some change in rainfall pattern.

REFERENCES

Allard, V., Newton, P.C.D., Lieffering, M., et al. (2003) Nitrogen cycling in grazed pastures at elevated CO_2: returns by animals. *Global Change Biology,* 9, 1731–1742.

Allard, V., Newton, P.C.D., Lieffering, M., Soussana, J-F., Carran, R.A., and Matthew, C. (2005) Increased quantity and quality of coarse soil organic matter fraction at elevated CO_2 in a grazed grassland are a consequence of enhanced root growth and turnover. *Plant and Soil*, 276, 49–60.

Allard, V., Robin, C., Newton, P.C.D., Lieffering, M., and Soussana, J-F. (2006) Short-term and long-term effects of elevated CO_2 on *Lolium perenne* root exudation and its consequences on soil organic matter turnover and plant N yield. *Journal Soil Biology and Biochemistry.* In press.

Ball, P.R. and Ryden, J.C. (1984) Nitrogen relationships in intensively managed temperate pastures. *Plant and Soil,* 76, 23–33.

Bolan, N.S., Saggar, S., Luo, J., Bhandral, R., and Singh, J. (2004) Gaseous emission from grazed pastures: processes, measurements and modelling. Environmental implications and mitigation. *Advances in Agronomy,* 84, 37–120.

Carran, R.A. and Theobald, P.W., (2000) Effects of excreta return on properties of a grazed pasture soil. *Nutrient Cycling in Agroecosystems,* 56, 79–85.

Daepp, M., Nösberger, J., and Lüscher, A. (2001) Nitrogen fertilisation and developmental stage affect the response of yield, biomass partitioning and morphology of *Lolium perenne* L. swards to elevated pCO_2. *New Phytologist,* 150, 347–358.

DeLuca, T.H. and Keeney, D.R. (1994) Soluble carbon and nitrogen pools of prairie and cultivated soils: seasonal variation. *Soil Science Society of America Journal,* 58, 835–840.

Diaz, S., Grime, J.P., Harris, J., et al. (1993) Evidence of a feedback mechanism limiting plant response to elevated atmospheric carbon dioxide. *Nature,* 364, 616–617.

Hadley, M, (1993) Grasslands for sustainable ecosystems. In *Proceedings of the XVII International Grassland Congress,* New Zealand Grassland Association, New Zealand, 21–28.

Haynes, R.J. and Williams, P.H. (1993) Nutrient cycling and soil fertility in the grazed grassland ecosystem. *Advances in Agronomy,* 49, 119–190.

Hungate, B.A., Stiling, P.D., and Dijkstra, P. (2004) CO_2 elicits long term decline in nitrogen fixation. *Science,* 304, 1291.

Hutton, J.B., Jury, K.E., and Davies, E.B. (1967) Studies of the nutritive value of New Zealand dairy pastures. V. The intake and utilisation of potassium, sodium, calcium, phosphorus, and nitrogen in pasture herbage by lactating dairy cattle. *New Zealand Journal of Agricultural Research,* 10, 367–388.

Ineson, P., Coward, P.A., and Hartwig, U.A. (1998) Soil gas fluxes of N_2O, CH_4 and CO_2 beneath *Lolium perenne* under elevatedCO_2: the Swiss free air carbon dioxide enrichment experiment. *Plant and Soil,* 198, 89–95.

IPCC (Intergovernmental Panel on Climate Change) (2000) *Good Practice Guidance and Uncertainty Management in National Greenhouse Gas Inventories,* Penman, J., Kruger, D., Galbally, I., et al., Eds., Technical Support Unit, Kanagawa, Japan.

Jenny, H. (1941) *Factors of Soil Formation,* McGraw-Hill, New York, 281.

Kadzere, C.T., Murphy, M.R., Silanikove, N., and Matz, E., (2002) Heat stress in lactating cows: a review. *Livestock Production Science,* 77, 59–99.

Luo, Y., Su, B., Currie, W.S., et al. (2004) Progressive nitrogen limitation of ecosystem responses to rising atmospheric carbon dioxide. *Bioscience,* 54, 731–739.

Newton, P.C.D., Carran, R.A., and Lawrence, E.J. (2003) Reduced water repellency of a grassland soil under elevated atmospheric CO_2. *Global Change Biology,* 10, 1–4.

Newton, P.C.D., Allard, V., Carran, R.A., and Lieffering, M. (2006) Impacts of elevated CO_2 on grassland grazed by sheep: the New Zealand FACE experiment. In *Managed Ecosystems and CO_2: case Studies, Processes and Perspectives,* Nösberger, J., Long, S.P., Hendrey, J.R., Stitt, M., Norby, R.J., and Blum, H., Eds., Ecological Studies Series. Springer-Verlag, Heidelberg, Germany.

Newton, P.C.D., Clark, H., Edwards, G.R., and Ross, D.J. (2001) Experimental confirmation of ecosystem model predictions comparing transient and equilibrium plant responses to elevated atmospheric CO_2. *Ecology Letters,* 4, 344–347.

Niklaus, P.A., Alphei, J., Ebersberger, D., Kampichler, C., Kandeler, E., and Tscherko, D. (2003) Six years of *in situ* CO_2 enrichment evoke changes in soil structure and soil biota of nutrient-poor grassland. *Global Change Biology,* 9, 585–600.

Niklaus, P.A., Glöckler, E., Siegwolf, R., and Körner, C. (2001) Carbon allocation in calcareous grassland under elevated CO_2: a combined ^{13}C pulse-labelling/soil physical fractionation study. *Functional Ecology*, 15, 43–50.

Niklaus, P.A. and Körner, Ch. (2004) Synthesis of a six-year study of calcareous grassland responses to *in situ* CO_2 enrichment. *Ecological Monographs*, 74, 491–511.

Orr, R.J., Penning, P.D., Parsons, A.J., et al. (1995) Herbage intake and N excretion by sheep grazing monocultures or a mixture of grass and clover. *Grass and Forage Science*, 50, 31–40.

Rillig, M.C., Wright, S.F., Allen, M.F., and Field, C.B. (1999) Rise in carbon dioxide changes soil structure. *Nature*, 400, 628.

Schwinning, S. and Parsons, A.J. (1996) Analysis of the coexistence mechanisms for grasses and legumes in grazing systems. *Journal of Ecology*, 84, 799–813.

Shaw, M.R., Zavaleta, E.S., Chiariello, N.R., Cleland, E.E., Mooney, H.A., and Field, C.B. (2002) Grassland responses to global environmental changes suppressed by elevated atmospheric CO_2. *Science*, 298, 1987–1990.

Six, J., Carpentier, A., van Kessel, C., Merckx, R., Harris, D., Horwath, W.R., and Lüscher, A. (2001) Impact of elevated CO_2 on soil organic matter dynamics as related to changes in aggregate turnover and residue quality. *Plant and Soil*, 234, 27–36.

Smith, K.A. and Frost, J.P. (2000) Nitrogen excretion by farm livestock with respect to land spreading requirements and controlling nitrogen losses to ground and surface waters. Part 1: Cattle and sheep. *Bioresource Technology*, 71, 173–181.

Steele, K.W. and Brock, J.L., (1985) Nitrogen cycling in legume based forage production systems in New Zealand. In *Forage Legumes for Energy-Efficient Animal Production*, Barnes, R.F., Ed., U.S. Department of Agriculture, Agricultural Research Service, Washington, D.C., 171–176.

Theobald, P.W. and Carran R.A. (2000) Urine powers pasture production. In *Soil Research: A Knowledge Industry for Land Based Exporters*, Currie, L.D. and Loganathan, P., Eds., Occasional report no. 13. Fertilizer and Lime Research Centre, Massey University, Palmerston North, Australia, 193–196.

Thomas, R.B., van Bloem, S.J., and Schlesinger, W.H. (2006) Climate change and symbiotic nitrogen fixation in agroecosystems. In *Agroecosystems in a Changing* Environment, Newton, P.C.D., Carran, R.A., Edwards, G.R., and Niklaus, P.A., Eds.. CRC Press, Boca Raton, FL, 87–112.

Zanetti, S.U.A., Hartwig, A., and Luscher, T. (1996) Stimulation of symbiotic nitrogen fixation in *Trifolium repens* under elevated atmospheric pCO_2 in a grassland ecosystem. *Plant Physiology*, 112, 575–583.

7 Sustainability of Crop Production Systems under Climate Change

Jürg Fuhrer

CONTENTS

ABBREVIATIONS

D	Drainage
ET	Evapotranspiration
ET_l	Single leaf transpiration
ET_c	Canopy evapotranspiration
ET_o	Reference ET for alfalfa
ET_p	Whole-plant transpiration
ET_r	Reference ET for grass
FC	Field capacity
g_s	Leaf conductance
I	Interception of precipitation

IPCC	Intergovernmental Panel on Climate Change
K_c	Crop specific coefficient
K_s	soil dryness coefficient
LAI	Leaf area index
NUE	Nitrogen use efficiency
PET	Potential evapotranspiration
P	Precipitation
R	Runoff
r_a	Aerodynamic resistance
r_b	Boundary layer resistance
r_c	Canopy resistance
r_s	Stomatal resistance
Rn	Net radiation
SWC	Soil water content
T	Temperature
vp	Water vapour pressure
vpd	Water vapour pressure deficit
Wt	Capillary raise

7.1 INTRODUCTION

Agriculture makes increasing use of soil, water, and air to produce food for an ever-growing world population, and agricultural activities often harm the environment by changing the quality or the quantity of the locally available natural resources, which are also the foundations of natural habitats, biodiversity, and landscapes, and also of human health. Today's challenge is to elaborate and implement ways to efficiently and profitably produce sufficient safe food, while avoiding degradation of resources and harming the environment under rapidly changing climatic conditions.

Crop production systems can be regarded as environmentally sustainable if their use of resources such as water and nutrients is efficient, and environmental impacts are minimal relative to the production of goods. The efficiency of resource usage is influenced by the thermal and hydrological regimes, and thus by the local weather pattern, which varies from year to year within the boundaries of the respective climatic zone. Consequently, any substantial shift in these boundaries, as it is expected to occur during the next 100 years (see below), will affect the sustainability of cropping systems, and their maintenance through management, and will force farmers to adapt their practices. Moreover, the increase in atmospheric CO_2 concentration (pCO_2), which is a main driver of climate change, can affect plant–water relations and nutrient cycles in soil–plant systems. Such CO_2-induced effects interact with those of changes in temperature and precipitation (cf. Fuhrer 2003). They also depend on acclimation at the plant level (see Hovenden 2006, Chapter 12, this volume). The resulting complexity makes predictions of net effects of global change difficult and uncertain.

The focus of this contribution is to review results from both experimental and modeling studies on water and nitrogen (N) use of cropping systems in relation to climatic change, and to discuss implications for future irrigation and fertilization needs.

7.2 PROJECTIONS FOR ATMOSPHERIC $p\mathrm{CO}_2$ AND CLIMATE

Over the past 2 decades, $p\mathrm{CO}_2$ increased at an average rate of about 1.5 ppm (0.4%) per year, and it may rise up to between 540 and 970 ppm by 2100, depending on global socioeconomic development and related emission patterns (IPCC [Intergovernmental Panel on Climate Change] 2001). In the case of the Special Report on Emissions Scenarios (SRES) A1B emission scenario, $p\mathrm{CO}_2$ would reach ~500 ppm by 2050 and ~700 ppm by 2100, which would be equivalent to an approximate doubling of the current level.

The increase in $p\mathrm{CO}_2$ and in other greenhouse gas concentrations (mainly methane and nitrous oxide) forces an increase in the annual mean surface air temperature of 1.4 to 5.8°C by 2100, with distinct regional differences (IPCC 2001). Maximum warming is expected in the high latitudes of the Northern Hemisphere and minimum warming in the southern ocean. Nearly all land areas warm more rapidly than the global average. On top of an increase in mean annual temperature, an increase in the interannual variability and a larger frequency of extremes is expected; for the area of central Europe, Schär et al. (2004) found that by 2071 to 2100, year-to-year variability increases by up to 100%, on top of the increase in mean temperature; this scenario implies much more frequent heat waves and droughts during the growing season, which is likely to be of higher ecological relevance than just a higher mean temperature.

For precipitation, trend projections into the future are much more uncertain. Changes in total seasonal precipitation or its seasonal distribution, are both important. Globally, averaged precipitation is expected to increase (IPCC 2001), but the magnitude of regional precipitation changes varies considerably among models, with the typical range being 0 to 50% where the direction of change is strongly indicated, and around 30 to +30% where it is not. A projected increase is consistent in northern high-latitude regions in winter, whereas reductions are most likely in subtropical latitudes. Increased wintertime precipitation is expected in the mid to high latitudes of the Northern Hemisphere and over Antarctica (Giorgi and Francisco 2000). As reviewed recently by Fuhrer et al. (2006), in Europe both rain-day frequency and intensity during winter exhibit an increase north of about 45°N, while the rain-day frequency decreases to the south. This is consistent with an increase of mean winter precipitation by 10 to 30% over most of central and northern Europe, and a decrease over the Mediterranean. In summer, the most remarkable change is a strong decrease in the frequency of wet days, for instance to about half in the Mediterranean, which goes along with a 20 to 50% decrease of mean summer precipitation. In the tropics, models consistently show an increase in Africa, a small increase in South America, but practically no change in Southeast Asia. A decrease in summer precipitation is expected in the Mediterranean basin and in regions of Central America and northwestern Europe.

7.3 SOIL WATER BALANCE AND CROP WATER USE

Soil water content (SWC) is an important variable in the climate system, and a key limiting factor for crop production. One of the earlier predictions of global warming by IPCC showed increasing midlatitude summer drought in important agricultural regions of the Northern Hemisphere (Manabe et al. 1981; Houghton et al. 1996; Watson et al. 1996). This has been confirmed by regional studies in central Europe based on IPCC grid-point scenarios (e.g., Jasper et al. 2004), while for other regions, for instance in parts of Canada, increasing SWC was predicted (McGinn et al. 2001). A change in the water balance and, thus, in the amount of plant-available water in the soil, can be crucial for the yield of rain-fed crops, and it determines the amount of supplemental water needed in irrigated crops. In grasslands, which are typically rain fed, 90% of the variance in primary production can be accounted for by variations in annual precipitation (see Campbell et al. 1997).

7.3.1 SOIL WATER BALANCE

SWC of nonirrigated systems depends on the balance between the inputs of water, mainly through precipitation (P), interception (I), and capillary rise from shallow water tables (Wt), and outputs via evapotranspiration (ET), runoff (R) from saturated soils, and drainage (D):

$$dSWC/dt = P + I + Wt - ET - R - D \qquad (7.1)$$

Consequently, SWC varies mainly with changes in P, Wt, and in the conditions affecting system-level ET. The latter is associated with the gradient of vapor pressure (vp) between the ground surface and the layer of the atmosphere receiving the evaporated water, and is driven by net radiation (Rn). Control over ET is exerted by the surface boundary conditions — by crop-specific characteristics such as leaf area index (LAI), row spacing, and so forth, as well as soil characteristics such as SWC and soil hydraulic properties. In the absence of any boundary layer limitation, ET equals potential evapotranspiration (PET) and varies with air humidity, temperature (T), and net radiation (Rn).

In relation to the soil water balance, and to estimated crop water demand, the concept of reference crop ET is commonly applied. This refers to ET from a uniform, well-defined green crop surface under well-watered conditions, typically alfalfa (ET_o) or grass (ET_r) (Jensen et al., 1990). The reference ET can be estimated by applying the Penman-Monteith equation with an assumed crop height of 12 cm, a fixed canopy resistance of 70 m s^{-1}, and an albedo of 0.23 (closely resembling ET from an extensive surface of green grass) (Doorenbos and Pruitt, 1977), or by using simpler approaches when data availability is limited (e.g., Trajkivic 2005). Values for annual ET_o range from below 500 mm in northern latitudes to around 1000 mm in mid latitudes, and highest values of >2000 mm in subtropical areas of Africa, South America, and Australia. From ET_o, estimates of actual ET of a specific crop can be derived as,

$$ET = ET_o \text{ (or } ET_r) \times K_c \times K_s, \tag{7.2}$$

with the crop-specific coefficient K_c, and a soil dryness coefficient K_s, which is 1 for well-watered conditions. Under the latter conditions, ET_o provides a good approximation of PET for a particular site, but it ignores physiological effects of increasing pCO_2.

7.3.2 EFFECTS OF ELEVATED pCO_2 ON ET

Both elevated pCO_2 and increasing T affect crop water use by changing all parameters of Equation 7.2 for ET. Elevated pCO_2 affects canopy resistance (r_c) through a reduction in stomatal diffusion (g_s). As discussed by Ghannoum et al. (Chapter 3, this volume), elevated pCO_2 decreases g_s, thereby causing a reduction in transpiration rates of single leaves (ET_l). At the whole-plant level, this effect is partly compensated for by increasing total leaf area, thus leading to a smaller reduction in plant ET (ET_p) than in ET_l, or even a smaller reduction in canopy ET (ET_c). The magnitude of the effect on ET_c depends on canopy structure, the degree of coupling of the canopy to the atmosphere, and microclimatic conditions. The better the coupling of the surface to the prevailing atmospheric conditions, the greater the control of transpiration by the plant. In the absence of soil moisture limitation crops and grasslands are not well coupled; hence, transpiration is only weakly determined by surface conditions, and effects of elevated pCO_2 on ET_c are small. In fact, in FACE (free air CO_2 enrichment) experiments providing a close-to-natural low degree of coupling, measured and simulated reductions in cumulative ET_c at elevated pCO_2 were typically below <10%, and much less than reductions in ET_l. In open-top chamber experiments, measured reductions in ET_c at elevated pCO_2 were often larger because of the much stronger coupling due to forced turbulence by ventilation. The effect of elevated pCO_2 is further limited by increased T and decreased humidity in the air surrounding the leaves. Moreover, only marginal compensation through a small effect of elevated pCO_2 on crop LAI was found (Bunce 2004). Nevertheless, even under FACE conditions, a small effect of pCO_2 on ET_c can lead to increased SWC, as observed in sorghum (Wall et al. 2001), wheat, cotton, and maize (*Zea mays* L.) (Kang et al. 2002), and in tallgrass prairie (Owensby et al. 1999) (cf. Kimball et al. 2002). In turn, crop productivity may benefit from improved SWC, which is more important under water-limited than under well-watered conditions (see Ghannoum et al. Chapter 3, this volume). This benefit was also observed for seminatural and natural grasslands (Morgan et al. 2004). Under dry soil conditions, the coupling is much stronger than under wet soil conditions because stomatal resistance (r_s) is large relative to atmospheric resistance (r_a). Hence, under dry conditions, the relative effect of elevated pCO_2 on transpiration is more important than under wet conditions (Lockwood 1999). In C_4 plants, pronounced drying of soils may increase r_s, and consequently lower leaf-internal CO_2 concentrations (C_i) to levels over which photosynthesis may become sensitive to CO_2, as suggested by Polley (2002). Thus, increasing pCO_2 may have a stronger positive effect on the yield of C_4 plants under dry conditions, in contrast to wet conditions. However, this may be insignificant, as discussed by Ghannoum et al. (2003), and should be viewed together with other factors such as increased leaf temperature (Siebke et al. 2003), and improved shoot and leaf water relations due to increased r_s (Seneweera et al. 1998).

Overall, increasing pCO_2 affects the water balance of cropping systems positively, although the effect of a doubling of pCO_2 appears to be small. Of course, these projections ignore any crop acclimation to elevated pCO_2, and the replacement of crops and genotypes in cropping systems as a farmer's most likely adaptive responses to climate change (Hovenden Chapter 12, this volume).

7.3.3 EFFECTS OF TEMPERATURE ON *ET*

Theoretically, an increase in T should act opposite to the effect of elevated pCO_2. Simulations using a soil–crop–climate model indicated that an increase by 3°C induces an increase in *PET* of 14% (Ramirez and Finnerty 1996). Similarly, regional climate change scenarios with a mean annual increase by 3°C resulted in an increase in ET_c under temperate central European conditions of 16.7% (Jasper et al. 2004), and of up to 8% per 1°C warming over central Asia (Chattopadhyay and Hulme 1997). Higher leaf temperature (T_l) may offset positive effects of elevated pCO_2, as discussed by Ghannoum et al. (Chapter 3, this volume), because the latter causes larger leaf-to-air gradients in vapor pressure deficit (*vpd*). Calculations using the Penman–Monteith equation predict that *PET* increases by about 2 to 3% for each 1°C rise in T; the relative increase is slightly less than expected from changes in *vpd* because ET_c is also controlled by the surface net radiation balance, which declines with increasing T (Lockwood 1999).

The proposed increase in ET_c with increasing air temperature is consistent with simulated changes for the conterminous U.S. of +4.3% over the period 1962 through 1988 (Hobbins et al. 2004). However, these results are in conflict with decreasing trends in *ET* observed over the past decades using data from pan evaporimeters. An analysis in Italy revealed a significant decrease during the past 122 years (Moonen et al. 2002), in agreement with a survey of pan *ET* data collected at a large number of stations in the former Soviet Union and the United States (Peterson et al. 1995), in India (Chattopadhya and Hulme 1997), and in Australia (Roderick and Farquhar 2004). The authors of the Italian study argued that the decrease might be caused by the asymmetry of increasing T, that is, a strong increase in minimum (nighttime) T, and slightly decreasing maximum T. This asymmetric warming is characteristic of a climate influenced by increasing cloudiness that reduces solar irradiance (which is often referred to as *global dimming*) and decreases *ET* from pans (Roderick and Farquhar 2002). However, pan evaporimeter data do not reflect terrestrial *ET*. In fact, after recalibration of pan *ET* data for sites in the United States and Russia to make them more representative of actual *ET*, the trend with time became positive (Golubev et al. 2001). While increasing cloud coverage and aerosol loading may be responsible for the past decrease in pan *ET*, it leads to more scattering of light and less shade inside plant canopies, resulting in increasing diffuse radiation with a positive effect on plant productivity and terrestrial *ET* (Roderick et al. 2001). Therefore, any projection of future trends in *ET* should consider not only effects of increasing T, but also of increasing diffuse radiation perceived by the vegetation. However, future changes in radiation are uncertain; recent observations suggest that the trend of global dimming has been reversed in most recent years because of declining sulfur aerosol loading of the atmosphere (Wild et al. 2005).

Higher T also affects K_s (see Equation 7.2), which describes the crop-specific modification of ET_0, and which depends on the stage of canopy development (Jensen et al. 1990). K_s varies in relation to canopy structure and length of individual growth stages. Under current conditions, K_s ranges from typically below 0.5 during initial growth to values well above 1 in many crops during mid-season, followed by a decline until maturity. If warming affects the growth cycle by accelerating crop development and shortening phenophases, this will affect K_s and may counteract the enhancement of ET_c by direct effects of warming and diffuse radiation. This effect was observed in simulations of the growth of wheat in Spain (Guerena et al. 2001), but observational data to quantify this indirect effect are scarce. For perennial temperate grasslands, simulations showed that effects of doubled pCO_2 and climate change on seasonal total ET_c were additive, and that the reduction by doubled pCO_2 was antagonized by the effect of warming, thus still leading to increased ET_c (Riedo et al. 1999). In other words, with increasing T, the positive effect of elevated pCO_2 on water use declines. In small plots of Mediterranean grassland, Zavaleta et al. (2003) observed a reduction in water loss due to an acceleration of plant senescence leading to a higher average SWC. In places where SWC is often limiting (i.e. $K_s <$ 1), this diminishing effect of elevated pCO_2 on crop water loss may positively feed back on growth through increased SWC.

In summary, future climate warming, possibly in combination with increasing diffuse radiation, stimulates crop ET in the absence of soil water limitation, but shifts in crop phenology and accelerated crop development could reduce ET summed over the life cycle of a determined annual crop. Quantitatively, the outcome of the interplay of various factors depicted in Figure 7.1, including effects of elevated pCO_2, remains uncertain because of the lack of data from well-designed factorial field experiments.

7.3.4 IMPLICATIONS FOR IRRIGATION

Irrigation is necessary in areas with a negative soil water balance, that is, where losses exceed natural inputs through P and capillary flow (see Equation 7.1). Today, about 20% of the world's cropland is irrigated, with a major fraction located in Asia, producing about 40% of the annual global crop yield. Irrigated agriculture is responsible for approximately 70% of all the freshwater withdrawn worldwide. More water and related energy will be required for irrigation in the future, as world food demand continues to grow. In view of this trend, with its well known ecological, economic, and social implications (cf. Brouwer and Falkenmark 1989), any additional change in irrigation requirements brought about by changing climatic conditions is of great importance.

Earlier reports have concluded that (1) the demand for irrigation water will rise in a warmer climate, (2) falling water tables and the resulting increase in the energy needed to pump water will make the practice of irrigation more expensive, (3) the area of irrigated land will change, or (4) water intensity (i.e., the amount of irrigation water needed per unit of irrigated land) may increase or decrease, depending on

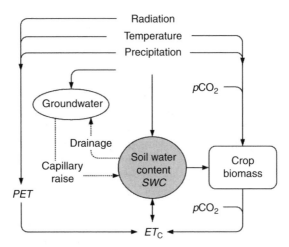

FIGURE 7.1 Simplified diagram of the interplay between climatic factors and elevated pCO_2, and potential evapotranspiration (*PET*), actual canopy evapotranspiration (ET_c), possible groundwater exchange (dotted lines), which determines the soil water content (*SWC*) in cropping systems either directly or indirectly via changing plant biomass.

location (Rosenzweig 1990; Adams et al. 1990). Peak irrigation demand could rise due to more frequent severe heat waves. Most of these notions are based on the fact that irrigation scheduling generally uses a water-balance approach (see above and Jensen et al. 1990). Critical conditions for crop growth occur during periods with an amount of *P* of less than about 50% of *ET* (Brouwer and Falkenmark 1989). Thus, it seems reasonable to assume that increased *ET* due to climate change will increase irrigation requirements, provided that *P* declines, remains unchanged, or increases less than *ET*. A recent study for the United Kingdom supported this assumption and projected that climate change impacts on the use of water for irrigation could increase by about 20% by the 2020s, and by around 30% by the 2050s, with regional differences (Downing et al. 2003). However, any projection of future irrigation requirements suffers from several sources of uncertainty:

- Uncertain projections of *P* amount and intensity, and in the seasonal and spatial distribution, due to uncertainties with respect to current general circulation models (GCM) and future greenhouse gas emission patterns. As an example, two different GCMs combined with crop models gave significantly different irrigation and yield trends in U.S. agriculture, both in magnitude and direction (Brumbelow and Georgakakos 2001).
- Uncertain changes in *ET* because of interactions between effects on *ET* of *T* and other factors including increased cloudiness during the daytime. Conversely, elevated pCO_2 may decrease *ET* especially in drier regions, but the effect depends on crop, canopy structure, and field management, and may be sensitive to acclimation (Hovenden, Chapter 12, this volume). Moreover, uncertain effects of both *T* and pCO_2 on phenology and *LAI*

affect the relationship between ET_o and actual ET, which in turn determines the estimated amount of irrigation water needed.

- Finally, uncertainty in the simulation of SWC at resolutions relevant for planers: Simple soil water balance models ignore subsurface flows, which can strongly influence root-zone SWC, and thus irrigation requirements and agricultural benefits (Ramirez and Finnerty 1996).

Simulations of future changes in SWC have been performed at different scales. Global simulations with a coupled ocean–atmosphere–land model suggested that by the middle of the 21st century, SWC may decline in many semiarid regions of the world, such as the southwestern region of North America, the northeastern region of China, the Mediterranean region of Europe, and the grasslands of Australia and Africa, with largest reductions during the dry season (Manabe et al. 2004). For middle to high latitudes, these and other simulations projected decreasing SWC during summer and increasing SWC during winter. In agreement, in simulations for a catchment in Switzerland with a hydrological model driven by weather data downscaled from GCM outputs, SWC declined significantly during the growing season even in the absence of large shifts in P, but with strongly increased ET, but it increased during wintertime (Jasper et al. 2004). This suggests that in the future, water limitation for crops may occur in regions where there is no need for irrigation under current conditions. As shown by Jasper et al. (2006), this applies in the first place to flat lowland areas where currently the groundwater is connected to the rooting zone during much of the growing season; this connection is lost earlier during a warmer growing season, thus leading to a more rapid drop in SWC. An increase in depth to the groundwater table may have larger effects on SWC than altered P (Ramirez and Finnerty 1996). In sloped terrain, interflow may play an equally important role. Thus, in any assessment of future irrigation requirements, modeling of SWC dynamics should not be based on single-column soil moisture schemes that largely ignore lateral fluxes and fluctuations in the depth to groundwater table.

A set of simulation models used to estimate climate change impacts on irrigated corn in the United States indicated that beneficial effects of an extended growing season and a larger water supply through higher amounts of P were generally overcome by negative impacts on crop yield of increased ET combined with reduced solar radiation during plant maturing stages (Tung and Haith 1998). Consequently, the irrigation requirement for corn was found to increase with respect to the baseline in 17 out of 18 test basins across the United States using a climate change scenario, which falls in the middle of a range given by the IPCC (Intergovernmental Panel on Climate Change) (Izaurralde et al. 2003). In contrast, results for Spain showed that irrigation requirements in wheat production could decrease in response to doubled pCO_2, GCM, and RCM (regional climate model) forcing because of the shorter crop duration and an earlier start of the growing season, which reduces cumulative ET during the life cycle of the crop (Guerena et al. 2001). The following crop could benefit from resulting soil water savings, provided that suitable field practices are applied to avoid water losses from the bare soil after harvest (e.g., mulching or no till). This underlines the need for linking hydrological models with crop growth models in order to capture the influence of shifts in crop phenology on crop water use.

These few examples indicate that in a warmer world, effects of increased *ET* may lead to either increased or decreased irrigation requirements. The direction of change in cumulative crop *ET* depends on the magnitude and direction of the concurrent change in *P* (which is highly uncertain); shifts in crop phenology, canopy structure, and rooting depth, and the influences of groundwater and interflow (which depends on soil texture); depth to the groundwater table; and terrain characteristics such as slope. The change in the need for irrigation may be only marginally affected by increasing pCO_2. Where the need for supplemental irrigation will increase, farmers will have to seek optimal water application techniques and timing; as shown by Tavakkoli and Oweis (2004) for the western Asia and North Africa region (with a Mediterranean climate), water use efficiency related to yield can be optimized by adjusting irrigation amounts and scheduling to specific crop water requirements during different phenological stages. Finally, management can support the sustainability of cropping systems through field practices that positively influence the soil water balance and counteract soil water depletion, for instance, during periods when no crop is present, or from the area of bare ground between rows of crops. Also, tillage systems and equipment have great impacts on water infiltration, storage, and plant efficiency (Sullivan 2002). Besides effects on soil physical traits, most important is the extent at which the surface cover is maintained. Figure 7.2 illustrates how *SWC* can be influenced positively by applying no-till to corn. This example shows that the positive effect of no-till is more evident in the top 15 cm than between 15 and 30 cm.

7.4 NITROGEN CYCLING AND N USE

Through fertilizer production and use, fossil fuel combustion, and cultivation of legumes, human activities add about 140 Tg N yr[1] to the global N cycle, a rate that exceeds natural N fixation (Vitousek et al. 1997). Intensive crop production requires N inputs, but the more N is introduced via fertilization or biological fixation, the greater the potential for losses of excess N to the environment. N losses from cropping systems are the result of excess inputs relative to crop demand, or of the asynchrony between rates at which N in the soil is made available to plant and patterns of N uptake. Sustainable management of cropping systems requires that the demand of a crop for N is closely synchronized with fertilizer addition and processes that regulate N availability in the soil, in order to avoid nitrate (NO_3) accumulation, and thus to minimize off-site environmental effects through leaching or denitrification. In the context of this review, the main question is how increasing pCO_2 and climatic change could influence soil-related processes and crop N use, with possible implications for fertilization.

7.4.1 EFFECT OF ELEVATED pCO_2 ON N CYCLING AND N USE

Ghannoum et al. (Chapter 3, this volume) concluded that elevated pCO_2 is unlikely to alter plant demand for N, whereas requirements for phosphorus (P) may increase.

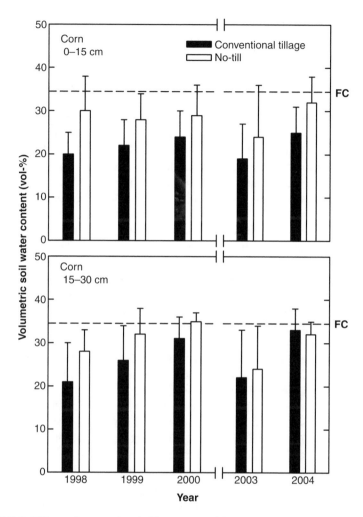

FIGURE 7.2 Effects of conventional tillage vs. no-till on seasonal mean volumetric water content of a sandy loam soil cultivated with corn (*Zea mays* L.) near Bern, Switzerland. FC = Field capacity. Means were calculated from repeated measures during the season. Vertical lines: Standard deviation of repeated measures. (Data are courtesy of U. Zihlmann and P. Weisskopf, Agroscope FAL Reckenholz, Zürich, Switzerland, unpublished.)

This view is supported by recent experiments; Torbert et al. (2004) found that in grain sorghum, a C_4 plant, total N uptake and N yield were not affected at elevated pCO_2, but the C:N ratio and N-use efficiency (*NUE*) increased by 20 to 30%. In spring wheat, a C_3 plant, Adamsen et al. (2005) observed that elevated pCO_2 did not result in major differences in either aboveground or belowground N when soil N fertility was adequate, but it increased early vegetative growth. At low N supply, elevated pCO_2 resulted in lower leaf N concentrations in wheat because the increase in plant biomass was not matched by increased N uptake (Sinclair et al. 2000). In general, increasing N uptake caused by elevated pCO_2 does not necessarily increase

N yield in cereals (Kimball et al. 2002) because plant N concentration decreases as a result of dilution due to more rapid growth, or due to changes in N allocation at the level of the photosynthetic apparatus, as discussed by Ghannoum et al. (Chapter 3, this volume). In soybean, increased yield with elevated pCO_2 was not accompanied by a change in the C:N ratio and N contents, but N yield increased due to increased N_2 fixation. In N_2-fixing crops, increasing N demand due to CO_2-stimulated growth is met by higher rates of N_2 fixation, provided that P is not limiting (Thomas et al., Chapter 4, this volume). Through N-rich plant residues, larger amounts of N are transferred to the organic soil N pool. If the extra N is not fixed in the soil or taken up by an accompanying intercrop or by following nonleguminous crops, it may be volatilized or leached. Hence, proper management practices to control the soil-available N pool in rotations with legumes may become more important in a high-CO_2 world.

Plant responses to environmental change are underpinned by mechanisms that control microbial processes in soils, and by nutrient-cycling feedbacks. Elevated pCO_2 increases plant biomass above and below the ground (Kimball et al. 2002), as discussed by Thomas et al. (Chapter 4, this volume). In turn, this stimulates the input of organic material derived from roots and litter, which may act as substrate for soil microbes. This can influence mineralization of soil organic matter, N transformation, and N fixation, and thus N availability in soil. The relevance of this input depends on the size of preexisting organic N pools.

In an experiment with seedlings of spring barley, concentrations of total inorganic N in the rhizosphere soil were not significantly affected by pCO_2, but the addition of N under enriched pCO_2 increased the amount of N recovered by the plant from both fertilizer and soil (Martin-Olmedo et al. 2002). Retrieval and utilization of N were also stimulated by elevated pCO_2 in field-grown cotton (Prior et al. 1998), and in wheat (Fangmeier et al. 1999). Increased recovery of soil N could be linked to higher *NUE*, which is a common response of C_3 plants to elevated pCO_2 (Drake et al. 1997), or to improved soil N exploitation by the root system. The latter would be consistent with the higher fraction of fertilizer-derived N in plants grown at elevated pCO_2 mentioned above (Martin-Olmedo et al. 2002), and with the decline in topsoil mineral N concentration under spring wheat exposed to elevated pCO_2, which was related to increased root biomass (Adamsen et al. 2005).

In contrast, N uptake remained unchanged in *Lolium perenne* (Hartwig et al. 2002) and grain sorghum (*Sorghum vulgare* Pers.) (Torbert et al. 2004). Torbert et al. (1996) reported that elevated pCO_2 decreased NO_3 concentrations below the rooting zone in both soybean and grain sorghum crops, thus indicating improvements in N retention in soil organic matter pools. In intact C_3/C_4 grassland in central Texas (U.S.), increasing pCO_2 decreased net N mineralization, which led to a strong, nonlinear decrease in N availability (Gill et al. 2002). In swards of *L. perenne* after 4 years with elevated pCO_2, N yield and *NUE* were reduced, while system N gain decreased at low, and increased at high N fertilization rates; an increase in N yield and *NUE* in clover was associated with higher system (i.e., plant and soil) N gains at both low and high N input rates (Hartwig et al. 2002). The larger system N gain was mainly due to higher retention of N in the soil. However, after 10 years of exposure, analysis of the same system revealed that N mobilization from soil, mainly from soil organic matter, increased and N limitation of plant growth decreased under elevated pCO_2 (Schneider et al. 2004).

7.4.2 EFFECT OF CLIMATE CHANGE ON N BALANCE

In contrast to elevated pCO_2, warming tends to have a negative effect on plant *NUE*. This can be due to increased sink limitation with increasing T, especially in plants with reproductive sinks (Reddy et al. 1991). At the system level, warmer conditions stimulate soil N availability through higher rates of mineralization, which can stimulate productivity (Parton et al. 1995), but potentially also higher system N losses, if N demand by the plant is not synchronized with N supply. Tillage strongly stimulates N mineralization; therefore, reduced tillage under warmer conditions may prevent N release during early parts of the season when N uptake by the crops is still small, especially in spring-sown crops.

In legume-based systems in temperate agriculture, N supply through symbiotic N_2 fixation may benefit from higher T because currently low temperatures during autumn and winter are a substantial impediment to legume growth and nodulation. On the other hand, high T associated with water deficits during spring and early summer terminates the growing season and thus limits N_2 fixation in annual legumes in Mediterranean-type climates, and in the tropics where high soil temperatures (< 50°C) has direct negative effects on N_2 fixation, nodulation, and plant growth. Hence, in some currently cooler regions, higher T without major water limitations could lead to an expansion of N_2-fixing crops, whereas in others, cultivation may decline unless current host plants or their associated strains of N_2-fixing bacteria are replaced by better adapted ones. Thomas et al. (Chapter 4, this volume) suggested that positive effects of elevated pCO_2 (see above) are likely to be greater at higher T, but further experimental data are needed to support this notion.

7.4.3 IMPLICATIONS FOR FERTILIZATION

The total amount of N fertilizer required by C_3 crops to support optimal productivity at elevated pCO_2 is unlikely to become larger, despite the increase in biomass (Poorter and Navas 2003). This view is supported by experimental data for wheat (Adamsen et al. 2005) and for sorghum (Torbert et al. 2004) showing that total N uptake was not affected, and that C:N ratios increased at elevated pCO_2, while no impact on fertilizer N was observed,. With soybean, the extra N in shoot and root residues returned to the soil may provide extra N credits to the following nonleguminous crops, which reduces the need for mineral N fertilizer inputs.

Shifts in phenology and growth due to warming will require adjustments in fertilization scheduling to maintain the synchrony between N inputs and N demand during individual crop stages. Higher T generally reduces grain yield in C_3 crops due to accelerated development, in particular shorter periods of grain filling. Whether or not this effect changes total crop N use and thus N fertilizer requirements remains uncertain. Warmer soil T increases N supply from organic pools, which could reduce the need for N additions for some time. Higher T generally favors C_4 crops, and

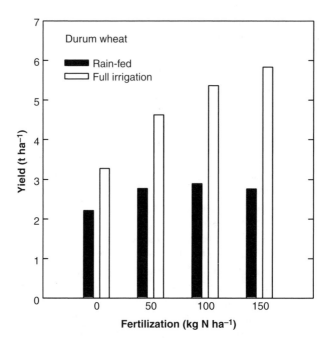

FIGURE 7.3 Mean effects of fertilizer N on durum wheat grain yield in the presence or absence of supplemental irrigation (data are for full irrigation) in a semiarid region. (Data from Oweis, T., Pala, M., and Ryan, J. [1999] *European Journal of Agronomy*, 11, 255–266.)

increased N inputs may be necessary to maintain critical levels of N concentration to maintain optimal productivity (see Ghannoum et al., Chapter 3, this volume).

More frequent drought will reduce soil N mineralization and N inputs through N_2 fixation in legume-based systems, and thus lower productivity. Under these limiting conditions, extra fertilizer input is of little use, giving the well-known relationship between N and soil moisture (Raming and Rhoades 1962). However, excessive input to dry soils enhances the potential for off-site environmental impacts when soils are rewetted, due to the loss of N, with volatile N gases (NH_3, N_2O) and through NO_3-leaching. Data for wheat cultivated under dry conditions in Syria illustrate how fertilization increases the yield in the presence of supplementary irrigation, but not in rain-fed crops where water limits production across the range of N additions (Figure 7.3). Hence, in regions where future conditions are characterized by higher T and more frequent drought, productivity of cropping systems may be maintained by additional irrigation and possibly through expansion of N_2-fixing crops. In the absence of irrigation, potential productivity is likely to decline, particularly in water-demanding crops such as corn, and additional fertilization or use of legumes will be of little benefit. In regions where water will not be limiting, an increase in T may increase crop productivity directly, and indirectly through additional N input via N_2 fixation, whereas mineral fertilizer use is likely to remain stable. These trends are not influenced substantially by increased pCO_2.

7.5 CONCLUSIONS

From this review, the following conclusions can be drawn with respect to future management of crop production systems:

- Global climate change will alter the conditions for agricultural production in all parts of the world, with regional differences in magnitude and direction. Higher mean T is likely affecting the currently cooler high latitude regions relatively more than warmer regions in the south, while increased frequency of heat waves may cause significant risks to regions in mid to lower latitudes. Where warming is associated with unchanged or reduced P, it is likely causing more frequent water limitation, whereas where P is increasing, productivity may benefit from higher T through more favorable growing conditions and expansion of the growing season.
- Key to any projection of future crop productivity is knowledge of shifts in SWC. Reduced SWC during sensitive stages of crop development limits productivity, whereas improved SWC resulting from reduced ET due to effects of elevated pCO_2 could increase productivity, especially in C_3 crops. The interaction between these two opposing mechanisms remains uncertain and depends on plant acclimation and farmers' adaptations in terms of crop and cultivar selection, planting procedures (timing, row spacing, and so forth), and soil tillage and plant residue management.
- In a warmer world, with possibly higher levels of diffuse radiation, PET and likely ET are higher. Where $ET{:}P$ becomes critical (< 0.4 to 0.5), irrigation demand will grow. However, cumulative ET during the crop's growth cycle may decline due to faster plant development, associated with lower yield. Thus, management practices aiming to preserve soil water for subsequent crops will be important.
- Where shifts in climate result in more favorable growing conditions, additional fertilizer input may increase actual yield levels. Where SWC is limiting, additional N fertilization will have no effect, but excess soil-available N will cause higher risks for off-site impacts. Increased pCO_2 may have only a minor effect on future fertilizer needs, but fertilization protocols may need to be adapted to altered growth patterns.
- Climate change will affect several constraints to crop production, and farmers' adaptive responses will need to focus on SWC conservation to avoid new investments for irrigation, or on optimal water application where irrigation equipment is installed. Fertilization will need to be adjusted to changes in crop phenology and soil conditions in order to match inputs with crop demand, and to avoid low NUE and the risk of N losses to the environment. These are some measures that will be necessary in order to reach maximum sustainability under new climatic conditions.
- Overall, water and fertilizer management at the plot and catchment scale will need to be adjusted to changing climatic conditions and to the needs of future, better-adapted crop species or varieties.

ACKNOWLEDGMENT

This review was produced in the framework of the National Centre for Competence in Research (NCCR) Climate Project titled *Climate Change and Food Production*, funded by the Swiss National Science Foundation.

REFERENCES

Adams, R.M., Rosenzweig, C., Peart, R.M., et al. (1990) Global climate change and US agriculture. *Nature*, 345, 219–224.

Adamsen, F.J., Wechsung, G., Wechsung, F., et al. (2005) Temporal changes in soil and biomass nitrogen for irrigated wheat grown under free-air carbon dioxide enrichment (FACE). *Agronomy Journal*, 97, 160–168.

Brouwer, F. and Falkenmark, M. (1989) Climate-induced water availability changes in Europe. *Environmental Monitoring and Assessment*, 13, 75–98.

Brumbelow, K. and Georgakakos, A. (2001) An assessment of irrigation needs and crop yield for the United States under potential climate changes. *Journal of Geophysical Research*, 106, D21, 27383–27405.

Bunce, J.A. (2004) Carbon dioxide effects on stomatal responses to the environment and water use by crops under field conditions. *Oecologia*, 140, 1–10.

Campbell, B.D., Stafford Smith, M.D., and MecKeon, G.M. (1997) Elevated CO_2 and water supply interactions in grasslands: a pasture and rangelands management perspective. *Global Change Biology*, 3, 177–187.

Chattopadhyay, N. and Hulme, M. (1997) Evaporation and potential evapotranspiration in India under conditions of recent and future climate change. *Agriculture and Forest Meteorology*, 87, 55–73.

Doorenbos, J. and Pruitt, W.O. (1977) Crop water requirements, irrigation and drainage. Paper No. 24, FAO, United Nations, Rome, Italy.

Downing, T.E., Butterfield, R.E., Edmonds, B., et al. (2003) *Climate Change and the Demand for Water*. Research Report, Stockholm Environment Institute Oxford Office, Oxford, U.K.

Drake, B.G., Gonzalez Meler, M.A., and Long, S.P. (1997) More efficient plants: a consequence of rising atmospheric CO_2? *Annual Revue of Plant Physiology and Plant Molecular Biology*, 48, 609–639.

Fangmeier, A., De Temmerman, L., Mortensen. L., et al. (1999) Effects on nutrients and on grain quality in spring wheat crops grown under elevated CO_2 concentrations and stress conditions in the European, multiple-site experiment "Espace-wheat." *European Journal of Agronomy*, 10, 215–229.

Fuhrer, J. (2003) Agroecosystem responses to combinations of elevated CO_2, ozone, and global climate change. *Agriculture, Ecosystems and Environment*, 97, 1–20.

Fuhrer, J., Beniston, M., Fischlin, A., et al. (2006) Climate risks and their impact on agriculture and forests in Switzerland. *Climatic Change*.

Ghannoum, O., Conroy, J.P., Driscoll, S.P., Paul, M.J., Foyer, C.H., and Lawlor, D.W. (2003) Nonstomatal limitations are responsible for drought-induced photosythetic inhibition in four C_4 grasses. *New Phytologist*, 159, 599–608.

Ghannoum, O., Searson, M.J., and Conroy, J.P. (2006) Nutrient and water demands of plants under global climate change. In *Agroecosystems in a Changing Environment*, Newton, P.C.D., Carran, R.A., Edwards, G.R., and Niklaus, P.A., Eds., CRC Press, Boca Raton, FL, 145–160.

Gill, R.A., Polley, H.W., Johnson, H.B., Anderson, L.J., et al. (2002) Nonlinear grassland response to past and future atmospheric CO_2. *Nature*, 417, 279–282.

Giorgi, F. and Francisco, R. (2000) Evaluating uncertainties in the prediction of regional climate change. *Geophysical Research Letters*, 27, 1295–1298.

Golubev, V.S., Lawrimore, J.H., Groisman, P.Y., et al. (2001) Evaporation changes over the contiguous United States and the former USSR: a reassessment. *Geophysical Research Letters,* 28, 2665–2668.

Guerena, A., Ruiz-Ramos, M., Diaz-Ambrona, C.H., et al. (2001) Assessment of climate change and agriculture in Spain using climate models. *Agronomy Journal*, 93, 237–249.

Hartwig, U.A., Lüscher, A., Nösberger, J., et al. (2002) Nitrogen-15 budget in model ecosystems of white clover and perennial ryegrass exposed for four years at elevated atmospheric pCO_2. *Global Change Biology*, 8, 194–202.

Hobbins, M.T., Ramirez, J.A., Brown, T.C. (2004) Trends in pan evaporation and actual evapotranspiration across the conterminous U.S.: paradoxical or complementary? *Geophysical Research Letters* 31, L13503.

Houghton, J.T., Meira Filho, L.G., Callander, B.A., et al., Eds. (1996) *Climate Change 1995— The Science of Climate Change*, Cambridge University Press, Cambridge, U.K.

Hovenden, M.J. (2006) Distinguishing between acclimation and adaptation. In *Agroecosystems in a Changing Environment*, Newton, P.C.D., Carran, R.A., Edwards, G.R., and Niklaus, P.A., Eds., CRC Press, Boca Raton, FL, 301–318.

IPCC (Intergovernmental Panel on Climate Change). (2001) *Climate Change 2001: The Scientific Basis*, Houghton, J.T., Ding, Y., Griggs, D.J. et al., Eds., Cambridge University Press, Cambridge, U.K.

Izaurralde, R.C., Rosenberg, N.J., Brown, R.A., et al. (2003) Integrated assessment of Hadley Center (HadCM2) climate-change impacts on agricultural productivity and irrigation supply in the conterminous United States. Part II. Regional agricultural production in 2030 and 2095. *Agriculture and Forest Meteorology*, 117, 97–122.

Jasper, K., Calanca, P.L., Gyalistras, D., et al. (2004) Differential impacts of climate change on the hydrology of two alpine river basins. *Climate Research*, 26, 113–129.

Jasper, K., Calanca, P.L., and Fuhrer, J. (2006) Changes in summertime soil water patterns in complex terrain due to climatic change. *Journal of Hydrology*.

Jensen, M.E., Burman, R.D., and Allen, R.G., Eds. (1990) *Evapotranspiration and Irrigation Water Requirements*, American Society of Civil Engineers, New York.

Kang, S.Z., Zhang, F.C., Hu, X.T., et al. (2002) Benefits of CO_2 enrichment on crop plants are modified by soil water status. *Plant and Soil*, 238, 69–77.

Kimball, B.A., Kobayashi, K., and Bindi, M. (2002) Responses of agricultural crops to free-air CO_2 enrichment. *Advances in Agronomy*, 77, 293–368.

Lockwood, J.G. (1999) Is potential evapotranspiration and its relationship with actual evapotranspiration sensitive to elevated atmospheric CO_2 levels? *Climatic Change*, 41, 193–212.

Manabe, S., Milly, P.C.D, and Wetherald, R. (2004) Simulated long-term changes in river discharge and soil moisture due to global warming. *Hydrological Sciences*, 49, 625–642.

Manabe, S., Wetherald, R.T., and Stouffer, R.J. (1981) Summer dryness due to increase of atmospheric CO_2 concentration. *Climatic Change*, 3, 347–384.

Martin-Olmedo, P., Rees, R.M., and Grace, J. (2002) The influence of plants grown under elevated CO_2. *Global Change Biology*, 8, 643–657.

McGinn, S.M., Shepard, A., and Akinremi, O.O. (2001) *Assessment of Climate Change Impacts on Soil Moisture and Drought on the Prairies.* Final Report for the Climate Change Fund (CCAF). Agriculture and Agri-Food Canada, Lethbridge Research Center, Lethbridge, Alberta, Canada.

Moonen, A.C., Ercoli, L., Mariotti, M., et al. (2002) Climate change in Italy indicated by agrometeorological indices over 122 years. *Agriculture and Forest Meteorology,* 111, 13–27.

Morgan, J.A., Pataki, D.E., Körner, C., et al. (2004) Water relations in grassland and desert ecosystems exposed to elevated CO_2. *Oecologia,* 140, 11–25.

Oweis, T., Pala, M., and Ryan, J. (1999) Management alternatives for improved durum wheat production under supplemental irrigation in Syria. *European Journal of Agronomy,* 11, 255–266.

Owensby, C.E., Ham, J.M., Knapp, A.K., et al. (1999) Biomass production and species composition change in a tallgrass prairie ecosystem after long-term exposure to elevated atmospheric CO_2. *Global Change Biology,* 5, 497–506.

Parton, W.J., Scurlock, J.M.O., Ojima, D.S., et al. (1995) Impact of climate change on grassland production and soil carbon worldwide. *Global Change Biology,* 1, 13–22.

Peterson, T.C., Golubev, V.S., and Groisman, P.Y. (1995) Evapotranspiration losing its strength. *Nature,* 377, 687–688.

Polley, H.W., Johnson, H.B., and Derner, J.D. (2002) Soil- and plant-water dynamics in a C_3/C_4 grassland exposed to subambient to superambient CO_2 gradient. *Global Change Biology,* 8, 1118–1129.

Poorter, H. and Navas, M.L. 2003 Plant growth and competition at elevated CO_2: on winners, losers and functional groups. *New Phytologist,* 157, 175–198.

Prior, S.A., Torbert, H.A., Runion, G.B., et al. (1998) Effects of carbon dioxide enrichment on cotton nutrient dynamics. *Journal of Plant Nutrition,* 21, 1407–1426.

Raming, R.E. and Rhoades, H.F. (1962) Inter-relationship of soil moisture level and nitrogen fertilization on winter wheat production. *Agronomy Journal,* 54, 123–127.

Ramirez, J.A. and Finnerty, B. (1996) CO_2 and temperature effects on evapotranspiration and irrigated agriculture. *Journal of Irrigation and Drainage Engineering,* 122, 155–163.

Reddy, V.R., Baker, D.N., and Hodges, H.F. (1991) Temperature effects on cotton canopy growth, photosynthesis, and respiration. *Agronomy Journal,* 83, 699–704.

Riedo, M., Gyalistras, D., Fischlin, A., et al. (1999) Using an ecosystem model linked to GCM-derived local weather scenarios to analyse effects of climate change and elevated CO_2 on dry matter production and partitioning, and water use in temperate managed grasslands. *Global Change Biology,* 5, 213–223.

Roderick, M.L. and Farquhar, G.D. (2002) The cause of decreased pan evaporation over the past 50 years. *Science,* 298, 1410–1411.

Roderick, M.R. and Farquhar, G.D. (2004) Changes in pan evaporation from 1970 to 2002. *International Journal of Climatology,* 24, 1077–1090.

Roderick, M.R., Farquhar, G.D., and Berry, S.L., et al. (2001) On the direct effect of clouds and atmospheric particles on the productivity and structure of vegetation. *Oecologia,* 129, 21–30.

Rosenzweig, C. (1990) Crop response to climate change in the southern Great Plains: a simulation study. *Professional Geographer,* 42, 20–37.

Schär, C., Vidale, P.L., Lüthi, D., et al. (2004) The role of increasing temperature variability for European summer heat waves. *Nature,* 427, 332–336.

Schneider, M.K., Lüscher, A., Richter, M., et al. (2004) Ten years of free-air CO_2 enrichment altered the mobilization of N from soil in *Lolium perenne* L. swards. *Global Change Biology,* 10, 1377–1388.

Seneweera, S.P., Ghannoum, O., and Conroy, J.P. (1998) High vapour pressure deficit and low soil water availability enhance shoot growth responses of a C_4 grass (*Panicum coloratum* cv. Bambatsi) to CO_2 enrichment. *Functional Plant Biology*, 25, 287–292.

Siebke, K., Ghannoum, O., Conroy, J.P., et al. (2003) Photosynthetic oxygen exchange in C_4 grasses: the role of oxygen as electron acceptor. *Plant, Cell and Environment*, 26, 1963–1972.

Sinclair, T.R., Pinter, P.J., Jr., Kimball, B.A., et al. (2000) Leaf nitrogen concentration of wheat subjected to elevated $[CO_2]$ and either water or N deficit. *Agriculture, Ecosystems and Environment*, 79, 53–60.

Sullivan, P. (2002) Drought resistant soils. *Agronomy Technical Note by Appropriate Technology Transfer for Rural Areas (ATTRA)*. Source: http://www.attra.ncat.org/attra-pub/drought.pdf/html, accessed July 13, 2005.

Tavakkoli, A.R. and Oweis, T.Y. (2004) The role of supplemental irrigation and nitrogen in producing bread wheat in the highlands of Iran. *Agricultural Water Management*, 65, 225–236.

Thomas, R.B., van Bloem, S.J., and Schlesinger, W.H. (2006) Climate change and symbiotic nitrogen fixation in agroecosystems. In *Agroecosystems in a Changing Environment*, Newton, P.C.D., Carran, R.A., Edwards, G.R., and Niklaus, P.A., Eds., CRC Press, Boca Raton, FL, 87–112.

Torbert, H.A., Prior, S.A., Rogers, H.H., et al. (1996) Elevated atmospheric carbon dioxide in agroecosystems affects groundwater quality. *Journal of Environmental Quality*, 25, 720–726.

Torbert, H.A., Prior, S.A., Rogers, H.H., et al. (2004) Elevated CO_2 effects on N fertilization in grain sorghum and soybean. *Field Crops Research*, 88, 47–57.

Trajkivic, S. (2005) Temperature-based approaches for estimating reference evapotranspiration. *Journal of Irrigation and Drainage Engineering*, 131, 316–323.

Tung, C.P. and Haith, D.A. (1998) Climate change, irrigation and crop response. *Journal of the American Water Research Association*, 34, 1071–1085.

Vitousek, P.M., Aber, J.D., Howarth, R.W., et al. (1997) Human alteration of the global nitrogen cycle: sources and consequences. *Ecological Applications*, 7, 737–750.

Wall, G.W., Brooks, T.J., Adam, R., et al. (2001) Elevated atmospheric CO_2 improved sorghum plant water status by ameliorating the adverse effects of drought. *New Phytologist*, 152, 231–248.

Watson, R.T., Zinyowera, M.C., and Moss, R.H., Eds. (1996) *Climate Change 1995—Impacts, Adaptations and Mitigation of Climate Change: Scientific-Technical Analyses*. Cambridge University Press, Cambridge, U.K.

Wild, M., Gilgen, H., Roesch, A., et al. (2005) From dimming to brightening: decadal changes in solar radiation at Earth's surface. *Science*, 308, 847–850.

Zavaleta, E.S., Thomas, B.D., Chiariello, N.R., et al. (2003) Plants reverse warming effect on ecosystem water balance. *Proceedings of the National Academy of Sciences USA*, 100, 9892–9893.

Zihlmann, U. and Weisskopf, P. Agroscope FAL Reckenholz, Zürich, Switzerland, unpublished.

Part II

Pests, Weeds, and Diseases

8 Plant Performance and Implications for Plant Population Dynamics and Species Composition in a Changing Climate

Grant R. Edwards and Paul C.D. Newton

CONTENTS

8.1 INTRODUCTION

The aim of this chapter is to review our knowledge of the effects of climate change on the dynamics and persistence of plant populations. In common with many other chapters in this book, we concentrate primarily on the effects of elevated $[CO_2]$ as this is the background against which responses to other environmental changes will occur, that is, it is interactions between $[CO_2]$ and temperature or precipitation changes that are relevant and we include these where they are known. While there is a strong theoretical case that plant physiological responses to elevated $[CO_2]$ should be modified by temperature (Morison and Lawlor 1999), there is very little evidence for $[CO_2]$–temperature interactions operating on higher-order responses

(Morison and Lawlor, 1999; Norby and Luo, 2004). For example, Volder et al. (2004), Newman et al. (2001), and Lilley et al. (2001) found no interactive effects of temperature and $[CO_2]$ on the dry matter yields of forage plants; similarly there were no interactive effects observed on the productivity of the tree species *Pinus sylvestris* (Peltola et al., 2002), *Pseudotsuga menzeisii* (Olszyk et al., 2003), or *Acer rubrum* (Norby et al., 2003). Species diversity effects also appear unresponsive (Zavaleta et al., 2003; Harmens et al., 2004). Despite limited evidence for frequent $[CO_2]$–temperature interactions, we cannot ignore the possibility that these may occur (Shaw et al., 2002), and the quantitative outcome of additive effects, of course, are also of interest.

In studying the dynamics of populations, we are concerned, in the current context, with the ways in which increased $[CO_2]$ in interaction with other biological and physical factors might alter birth, deaths, emigration, and immigration, so as to bring about changes in plant numbers, size, and age through time and space. In general, the experimental information, particularly that dealing with $[CO_2]$–temperature effects, deals with the outcome of population dynamics (e.g., productivity, Shaw et al. [2002] or community composition Zavaleta et al. [2003], rather than the processes that contribute to these changes.

The number, size, and distribution of plants is critically important for many other ecosystem processes such as plant production, nutrient cycling, and herbivore growth. Hence, changes in plant population dynamics under climate change are likely to have important consequences for ecosystem function. From an applied perspective, this means predicting the effects of climate change on plant population dynamics is fundamental to understanding how many agroecosystem practices will change under future climates (Fuhrer, 2003). For example, an understanding of plant population dynamics contributes to the design of weed and pest control strategies, crop and pasture management systems, and fertilizer and irrigation regimes.

8.2 DISTINGUISHING PERFORMANCE AND DYNAMICS

In considering the effects of elevated $[CO_2]$ on plant communities, there is a need to distinguish between the effects of elevated $[CO_2]$ on plant performance at different stages of the plant's life cycle and its effects on population dynamics. There is vast literature showing how elevated $[CO_2]$ can alter germination, seedling survival, ramet dynamics, and seed production, but this tells us little about the effects on plant population dynamics. The extent to which elevated-$[CO_2]$-induced changes in performance will affect plant population dynamics will depend critically on which factors are regulating populations. For example, while elevated $[CO_2]$ may alter seed production or quality, changes in seed production may not lead to altered plant recruitment in the next generation (see Bazzaz et al., 1992; Crawley, 1990; Edwards et al., 2001a). The impact will depend on the extent to which populations are limited by other factors such as the availability of microsites, intra- and interspecific competition, fungal pathogens, and how these change in the future. Indeed, an increase in seed number under elevated $[CO_2]$ might be accompanied by an increase in the size and growth of existing plants, so leading to increased competition with little net effect on seedling recruitment (Edwards et al., 2001a). Throughout this chapter, this will be a recurrent theme — the importance of elevated $[CO_2]$ and other climate change to plant population dynamics remains difficult to assess until the limits of plant recruitment are known.

In this chapter we consider each stage of a plant's life cycle in turn — germination, seedling growth and survival, plant development rate, seed production, and seed quality. For each stage, we then consider how elevated $[CO_2]$ and climatic change might modify these processes, and consider the consequences for plant population dynamics and community composition.

8.3 GERMINATION

A change in the proportions of species recruiting from the seed bank can result in change in the balance of plant species in a community. Elevated $[CO_2]$ could affect germination prior to dispersal, through indirect effects on seed quality, or postdispersal, through effects on the germination environment. In this section we consider the effects of elevated $[CO_2]$ that operate after seed dispersal. Indirect effects of elevated CO_2 that operates through changes in seed provisioning are considered later in this chapter in Section 8.7 on seed mass and seed quality.

Laboratory studies have shown that elevated $[CO_2]$ can both increase (Ballard, 1958, 1961) and decrease (Popay and Roberts,, 1970) germination, but it is not clear how these laboratory responses relate to the more complex field environment. The traditional view is that elevated $[CO_2]$ may have little direct impact on germination and emergence in the field, as $[CO_2]$ in the soil or at the soil surface is already very high and increases on the order of 350 ppm may be inconsequential. CO_2 concentrations of 1800 ppm within the soil and concentrations of 1400 ppm at the soil surface and of 500 ppm above the soil surface have been measured (Bazzaz and Williams, 1991; Schwartz and Bazzaz, 1973).

This prediction is supported by several field or mesocosm studies showing no impact of elevated $[CO_2]$ on the timing of emergence, germination rate or total germination (Bazzaz et al., 1992; Edwards et al., 2001b; Garbutt et al., 1990; Morse and Bazzaz, 1994). For example, Edwards et al. (2001b) showed no effects of elevated $[CO_2]$ on germination in six grassland species when either planted on bare soil or when sown into microsites in existing pasture canopies maintained at ambient and elevated CO_2. In contrast, Heichel and Jaynes (1974) found *Kalmia* seeds located on the soil surface emerged faster under conditions of elevated $[CO_2]$. Furthermore, Ziska and Bunce (1993) found in recently cultivated soil that elevated $[CO_2]$ significantly increased the total number of weed seedlings, although there was significant variation between species (Table 8.1). Significant increases in seedling number were observed for grasses and *Amaranthus* sp. but not for *Chenopodium album, Portulaca oleracea,* or *Lepidium virginicum*. The exact reason for the increase in germination was unclear, although it has been speculated that increased production of ethylene, a gas known to mediate germination (Eglee, 1980), may be involved (Esashi et al., 1986). However, the levels of CO_2 in these studies required for the stimulation of

TABLE 8.1
Average Number of Weed Seedlings Present in Open-Top Field Chambers at Three Different CO_2 Concentrations about 3 Weeks after CO_2 Fumigation

Species	CO_2 (ppm)			Significance
	350	575	700	
Amaranthus sp.	2	5	15	*
Chenopodium album	17	15	29	ns
Portulaca oleracea	60	64	112	ns
Lepidium viginicum	83	114	238	ns
Grasses	49	87	154	*
Total	211	285	547	*

Note: Grasses include all grasses not separated by species. Asterisk: Statistically significant ($P < 0.05$) using a t-test. ns: no significant difference.

Source: Ziska, L.H. and Bunce, J.A. (1993) *Field Crops Research*, 34, 147–157. (With permission from Elsevier.)

ethylene production (1 to 10%) are much higher than those expected with global change (Esashi et al., 1986).

From these data it is difficult to draw conclusions about the direction of response or the implications, other than that there is likely to be species-specific variation. As differences in competitive hierarchies are frequently produced by the timing and extent of emergence, particularly in annual communities, these species-specific effects are likely to be important and require further study. Studies must also consider how germination will be influenced by other resources that $[CO_2]$ regulates, such as soil water and nutrients (Morgan et al., 2004). Newton et al. (2004) noted reduced water repellency of a grassland soil under elevated $[CO_2]$, a factor that may alter the availability of germination sites and modify recruitment from seeds (Crabtree and Gilkes, 1999).

8.4 SEEDLING SURVIVAL

It has long been established that the mortality of plants following emergence is a critical stage in the determination of plant population density, stand structure, and community composition (Harper, 1977). Thus, understanding the effects of elevated $[CO_2]$ and climatic change on seedling survival remains a priority. It has been argued that some aspects of the effects of elevated $[CO_2]$ on plant populations may be predictable in terms of general theory regarding population-level responses to increasing resource levels (Bazzaz et al., 1996; Bazzaz and McConnaughay, 1992). For example, in addition to enhancing plant growth, higher resource levels generally intensify interactions among individuals. Self-thinning (density-dependent mortality), for instance, generally occurs at a greater rate in high-fertility environments (Harper, 1977). Increased resource levels may also have predictable effects on yield-density relationships; when population dynamics are driven by intraspecific density dependence (as in many annual plant populations), increased resource levels can operate to increase the amplitude of population fluctuations (Symonides et al., 1986; Thrall et al., 1989). From this logic emerge the population-level predictions that elevated [CO_2] may enhance self-thinning and increase the likelihood of local population extinction.

Bazzaz et al. (1992) considered these predictions by investigating the effects of elevated [CO_2] (350 and 700 ppm) and plant density (100, 500, 1500, and 4000 plants m^2) on density-dependent patterns of demography and reproduction in the annual plant *Abutilon theophrasti*, a species that had previously been used as a model in several detailed analyses of plant population dynamics (e.g., Pacala and Silander, 1990; Thrall et al., 1989). In the elevated-[CO_2] environment, survivorship was significantly reduced, particularly at plant densities of 500 and 1500 plants m^2, but the proportion of plants flowering and fruiting, and the number of fruiting individuals in each population all increased. Presumably the increased mortality due to CO_2 enrichment resulted from an initial increase in growth rates, leading to more rapid canopy closure and increased self-thinning. The cumulative result was a trend toward increased population-level fecundity under elevated CO_2 despite a lack of change in individual fecundity. The [CO_2] and density effects on survival, individual and stand levels of fecundity, and germination success were used to parameterise a simple difference equation model examining long-term population dynamics (Bazzaz et al., 1992). The model predicted that elevated-[CO_2]-grown stands were less stable, exhibiting rapid, transient increases in population size followed by extinction. Although the results may be specific to these data, they highlight that a lack of responsiveness of individual plants to elevated [CO_2] does not preclude responsiveness at the population level.

Other studies have also noted reduced seedling survival under elevated [CO_2] (Fischer et al., 1997; Morse and Bazzaz, 1994) and point toward increased competition being the mechanism leading to this response. For example, in calcareous grassland, Fischer et al. (1997) found survival of the rare annual *Gentiana cruciata* was reduced under elevated [CO_2]. Further analysis revealed this effect was mediated by increased competition with existing vegetation because survival was correlated with total plot biomass at the time of fastest growth in late spring.

The negative effects on survival are by no means universal, and elevated [CO_2] has also been shown to have no measurable effect (Edwards et al., 2001a; Edwards et al., 2001b) or to increase survival (Housman et al., 2003; Polley et al., 1996; Polley et al., 2003; Polley et al., 2002; Rochefort and Bazzaz, 1992; Wang et al., 1995). From these studies emerge the need to consider what factors (e.g., grazing, water, nutrients) are actually regulating survival, and how this is affected by CO_2 enrichment and climatic change. For example, in grasslands where low water availability reduces the establishment of the invasive shrub *Prosopis glandulosa*, elevated [CO_2] increased seedling survival of the shrub, with this effect assigned to CO_2 enrichment reducing soil water depletion by associated grasses (Polley et al., 2003).

Johnson et al. (2003) explored interactions between arbuscular mycorrhizal (AM) fungi and elevated [CO_2] and found AM fungi increased plant diversity at elevated CO_2, but decreased plant diversity at ambient [CO_2]. This was due to decreased mortality of several C_3 forbs under elevated [CO_2], and suggests that [CO_2] enrichment may ameliorate the carbon cost of some AM symbioses.

Another important consequence of differences in establishment and survival is the effect that this may have on the structure of populations—in particular size hierarchies. Models of plant–plant interactions suggest that for constant density, the mean mass of an individual within the population will increase as resources (other than light) increase (Weiner, 1985; Weiner and Thomas, 1986). However, variation among individuals is not predicted to change because nonpreemptable resources are used in proportion to an individual's size; hence, the prediction that as [CO_2] concentration increases there should be an increase in mean mass with little or no change in the coefficient of variation because [CO_2] acts as a gaseous resource and a regulator of another diffusible resource, water. Morse and Bazzaz (1994) investigated this prediction in *Abutilon theophrastri* (C_3) and *Amaranthus retroflexus* (C_4). For *A. retroflexus*, there was little impact of elevated CO_2 on the coefficient of variation in plant biomass. A similar response was found for *A. theophrastri*, at 18°C and 28°C, as there was little change in the coefficient of variation on plant size, despite an increase in mean mass. At the warmer temperature of 38°C, however, significant self-thinning had occurred within these stands, eliminating the smaller suppressed individuals from the populations, and the size structure of these stands was more equitable than low-[CO_2] stands grown at this temperature.

8.5 DEVELOPMENT RATE

A number of studies reveal that altered plant phenology (development rate) at both the leaf and whole-plant level can be a response to elevated [CO_2]. In a study with *Pinus sylvestris*, needle fall occurred earlier at elevated CO_2 than at ambient [CO_2], a result attributed to changes in translocation rate or earlier translocation of nutrients away from leaves (Jach and Ceulemans, 1999). *Nardus stricta* plants growing near a natural [CO_2] spring exhibited earlier leaf senescence than plants growing farther away from the spring (Cook et al., 1998). In young *Populus trichocarpa*, Sigurdsson (2001) found earlier and more pronounced leaf senescence at elevated [CO_2], with this effect more pronounced in low- (natural) than high-nutrient conditions. By contrast, Hardy and Havelka (1975) found delayed leaf senescence in *Glycine max* L. in elevated compared to ambient [CO_2] conditions, and Tissue and Oechel (1987) showed leaves of *Eriophorum vaginatum* L. maintained high photosynthetic rates until later in the growing season at elevated rather than ambient [CO_2] conditions. The variation in response indicates the effects of elevated CO_2 on leaf phenology may alter leaf-level carbon acquisition during the growing season in ways that are difficult to predict.

Changes in the timing of flowering of 1 to 2 weeks have been recorded in response to elevated [CO_2]. Delayed flowering in [CO_2]-enriched environments has been found in studies using both perennial (Carter and Peterson, 1983; Curtis et al., 1989a) and annual (Garbutt et al., 1990) graminoids. In contrast, [CO_2] enrichment of annual herbaceous plants usually accelerates flowering and senesce compared to plants grown at ambient $[CO_2]$ (Garbutt and Bazzaz, 1984; Garbutt et al., 1990; Stomer and Horvath, 1983). In more detailed work, (Reekie et al., 1994) documented a consistent pattern of altered flowering in short- and long-day plants, indicating that varying effects of elevated $[CO_2]$ on plant phenology may be predictable from species' physiological characteristics. There was a consistent pattern of delayed flowering in four short-day plants and of accelerated flowering in four long-day plants exposed to elevated $[CO_2]$.

Flowering phenology is also very responsive to potential climate changes such as temperature. In a 47-year study of 537 plant species in southcentral England, Fitter and Fitter (2002) were able to show that 16% of these species flowered significantly earlier (15 days on average) when the periods 1954 through 1990 and 1991 through 2000 were compared. A change in flowering date was more likely to occur in species that were insect rather than wind pollinated. Fitter and Fitter identified a strong correlation with the local temperature record and concluded that increasing mean temperature was resulting in changes in phenology. Of course, this approach does not confirm a causal relationship; other changes in the environment that might influence phenology also occurred during this period. For example, the United Kingdom experienced annual N deposition rates of between 5 and 35 kg ha^{-1} yr^{-1} (Stevens et al., 2004) and N is a strong influence on flowering time (Mirschel et al., 2005). As outlined above, $[CO_2]$ concentration can also alter phenology and the atmospheric concentration of CO_2 increased by 15% from 1954 to the mid 1990s. Consequently, the observed changes in phenology are likely to be the result of changes in both temperature and $[CO_2]$ concentration; at present we do not have the information to apportion responses between these drivers making prediction difficult.

What are the likely consequences of changes in flowering phenology? Changes in the timing of flowering may be of particular importance in insect-pollinated plants, as the reproductive output could be reduced in plants flowering too early or too late with respect to the availability of ephemeral pollinators (Garbutt and Bazzaz, 1984). There is little data on which to test this prediction, although Garbutt and Bazzaz (1984) noted that the acceleration of flowering of 10 days in *Phlox drummondii* would probably not be significant in terms of the life cycle of the *Lepidopteran* pollinators. Shifts in flowering phenology may be critically important for species with maturation times that approach the length of the growing season. In annual species where growth is terminated by frosts or drought, delays in flowering might increase the probability of mortality prior to successful reproduction. In modular plants (e.g., grasses) where the transition of ramets from vegetative to reproductive state, leads to the death of the ramet, then changes in the timing and extent of flowering may lead to altered ramet densities. Clark et al. (1995), for example, found fewer tillers of *Lolium perenne* at 700 than 360 ppm $[CO_2]$, an effect that was assigned in part to earlier and more extended flowering.

8.6 SEED PRODUCTION

From a population perspective, the number of seeds produced determines the number of potential colonists at establishment microsites, and so $[CO_2]$-induced changes in seed production may have important implications for the future composition of plant communities. In this context, there is increasing evidence of a relatively poor relationship between the response of vegetative traits and reproductive traits to elevated CO_2 (Ackerly and Bazzaz, 1995; Farnsworth and Bazzaz, 1995; Jablonski, 1997), thus making it difficult to predict plant population responses from changes in vegetative growth alone.

A meta-analysis was conducted using data from 79 crop and wild species to consider effects of elevated $[CO_2]$ on seed production, seed size, and seed quality (Jablonski et al., 2002). For the crop species, they found seed numbers produced per plant were greater in elevated- compared to ambient-$[CO_2]$-grown plants (+16%), although there was substantial variation among individual species (e.g., rice +42%, soybean +20%, wheat +15%, maize +5%). In contrast, the increase in seed production of wild species in response to elevated $[CO_2]$ was much less pronounced (+4%). Furthermore, there was substantial interspecific variation in the absolute magnitude of the $[CO_2]$ response, with wild species equally distributed between those that were stimulated and those that were inhibited by the $[CO_2]$ treatment (and no clear functional groupings). This confirms earlier studies, demonstrating variability even in closely related wild species (e.g., Farnsworth and Bazzaz, 1995) or for those occurring in the same communities (Ackerly and Bazzaz, 1995; Navas et al., 1997; Smith et al., 2000; Thurig et al., 2003).

The increased seed production of crops under elevated $[CO_2]$ obviously has important implications for crop yield, and these have been discussed elsewhere (Fuhrer, 2003; Kimball et al., 2002). What is less clear is whether effects of elevated $[CO_2]$ on seed production translate into differences in plant population dynamics and plant species composition within communities. The effect will critically depend on the extent to which plant recruitment is seed limited (i.e., do the changes in seed production lead to changes in plant recruitment?). If recruitment is not seed limited, then changes in seed production that cause moderate increases or decreases in fecundity may have no measurable impact on plant abundance or population stability. It is curious, given the many studies exploring the effect of elevated $[CO_2]$ on reproduction, that the relatively simple experiment of sowing extra seeds into elevated and ambient $[CO_2]$ sites and recording recruitment has been carried out so infrequently (Edwards et al., 2001a). If a population is not seed limited, then recruitment may be limited by microsite availability, by competition with mature plants of the same or different species, or by seed and seedling predators.

Early suggestions were that seed-limited recruitment was uncommon, particularly in undisturbed perennial vegetation where seedlings were rarely found (Crawley, 1990). However, seed limitation and its effect on community composition have been documented in many recent seed addition studies, including experiments conducted with perennial plants (Edwards and Crawley, 1999a, b; Jakobsson and Eriksson, 2000; Zobel et al., 2000). A recent review shows that the extent of seed limitation appears to depend on the kind of habitat and upon the growth form of the plant species (Turnbull et al., 2000). Overall, approximately 50% of seed sowing studies (where seeds were added to existing populations) showed evidence of seed limitation. The likelihood of seed limitation declined in the following order: Annuals and biennials > perennial nonwoody species > perennial woody species. This agrees with the traditional view that annual plants are seed limited because they need to reproduce each year, and that herbaceous perennials are not seed limited, because seedlings are so rarely found in undisturbed perennial vegetation (Barrett and Silander, 1992; Turkington et al., 1979). Within perennial nonwoody species, legumes showed a particularly high incidence of seed limitation compared with grasses and nonleguminous forb species. Seed limitation also depended strongly on habitat: Sand dunes > woodland > grassland. Within grasslands, seed limitation tended to occur more frequently in early successful habitats (newly ploughed > arid > mesic grassland). These patterns appear to be correlated with the availability of bare earth (for example, greater in sand dunes that in grassland) and, hence, the number of competition-free microsites.

Due to the high incidence of seed limitation (Turnbull et al., 2000) interspecific differences in seed number under elevated CO_2 are likely to be important in driving changes in recruitment and community composition in a wide range of communities. However, the broad patterns of seed limitation discussed above (Turnbull et al., 2000) would indicate that effects are most likely in communities or at times when there are adequate establishment microsites (e.g., following drought) and to be more prominent in early successional species.

Edwards et al. (2001a, b) investigated these general patterns in a 2-year study of the effects of elevated $[CO_2]$ on flowering, seed production, and plant recruitment in a dry pasture on sandy soil in New Zealand (Table 8.2). In both years, elevated $[CO_2]$ increased seed production of the grasses *Anthoxanthum odoratum*, *Lolium perenne*, and *Poa pratensis*; the legumes *Trifolium repens* and *Trifolium subterraneum*; and the herbs *Hypochaeris radicata* and *Leontodon saxatilis*. In the first year, where summer drought opened up the pasture and created recruitment microsites, the increased seed production resulted in greater seedling recruitment for all species except *P. pratensis*. Seed limitation was confirmed by seed-sowing studies (i.e., more seedlings following seed addition) in all species except *P. pratensis*. However, in the second year, where the summer drought was not as severe, no effect on recruitment was found for *A. odoratum*, *L. perenne*, or *P. pratensis* again. Seed-sowing experiments confirmed that in the second year, *A. odoratum*, *L. perenne*, and *P. pratensis* were not seed limited. Precision sowing experiments provided no evidence that microsite conditions were more conducive to seedling emergence under CO_2 enrichment. Moreover, despite higher seedling numbers under elevated CO_2, there was no difference in seedling survival. Presumably, seedling densities were sufficiently low to avoid strong seedling competition, while sheep grazing (which regularly removed biomass) ensured that the biomass of the existing vegetation was never sufficient to affect survival. Additional measurements showed that the size of individual seedlings and mature plants of *H. radicata*, *L. saxatilis*, *T. repens*, and *T. subterraneum* was unaffected by elevated $[CO_2]$. As all four species showed increased abundance in the biomass under elevated $[CO_2]$, results indicate an important way in which elevated $[CO_2]$ influenced plant species composition in this pasture was through changes in the pattern of seedling recruitment. The importance of this mechanism to community dynamics was highlighted 4 years later when seedling establishment following a severe summer drought was dominated at elevated $[CO_2]$ by the seed-limited species *T. repens*, *T. subterraneum* and *H. radicata* (Figure 8.1).

TABLE 8.2
The Number of Dispersed Seeds in Summer, Seedlings Surviving at the End of Spring, and Extra Seedlings That Emerged from Seed Sowing in Ambient (360 ppm) and Elevated (475 ppm) Atmospheric CO_2 Rings in 1998 and 1999

	Seeds Dispersed m^2			Seedlings Surviving m^2			Seed Sowing (seedlings m^2)			
1998	360	475		360	475		360		475	
Agrostis capillaris	746	596		10.0	8.7		4.0		2.0	
Anthoxanthum odoratum	432	740	*	14.0	33.3	*	10.6	*	11.3	*
Cerastium glomeratum				40.2	42.2					
Hypochaeris radicata	183	386	**	3.3	10.0	*	11.3	*	14.6	**
Leontodon saxatilis	104	373	**	2.0	10.0	*	16.0	*	11.4	*
Lolium perenne	137	281	**	8.0	17.3	*	16.0	*	19.3	*
Poa pratensis	144	320	*	12.0	14.7		3.3		4.6	
Trifolium repens	275	713	**	12.0	50.3	**	24.6	*	19.3	*
Trifolium subterraneum				16.0	60.0	**				
1999										
Agrostis capillaris	576	550		6.7	7.3		4.6		4.0	
Anthoxanthum odoratum	589	779		10.0	11.3		6.0		4.6	
Cerastium glomeratum[c]				36.7	34.0					
Hypochaeris radicata	157	386	*	5.3	15.3	*	18.0	*	14.7	**
Leontodon saxatilis	104	471	**	5.3	12.7	*	17.3	**	15.3	*
Lolium perenne	184	301	*	7.0	9.3		6.7		6.0	
Poa pratensis	203	340		8.0	9.3		4.6		4.0	
Trifolium repens	184	569	*	10.7	27.3	*	13.3	*	16.6	*
Trifolium subterraneum	56	148	*	10.6	51.3	*				

Note: Level of significance: **$P < 0.05$, **$P < 0.01$, all others nonsignificant.

[a] Seed sowing values are the number of extra seedlings at the end of spring in sown compared to unsown plots. Asterisks for seed sowing indicate that the difference in seedling numbers between sown and unsown plots is greater than zero within each CO_2 treatment.

Source: Edwards et al. (2001a) *Oecologia*, 127, 383–394. (With permission from Springer-Verlag.)

In a further example in nutrient-poor calcareous grassland, Thurig et al. (2003) examined how seed production and plant community composition respond to elevated [CO_2]. After 5 years of [CO_2] enrichment, 3 out of 15 abundant species (*Carex flacca, Carex caryophyllea*, and *Lotus corniculatus*) showed increased seed production and tiller densities, suggesting that elevated [CO_2] caused a shift in the species composition of the grassland. They concluded that because *L. corniculatus* does not reproduce clonally, the increase in abundance was mostly likely due to increased

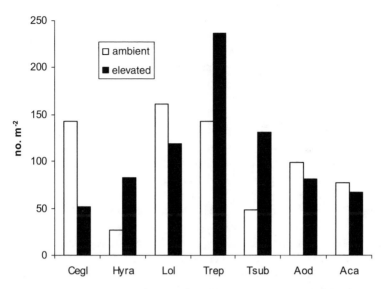

FIGURE 8.1 Seedling establishment measured in the winter 3 months after the end of an extended summer drought. (From Newton, P.C.D., unpublished). Cegl = *Cerastium glomeratum*; Hyra = *Hypochaeris radicata*; Lol = *Lolium perenne*; Trep = *Trifolium repens*; Tsub = *Trifolium subterraneum*; Aod = *Anthoxanthum odoratum*; Aca = *Agrostis capillaris*.

seed production and enhanced establishment of seedlings in previous years. In contrast, because recruitment in *C. flacca* and *C. caryophyllea* is rare, the increased biomass of these species under elevated [CO_2] was assigned to enhanced tillering.

In the Mojave Desert, Smith et al. (2000) found biomass and seed rain of the exotic annual grass *Bromus madritensis* increased more under elevated [CO_2] than several native annual species (*Erigonum trichopes*, *Lepidium lasiocarpum*, and *Vulpia octoflora*). The greater seed production response of the exotic vs. native species could enhance the success of the highly invasive species in arid North America, and so potentially expose these deserts to an accelerated fire cycle to which they are not adapted.

8.7 SEED MASS AND SEED QUALITY

The nutrient status of plants can affect seed traits at several stages, with the size and number of seeds largely determined by the time of flower bud initiation (Roach and Wulff, 1987; Wilson, 1983). Following this, however, subsequent environmental conditions, in particular nutrient supply and the extent of carbon assimilation, can affect seed filling (Egli, 1998; Fenner, 1992). Consequently, it might be expected that the capacity for seed filling, and so seed mass, may be greater under elevated [CO_2]. In their meta-analysis, Jablonski et al. (2002) found a small (+4%) but significant increase in individual seed mass, with the response greatest in legumes (+8%), less for nonlegume C_3 plants (+3%), and absent in C_4 plants (2%). This small response contrasts with the 5 to 25% increases reported in response to additions of mineral nutrients (Fenner, 1992). The exact reason for the small response to elevated

[CO_2] is unclear, but may be limited by breeding for uniform seed size in crop plants (Almekinders and Louwaars, 1999) and selection pressures for optimal seed size in wild species (Leishman et al., 2000).

Effects of elevated CO_2 on the chemical composition of vegetation are well studied. These generally show an increase in the C:N ratio of vegetative tissues (Chapter 6), and the implications of this to decomposition and herbivore feeding are well documented (Chapter 10). The effects of elevated [CO_2] on seed quality are less well studied, but may be important, due to effects of seed quality on early seedling growth. Seed nitrogen concentration was not affected by high [CO_2] concentrations in legumes, but declined significantly in most nonlegumes (15%), although there was significant variation among taxa in this latter group (Jablonski et al., 2002). The reductions in seed nitrogen are most likely due to a carbon dilution effect (e.g., as shown by the increased individual seed mass) or by the diversion of internal nitrogen resources to increased seed production (e.g., as shown by the increased seed numbers). The failure of legume N to respond may reflect that legumes are able to use the increased carbon gain under elevated [CO_2] for increased N_2 fixation (e.g., Allen and Boote, 2000) and so may be able to increase seed number and mass without any loss in seed nitrogen concentration.

How likely is it that any effect of elevated [CO_2] on seed mass and quality will translate into altered plant population dynamics? Increased seed weight may have important implications for seedling establishment, with seedlings from heavier seeds generally better able to survive hazards such as drought, shade, and competition (Westoby et al., 1992; Westoby et al., 1996; Wulff and Alexander, 1985). Changes in C:N ratios may have important implications for the competitive ability of seedlings. Parrish and Bazzaz (1985) noted seedlings from *Abuliton theophrasti* seeds with higher N concentrations had improved performance, particularly under competitive conditions. The increased C:N ratios in seeds may also have important implications for seed predation (Lincoln et al., 1993; Lindroth et al., 1995), with higher C:N ratios being less nutritious for predators, so resulting in increased total seed consumption.

The data on seed quality and seed mass with [CO_2] enrichment may indicate an advantage to legumes over nonlegumes in terms of effects on seedling performance; in legumes seed mass was increased without the drop-off in seed quality observed in most nonlegumes. However, there is relatively little data on the performance of progeny grown at ambient and elevated [CO_2] to confirm this prediction. In *Bromus rubens*, Huxman et al. (1998) observed an increase in C:N ratio and a decline in seed mass for seeds collected from plants that were grown at elevated [CO_2]. As might be expected from the above arguments, seedlings from seeds developed at elevated [CO_2] exhibited a reduced relative growth rate and achieved smaller final mass over the same period. In contrast, in the congeneric grass *Bromus erectus*, no effect of parental growth environment on germination was found, despite seed mass being higher for seeds developed on parents growing at elevated CO_2 (Steinger et al., 2000). It seems that for this species, the CO_2-induced seed mass was entirely offset by an increase in C:N ratio observed in these seeds, resulting in equal performance of seedlings between different parental [CO_2] treatments. In a study of the growth and reproduction of nine herbaceous annual species grown at elevated [CO_2],

Farnsworth and Bazzaz (1995) found little impact of elevated $[CO_2]$ on seed mass, with only *Ipomoea purpurea* showing a significant decline in seed mass under elevated $[CO_2]$. They found that in *Ipomoea purpurea*, germination percentages (measured at ambient CO_2) were lower in seeds collected from elevated-$[CO_2]$ parents, whereas in *I. hederacea* seeds, germination percentages were increased. This last result of increased germination without a change in seed mass confirms that ultimate fitness cannot be inferred from single measures of apparent seed provisioning (Farnsworth and Bazzaz, 1995).

A further interesting result that emerges from recent studies is how seed mass and CO_2 concentration during early seedling growth interact to determine seedling responses. Steinger et al. (2000) showed that large seeds of *Bromus erectus* were more strongly stimulated by elevated $[CO_2]$ than small seeds (regardless of the $[CO_2]$ concentration at which they were grown). Similar results were observed for *Quercus rubra* at the intraspecific level (Miao, 1995) and in six deciduous tree species at the interspecific level (Bazzaz and Miao, 1993); in each case $[CO_2]$ stimulation of seedling growth was positively correlated with seed mass. In a parallel study, Bezemer et al. (1998) also observed that plants of the annual grass *Poa annua* exhibited a much higher responsiveness to elevated $[CO_2]$ when the seeds from which they were grown originated from plants reared at elevated $[CO_2]$, although no data was presented on seed mass (see also Edwards et al., 2001b). The interaction between seed mass and elevated $[CO_2]$ is most likely explained by the larger amount of resources stored in large seeds to sustain rapid seedling growth immediately after germination. Overall, these results would indicate that the relative seedling growth advantage of large seeds with high nutrient reserves over small seeds is likely to be enhanced under future climate conditions.

8.8 COMPETITIVE ABILITY

Variation in response to elevated $[CO_2]$ among species, leading to altered competitive ability, is thought to be of paramount importance in determining shifts in community composition and altering plant population dynamics (Bazzaz and McConnaughay, 1992). The results of experiments with individual species indicate that in general, C_3 species show greater yield gains than C_4 species (Wand et al., 1999). Consequently, it is not surprising to find that in many studies of mixtures of C_3 and C_4 plants, C_3 plants become more competitive as $[CO_2]$ increases. However, these results are by no means universal, with elevated CO_2 favouring C_4 species in some studies (Owensby et al., 1993). Moreover, studies are suggesting a strong dependence on the ecosystem under consideration (see Wand et al., 1999). For example, in an estuarine marsh system, where C_3 and C_4 species occur in close proximity, elevated CO_2 altered species composition in favour of C_3 species (Curtis et al., 1989b), whereas in a dry, tallgrass prairie system, the C_4 species were enhanced over the C_3 species because of their higher tolerance of drought (Owensby et al., 1993). Many studies have demonstrated that legumes have an advantage over nonlegumes (Clark et al., 1995, 1997; Hebeisen et al., 1997; Nosberger et al., 1998) although this may not occur under nutrient-poor (low P) conditions (Navas et al., 1995; Niklaus et al., 2001). What is not clear from these studies is the longer-term consequences of the increased legume abundance to the plant community. With time, a portion of the additional nitrogen fixed by legumes should become available to neighbouring grass species, which will likely utilize it to their own advantage, thereby increasing in abundance.

8.9 CLONAL GROWTH

Clonal plants are capable of proliferating through the production of potentially independent vegetative parts. These parts are generically described as ramets and include such structures as tillers, rhizomes, and tubers. Clonal growth is an important aspect of population dynamics in a wide variety of ecosystem types (Klimeš et al., 1997), and is a key component of yield in many agricultural crops; for example, tillering in cereal plants. Because the production of ramets is determined by the allocation of resources and the activity of meristems, clonal growth is potentially susceptible to changes in elevated CO_2 (which can modify carbohydrate acquisition and use) and temperature (which is a powerful modifier of developmental processes). Experiments with crop species suggest that clonal growth is likely to be stimulated by elevated CO_2 resulting in increased tillering of wheat where this is not constrained by cultivar characteristics (Ziska et al., 2004), and substantially increased tuber production of potatoes (Kimball et al., 2002), for example. In fact, these $[CO_2]$ responses can go some way to reducing the negative temperature effects in determinate crops that can result in faster organ development, but penalise yield through a shortening of the duration of development (Morison and Lawlor 1999).

Studies indicate that species with indeterminate growth, and a large capacity to branch, often respond more strongly to elevated CO_2 (Poorter and Navas 2003), probably because they have a strong sink for increased assimilate supply. This is likely to be the basis for the interesting finding that older cultivars of wheat, which have a greater capacity to tiller, can outperform more modern cultivars in current and future concentrations of $[CO_2]$ (Ziska et al., 2004). This characteristic may also enable plants to avoid the penalties associated with increased rates of development at higher temperatures, which can effectively shorten the maturation time (Morison and Lawlor, 1999).

In more natural communities, such as mixed-species grassland, there is evidence that changes in abundance can be driven through clonal growth mechanisms. In a mixed species pasture, detailed measurements were made over a year on marked tillers of *L. perenne* and *Paspalum dilatatum*, and growing points of *T. repens* grown at 360, 525, and 700 ppm $[CO_2]$. For *T. repens*, a species that increased in abundance, growing point densities at 700 ppm were almost double those found at 350 ppm. A contrasting response was found for *P. dilatatum* and *L. perenne*, which showed no change in abundance under elevated CO_2. For *P. dilatatum*, tiller densities were highest at ambient CO_2, while *L. perenne* showed reduced tiller densities at 525 ppm compared to 360 and 700 ppm (Clark et al., 1997). The difference in response among these clonal species cautions us from suggesting that all clonal plants will have enhanced growth at elevated $[CO_2]$, and indicates that clonal growth is determined by more complex factors than simply carbohydrate supply. However, on the whole it may be said that clonal plants have a large sink potential fulfilling one of the criteria for positive response to elevated CO_2 identified by Oechel and Strain (1985).

8.10 COMMUNITY DYNAMICS AND CONSEQUENCES

Because global change can differentially alter plant population dynamics, it is to be expected that changes will occur in plant community composition; indeed, experiments confirm that changes in community structure are highly likely, including changes in diversity (both richness, evenness) and the abundance of particular species (e.g., Harte and Shaw, 1995; Niklaus et al., 2001). Because of inherent differences in the biology of species (and perhaps genotypes within species; see e.g., Whitham et al., 2003 for examples), it is quite conceivable that climate change can alter ecosystem processes indirectly through changes in community structure. Some recent examples can be found in Allard et al. (2004), where increased rates of mass loss from litter in pastures exposed to elevated CO_2 could be traced to an increasing abundance of the more decomposable forb and legume components of the community, and in Shaw and Harte (2001), where a greater mass loss from forb litter compared to shrub litter was identified as leading to the prediction that the increased abundance of shrubs found in response to increased temperature would result in a lower rate of ecosystem decomposition. Clearly then, to predict the consequences of climate change, it is important to be able to identify changes in community composition. This is a difficult task, indeed, as after 2 decades of intensive research on the effects of elevated [CO_2] few generalisations are possible (Poorter and Navas, 2003). The same difficulty arises with warming; for example, in subalpine meadows, forbs were disadvantaged in a warmer environment (Harte and Shaw, 1995) while the reverse was true in an annual Californian grassland (Zavaleta et al., 2003).

Clearly predictions become even more difficult when more than one driver is involved — for example, elevated [CO_2] and temperature — and as we have seen in this chapter, there are very few data from long-term experiments involving multiple drivers. New initiatives to support such experiments are in train, but the experimental difficulties (for example, the choice of temperature increase to match a particular [CO_2] concentration) are formidable (Norby and Luo, 2004). In addition, in many cases the mechanisms driving community change are not identified, and this restricts our ability to look for general principles of response. In this chapter, we have identified some of the key stages in the population biology of plants and drawn together our understanding about how these might be modified by their changing environment. Often the responses are extremely variable, making prediction hazardous, and some of the reasons for this variability are discussed in Chapter 13. Nevertheless, we feel it is important to identify consistent trends where possible, and this leads us to the following list of attributes that should result in a competitive advantage for species in a world with elevated CO_2 concentration and a warmer climate.

1. Plants that have an indeterminate growth form often have the potential to respond strongly to elevated [CO_2].
2. Plants in which recruitment is seed limited will have a greater opportunity to benefit from any stimulation of seed production.
3. Plants that are able to maintain the concentration of N in their seeds at elevated [CO_2], such as legumes, may reduce any negative consequences of a reduction in seed quality on germination and seedling growth.
4. Plants that have large seeds stand to benefit disproportionately from any [CO_2] stimulation of growth.
5. Plants that are wind pollinated may suffer less disruption than insect pollinated species if times of flowering are altered.

Two further points need to be made here. First, it is interesting to note that several of the characteristics listed match those that are used to describe typical characteristics of invasive or weedy species. For example, wind pollination and an ability to reproduce vegetatively are all considered to be traits associated with a "weedy" habit (Baker, 1965). Similarly, wind pollination (Daehler, 1998) and vegetative reproduction (Reichard and Hamilton, 1997) are traits that are identified with successful invaders. Second, not all of the characteristics described above are traits in the sense that they are amenable to selection either through natural selection pressure or by plant breeders. At present, rather little is known about the traits that might lead to enhanced performance under climate change (Thomas and Jasieski, 1996; Ward and Kelly, 2004) although variation in cultivar (Ziska et al., 2004) and genotype performance (Ward and Kelly, 2004) are encouraging evidence of the untapped potential that exists (see Chapter 13).

REFERENCES

Ackerly, D.D. and Bazzaz, F.A. (1995) Plant-growth and reproduction along CO_2 gradients—nonlinear responses and implications for community change. *Global Change Biology*, 1, 199–207.

Allard, V., Newton, P.C.D., Lieffering, M., Soussana, J-F., Grieu, P., and Matthew, C. (2004) Elevated CO_2 effects on decomposition processes in a grazed grassland. *Global Change Biology*, 10, 1553–1564.

Allen, L.H. and Boote, K.J. (2000). Crop ecosystem responses to climatic change: soybean. In *Climate Change and Global Crop Productivity*, Reddy, K.R. and Hodges, H.F., Eds., CABI Publishing, New York, 133–160.

Almekinders, C.J.M. and Louwaars, N.P. (1999) *Farmers' Seed Production—New Approaches and Practices*. IT Publications, London.

Baker, H.G. (1965) Characteristics and modes of origins of weeds. In *The Genetics of Colonizing Species*, Baker, H.G. and Stebbins, G.L., Eds., Academic Press, London, 141–172.

Ballard, L.A.T. (1958) Studies of dormancy in the seeds of subterranean clover. I. Breaking of dormancy by carbon dioxide and by activated carbon. *Australian Journal of Biological Sciences*, 11, 246–260.

Ballard, L.A.T. (1961) Studies of the dormancy in the seeds of subterranean clover. *Australian Journal of Biological Sciences*, 14, 173–186.

Barrett, J.P. and Silander, J.A. (1992) Seedling recruitment limitation in white clover (*Trifolium-repens*, Leguminosae). *American Journal of Botany*, 79, 643–649.

Bazzaz, F.A., Ackerly, D.D., Woodward, F.I., and Rochefort, L. (1992) CO_2 enrichment and dependence of reproduction on density in an annual plant [*Abutilon theophrasti*] and a simulation of its population dynamics. *Journal of Ecology*, 80, 643–651.

Bazzaz, F.A., Bassow, S.L., Bernsten, G.M., and Thomas, S.C. (1996). Elevated CO_2 and terrestrial vegetation: implications for and beyond the global carbon budget. In *Global Change and Terrestrial Ecosystems*, Walker, B.H. and Steffen, W.L., Eds., Cambridge University Press, Cambridge, U.K., 43–76.

Bazzaz, F.A. and McConnaughay, K.D.M. (1992) Plant-plant interactions in elevated CO_2 environments. *Australian Journal of Botany*, 40, 547–563.

Bazazz, F.A. and Miao, S.L. (1993) Successional status, seed size, and responses of tree seedlings to CO_2, light, and nutrients. *Ecology*, 74, 104–112.

Bazzaz, F.A. and Williams, W.E. (1991) Atmospheric CO_2 concentrations within a mixed forest—implications for seedling growth. *Ecology*, 72, 12–16.

Bezemer, T.M., Thompson, L.J., and Jones, T.H. (1998) *Poa annua* shows inter-generational differences in response to elevated CO_2. *Global Change Biology*, 4, 687–691.

Carter, D.R. and Peterson, K.M. (1983) Effects of a CO_2-enriched atmosphere on the growth and competitive interaction of a C3 and a C4 Grass. *Oecologia*, 58, 188–193.

Clark, H., Newton, P.C.D., Bell, C.C., and Glasgow, E.M. (1995) The influence of elevated CO_2 and simulated seasonal changes in temperature on tissue turnover in pasture turves dominated by perennial ryegrass (*Lolium perenne*) and white clover (*Trifolium repens*). *Journal of Applied Ecology*, 32, 128–136.

Clark, H., Newton, P.C.D., Bell, C.C., and Glasgow, E.M. (1997) Dry matter yield, leaf growth and population dynamics in *Lolium perenne/Trifolium repens*-dominated pasture turves exposed to two levels of elevated CO_2. *Journal of Applied Ecology*, 34, 304–316.

Cook, A.C., Tissue, D.T., Roberts, S.W., and Oechel, W.C. (1998) Effects of long-term elevated [CO_2] from natural CO_2 springs on *Nardus stricta*: photosynthesis, biochemistry, growth and phenology. *Plant, Cell and Environment*, 21, 417–425.

Crabtree, W.L. and Gilkes, R.J. (1999) Improved pasture establishment and production on water-repellent soils. *Agronomy Journal*, 91, 467–470.

Crawley, M.J. (1990) The population-dynamics of plants. *Philosophical Transactions of the Royal Society of London Series B-Biological Sciences*, 330, 125–140.

Curtis, P.S., Drake, B.G., Leadley, P.W., Arp, W.J., and Whigham, D.F. (1989a) Growth and senescence in plant communities exposed to elevated CO_2 concentrations on an estuarine marsh. *Oecologia*, 78, 20–26.

Curtis, P.S., Drake, B.G., and Whigham, D.F. (1989b) Nitrogen and carbon dynamics in C_3 and C_4 estuarine marsh plants grown under elevated CO_2 *in situ*. *Oecologia*, 78, 297–301.

Daehler, C.C. (1998) The taxonomic distribution of invasive angiosperm plants: ecological insights and comparisons to agricultural weeds. *Biological Conservation*, 84, 167–180.

Edwards, G.R. and Crawley, M.J. (1999a) Effects of disturbance and rabbit grazing on seedling recruitment of six mesic grassland species. *Seed Science Research*, 9, 145–156.

Edwards, G.R. and Crawley, M.J. (1999b) Rodent seed predation and seedling recruitment in mesic grassland. *Oecologia*, 118, 288–296.

Edwards, G.R., Clark, H., and Newton, P.C.D. (2001a) The effects of elevated CO_2 on seed production and seedling recruitment in a sheep-grazed pasture. *Oecologia*, 127, 383–394.

Edwards, G.R., Newton, P.C.D., Tilbrook, J.C., and Clark, H. (2001b) Seedling performance of pasture species under elevated CO_2. *New Phytologist*, 150, 359–369.

Eglee, G.H. (1980) Stimulation of common cocklebur (*Xanthium pennsylvanicum*) and redroot (*Amaranthus retroflexus*) seed germination by injections of ethylene into the soil. *Weed Science*, 28, 510–514.

Egli, D. (1998) *Seed Biology and the Yield of Grain Crops*. CABI International, New York.

Esashi, Y., Ooshima, Y., Abe, M., Kurota, A., and Satoh, S. (1986) CO_2-enhanced C_2H_4 production in tissues of imbibed cocklebur seeds. *Australian Journal of Plant Physiology*, 13, 417–429.

Farnsworth, E.J. and Bazzaz, F.A. (1995) Inter- and intra-generic differences in growth, reproduction, and fitness of nine herbaceous annual species grown in elevated CO_2 environments. *Oecologia*, 104, 454–466.

Fenner, M. (1992) Environmental influences on seed size and composition. *Horticultural Reviews*, 13, 183–213.

Fischer, M., Matthies, D., and Schmid, B. (1997) Responses of rare calcareous grassland plants to elevated CO_2: a field experiment with *Gentianella germanica* and *Gentiana cruciata*. *Journal of Ecology*, 85, 681–691.

Fitter, A.H. and Fitter, R.S.R. (2002) Rapid changes in flowering times in British plants. *Science*, 296, 1689–1691.

Fuhrer, J. (2003) Agroecosystem responses to combinations of elevated CO_2, ozone, and global climate change. *Agriculture, Ecosystems and Environment*, 97, 1–20.

Garbutt, K. and Bazzaz, F.A. (1984) The effects of elevated CO_2 on plants. III. Flower, fruit and seed production and abortion. *New Phytologist*, 98, 433–446.

Garbutt, K., Williams, W.E., and Bazzaz, F.A. (1990) Analysis of the differential response of five annuals to elevated CO_2 during growth. *Ecology*, 71, 1185–1194.

Hardy, R.W.F. and Havelka, U.D. (1975) Nitrogen fixation research: a key to world food? *Science*, 188, 633–643.

Harmens, H., Williams, P.D., Peters, S.L., Bambrick, M.T., Hopkins, A., and Ashenden, T.W. (2004) Impacts of elevated atmospheric CO_2 and temperature on plant community structure of a temperate grassland are modulated by cutting frequency. *Grass and Forage Science*, 59, 144–156.

Harper, J.L. (1977) *Population Biology of Plants*, Academic Press, London.

Harte, J. and Shaw, R. (1995) Shifting dominance within a montane vegetation community: Results from a climate-warming experiment. *Science*, 267, 876–880.

Hebeisen, T., Luscher, A., Zanetti, S., Fischer, B.U., Hartwig, U.A., Frehner, M., Hendrey, G.R., Blum, H., and Nosberger, J. (1997) Growth response of *Trifolium repens* L. and *Lolium perenne* L. as monocultures and bi-species mixture to free air CO_2 enrichment and management. *Global Change Biology*, 3, 149–160.

Heichel, G.H. and Jaynes, R.A. (1974) Stimulating emergence and growth of *Kalmia* genotypes with CO_2. *Horticultural Science*, 9, 60–62.

Housman, D.C., Zitzer, S.F., Huxman, T.E., and Smith, S.D. (2003) Functional ecology of shrub seedlings after a natural recruitment event at the Nevada Desert FACE Facility. *Global Change Biology*, 9, 718–728.

Huxman, T.E., Hamerlynck, E.P., Jordan, D.N., Salsman, K.J., and Smith, S.D. (1998) The effects of parental CO_2 environment on seed quality and subsequent seedling performance in *Bromus rubens*. *Oecologia*, 114, 202–208.

Jablonski, L.M. (1997) Responses of vegetative and reproductive traits to elevated CO_2 and nitrogen in *Raphanus* varieties. *Canadian Journal of Botany*, 75, 533–545.

Jablonski, L.M., Wang, Z., and Curtis, P.S. (2002) Plant reproduction under elevated CO_2 conditions: a meta-analysis of reports on 79 crop and wild species. *New Phytologist*, 156, 9–26.

Jach, M.E. and Ceulemans, R. (1999) Effects of elevated atmospheric CO_2 on phenology, growth and crown structure of Scots pine (*Pinus sylvestris*) seedlings after two years of exposure in the field. *Tree Physiology*, 19, 289–300.

Jakobsson, A. and Eriksson, O. (2000) A comparative study of seed number, seed size, seedling size and recruitment in grassland plants. *Oikos*, 88, 494–502.

Johnson, N.C., Wolf, J., and Koch, G.W. (2003) Interactions among mycorrhizae, atmospheric CO_2 and soil N impact plant community composition. *Ecology Letters*, 6, 532–540.

Kimball, B.A., Kobayashi, K., and Bindi, M. (2002) Responses of agricultural crops to free-air CO_2 enrichment. *Advances in Agronomy,* 77, 293–368.

Klimeš, L., Klimešová, J., Hendriks, R., and Van Groenendael, J. (1997) Clonal plant architecture: a comparative analysis of forms and function. In *The Ecology and Evolution of Clonal Plants*, de Kroon, H. and Van Groenendael, G., Eds., Backhuys Publishers, Leiden, 1–29.

Leishman, M.R., Wright, I.J., Moles, A.T., and Westoby, M. (2000) The evolutionary ecology of seed size. In *Seeds, the Ecology of Regeneration in Plant Communities*, Fenner, M., Ed., CABI Publishing, New York, 31–58.

Lilley, J.M., Bolger, T.P., and Gifford, R.M. (2001) Productivity of *Trifolium subterraneum* and *Phalaris aquatica* under warmer, high CO_2 conditions. *New Phytologist*, 150, 371–383.

Lincoln, D.E., Fajer, E.D., and Johnson, R.H. (1993) Plant-insect herbivore interactions in elevated CO_2 environments. *Trends in Ecology and Evolution*, 8, 64–68.

Lindroth, R.L., Arteel, G.E., and Kinney, K.K. (1995) Responses of three saturniid species to paper birch grown under enriched CO_2 atmospheres. *Functional Ecology*, 9, 306–311.

Miao, S.L. (1995) Acorn mass and seedling growth in *Quercus rubra* in response to elevated CO_2. *Journal of Vegetation Science*, 6, 697–700.

Mirschel, W., Wenkel, K.O., Schultz, A., Pommerening, E., and Verch, G. (2005) Dynamic phenological model for winter rye and winter barley. *European Journal of Agronomy*, 23, 123–135.

Morgan, J.A., Pataki, D.E., Korner, C., Clark, H., Del Grosso, S.J., Grunzweig, J.M., Knapp, A.K., Mosier, A.R., Newton, P.C.D., Niklaus, P.A., Nippert, J.B., Nowak, R.S., Parton, W.J., Polley, H.W., and Shaw, M.R. (2004) Water relations in grassland and desert ecosystem responses to rising atmospheric CO_2. *Oecologia*, 140, 11–25.

Morison, J.I.L. and Lawlor, D.W. (1999) Interactions between increasing CO_2 concentration and temperature on plant growth. *Plant, Cell and Environment*, 22, 659–682.

Morse, S.R. and Bazzaz, F.A. (1994) Elevated CO_2 and temperature alter recruitment and size hierarchies in C_3 and C_4 annuals. *Ecology*, 75, 966–975.

Navas, M.L., Guillerm, J.L., Fabreguettes, J., and Roy, J. (1995) The influence of elevated CO_2 on community structure, biomass and carbon balance of Mediterranean old-field microcosms. *Global Change Biology*, 1, 325–335.

Navas, M.L., Sonie, L., Richarte, J., and Roy, J. (1997) The influence of elevated CO_2 on species phenology, growth and reproduction in a Mediterranean old-field community. *Global Change Biology*, 3, 523–530.

Newman, Y.C., Sollenberger, L.E., Boote, K.J., Allen, L.H., and Little, R.C. (2001) Carbon dioxide and temperature effects on forage dry matter production. *Crop Science*, 41, 399–406.

Newton, P.C.D., Carran, R.A., and Lawrence, E.J. (2004) Reduced water repellency of a grassland soil under elevated atmospheric CO_2. *Global Change Biology*, 10, 1–4.

Niklaus, P.A., Leadley, P.W., Schmid, B., and Korner, C. (2001) A long-term field study on biodiversity x elevated CO_2 interactions in grassland. *Ecological Monographs*, 71, 341–356.

Norby, R.J., Hartz-Rubin, J., and Verbrugge, M.J. (2003) Phenological responses in maple to experimental atmospheric warming and CO_2 enrichment. *Global Change Biology*, 9, 1792–1801.

Norby, R.J. and Luo, Y. (2004) Evaluating ecosystem responses to rising atmospheric CO_2 and global warming in a multi-factor world. *New Phytologist*, 162, 281–293.

Nosberger, J., Hebeisen, T., Luscher, A., Blum, H., Frehner, M., Hartwig, U.A., Boller, B., and Stadelmann, F.J. (1998) Effects of elevated CO_2 on managed grassland. In *Breeding for a Multifunctional Agriculture*, Kartause Ittingen, Switzerland, 9–12 September 1997, 179–182.

Oechel, W.C. and Strain, B.R. (1985) Native species responses to increased carbon dioxide concentration. In *Direct Effects of Carbon Dioxide on Vegetation*, Strain, B.R. and Cure, J.D., Eds., U.S. Department of Commerce, Springfield, Virginia, 117–154.

Olszyk, D.M., Johnson, M.G., Tingey, D.T., et al. (2003) Whole-seedling biomass allocation, leaf area, and tissue chemistry for Douglas-fir exposed to elevated CO_2 and temperature for 4 years. *Canadian Journal of Forest Research*, 33, 269–278.

Owensby, C.E., Coyne, P.I., Ham, J.M., Auen, L.M., and Knapp, A.K. (1993) Biomass production in a tallgrass prairie ecosystem exposed to ambient and elevated CO_2. *Ecological Applications*, 3, 644–653.

Pacala, S.W. and Silander, J.A. (1990) Field-tests of neighborhood population-dynamic models of two annual weed species. *Ecological Monographs*, 60, 113–134.

Parrish, J.A.D. and Bazzaz, F.A. (1985) Nutrient content of *Abutilon theophrasti* seeds and the competitive ability of the resulting plants. *Oecologia*, 65, 247–251.

Peltola, H., Kilpelainen, A., and Kellomaki, S. (2002) Diameter growth of Scots pine (*Pinus sylvestris*) trees grown at elevated temperature and carbon dioxide concentration under boreal conditions. *Tree Physiology*, 22, 963–972.

Polley, H.W., Johnson, H.B., Mayeux, H.S., Tischler, C.R., and Brown, D.A. (1996) Carbon dioxide enrichment improves growth, water relations and survival of droughted honey mesquite (*Prosopis glandulosa*) seedlings. *Tree Physiology*, 16, 817–823.

Polley, H.W., Johnson, H.B., and Tischler, C.R. (2003) Woody invasion of grasslands: evidence that CO_2 enrichment indirectly promotes establishment of *Prosopis glandulosa*. *Plant Ecology*, 164, 85–94.

Polley, H.W., Tischler, C.R., Johnson, H.B., and Derner, J.D. (2002) Growth rate and survivorship of drought: CO_2 effects on the presumed tradeoff in seedlings of five woody legumes. *Tree Physiology*, 22, 383–391.

Poorter, H. and Navas, M-L. (2003) Plant growth and competition at elevated CO_2: on winners, losers and functional groups. *New Phytologist* 157, 175–198.

Popay, A.I. and Roberts, E.H. (1970) Factors involved in the dormancy and germination of *Capsella bursa pastoris* (L.) Medik. and *Senecio vulgaris* L. *Journal of Ecology*, 58, 103–122.

Reekie, J.Y.C., Hicklenton, P.R., and Reekie, E.G. (1994) Effects of elevated CO_2 on time of flowering in four short-day and four long-day species. *Canadian Journal of Botany*, 72, 533–538.

Reichard, S.H. and Hamilton, C.W. (1997) Predicting invasions of woody plants introduced into North America. *Conservation Biology*, 11, 193–203.

Roach, D.A. and Wulff, R.D. (1987) Maternal effects in plants. *Annual Review of Ecology and Systematics*, 18, 209–235.

Rochefort, L. and Bazzaz, F.A. (1992) Growth response to elevated CO_2 in seedlings of four co-occuring birch species. *Canadian Journal of Forest Research*, 22, 1583–1587.

Schwartz, D.M. and Bazzaz, F.A. (1973) *In situ* measurements of carbon dioxide gradients in a soil-plant-atmosphere system. *Oecologia*, 12, 161–167.

Shaw, M.R. and Harte, J. (2001) Control of litter decomposition in a subalpine meadow-sagebrush steppe ecotone under climate change. *Ecological Applications*, 11, 1206–1223.

Shaw, M.R., Zavaleta, E.S., Chiariello, N.R., Cleland, E.E., Mooney, H.A., and Field, C.B. (2002) Grassland responses to global environmental changes suppressed by elevated CO_2. *Science*, 298, 1987–1990.

Sigurdsson, B.D. (2001) Elevated CO_2 and nutrient status modified leaf phenology and growth rhythm of young *Populus trichocarpa* trees in a 3-year field study. *Trees-Structure and Function*, 15, 403–413.

Smith, S.D., Huxman, T.E., Zitzer, S.F., Charlet, T.N., Housman, D.C., Coleman, J.S., Fenstermaker, L.K., Seemann, J.R., and Nowak, R.S. (2000) Elevated CO_2 increases productivity and invasive species success in an arid ecosystem. *Nature*, 408, 79–82.

Steinger, T., Gall, R., and Schmid, B. (2000) Maternal and direct effects of elevated CO_2 on seed provisioning, germination and seedling growth in *Bromus erectus*. *Oecologia*, 123, 475–480.

Stevens, C.J., Dise, N.B., Mountford, J.O., and Gowing, D.J. (2004) Impact of nitrogen deposition on the species richness of grasslands. *Science*, 303, 1876–1879.

Stomer, L. and Horvath, S.M. (1983) Elevated carbon-dioxide concentrations and whole plant senescence. *Ecology*, 64, 1311–1314.

Symonides, E., Silvertown, J., and Andreasen, V. (1986) Population-cycles caused by over-compensating density-dependence in an annual plant. *Oecologia*, 71, 156–158.

Thomas, S.C. and Jasieski, M. (1996) Genetic variability and the nature of microevolutionary responses to elevated CO_2. In *Carbon Dioxide, Populations and Communities*, Körner, C. and Bazzaz, F.A., Eds., Academic Press, San Diego, CA, 51–81.

Thrall, P.H., Pacala, S.W., and Silander, J.A. (1989) Oscillatory dynamics in populations of an annual weed species *Abutilon-theophrasti*. *Journal of Ecology*, 77, 1135–1149.

Thurig, B., Korner, C., and Stocklin, J. (2003) Seed production and seed quality in a calcareous grassland in elevated CO_2. *Global Change Biology*, 9, 873–884.

Tissue, D.T. and Oechel, W.C. (1987) Response of *Eriophorum-vaginatum* to elevated CO_2 and temperature in the Alaskan tussock tundra. *Ecology*, 68, 401–410.

Turkington, R., Cahn, M.A., Vardy, A., and Harper, J.L. (1979) The growth, distribution and neighbor relationships of *Trifolium repens* in permanent pasture. III. The establishment and growth of *Trifolium repens* on natural and perturbed sites. *Journal of Ecology* 67, 231–243.

Turnbull, L.A., Crawley, M.J., and Rees, M. (2000) Are plant populations seed-limited? A review of seed sowing experiments. *Oikos*, 88, 225–238.

Volder, A., Edwards, E.J., Evans, J.R., Robertson, B.C., Schortemeyer, M., and Gifford, R.M. (2004) Does greater night-time, rather than constant warming alter growth of managed pasture under ambient and elevated atmospheric CO_2? *New Phytologist*, 162, 397–411.

Wand, S.J.E., Midgley, G.F., Jones, M.H., and Curtis, P.S. (1999) Responses of wild C_4 and C_3 grass (Poaceae) species to elevated atmospheric CO_2 concentration: a meta-analytic test of current theories and perceptions. *Global Change Biology*, 5, 723–741.

Wang, Z.M., Lechowicz, M.J., and Potvin, C. (1995) Responses of black spruce seedlings to simulated present versus future seedbed environments. *Canadian Journal of Forest Research*, 25, 545–554.

Ward, J.K. and Kelly, J.K. (2004) Scaling up evolutionary responses to elevated CO_2: lessons from *Arabidopsis*. *Ecology Letters* 7, 427–440.

Weiner, J. (1985) Size hierarchies in experimental populations on annual plants. *Ecology*, 66, 743–752.

Weiner, J. and Thomas, S.C. (1986) Size variability and competition in plant monocultures. *Oikos*, 47, 211–222.

Westoby, M., Jurado, E., and Leishman, M. (1992) Comparative evolutionary ecology of seed size. *Trends in Ecology & Evolution*, 7, 368–372.

Westoby, M., Leishman, M., and Lord, J. (1996) Comparative ecology of seed size and dispersal. *Philosophical Transactions of the Royal Society of London Series B-Biological Sciences*, 351, 1309–1317.

Whitham, T.G., Young, W.P., Martinsen, G.D., et al. (2003) Community and ecosystem genetics: a consequence of the extended phentype. *Ecology*, 84, 559–573.

Wilson, M.F. (1983) *Plant Reproductive Ecology*, John Wiley & Sons, New York.

Wulff, R.D. and Alexander, H.M. (1985) Intraspecific variation in the response to CO_2 enrichment in seeds and seedlings of *Plantago lanceolata* L. *Oecologia*, 66, 458–460.

Zavaleta, E.S., Shaw, M.R., Chiariello, N.R., Thomas, B.D., Cleland, E.E., Field, C.B., and Mooney, H.A. (2003) Grassland responses to three years of elevated temperature, CO_2, precipitation, and N deposition. *Ecological Monographs*, 73, 585–604.

Ziska, L.H. and Bunce, J.A. (1993) The influence of elevated CO_2 and temperature on seed germination and emergence from soil. *Field Crops Research*, 34, 147–157.

Ziska, L.H., Morris, C.F., and Goins, E.W. (2004) Quantitative and qualitative evaluation of selected wheat varieties released since 1903 to increasing atmospheric carbon dioxide: can yield sensitivity to carbon dioxide be a factor in wheat performance? *Global Change Biology*, 10, 1810–1819.

Zobel, M., Otsus, M., Liira, J., Moora, M., and Mols, T. (2000) Is small-scale species richness limited by seed availability or microsite availability? *Ecology*, 81, 3274–3282.

9 Climate Change Effects on Fungi in Agroecosystems

Matthias C. Rillig

CONTENTS

9.1 GENERAL CONSIDERATIONS

9.1.1 DEFINITION OF FUNGI

Fungi are heterotrophic, eukaryotic organisms with, generally, a hyphal growth form, reproducing by means of spores (Kendrick 2000). Fungi are not a monophyletic group. The kingdom Eumycota ("true" fungi) comprises several fungal phyla,

Basidiomycota, Ascomycota, Zygomycota, Chytridiomycota, and a recently defined phylum, Glomeromycota (formerly part of the Zygomycota). In addition there are also several fungal phyla that are not related to eumycotan fungi, the most germane of which to the current discussion is the phylum Oomycota. There are several members of the Oomycota that are important pathogens of crop plants (e.g., *Phytophthora* spp.). There are some important phylum-level characteristics that could be used to make general predictions about responses of different groups to forcers of climate change. For example, members of the phylum Oomycota, which possess flagellated zoospores, may be more sensitive to changes in precipitation than members of the Basidiomycota. However, there are also vast ecophysiological differences within phyla or even narrowly defined fungal groups that make sweeping generalizations difficult.

9.1.2 FUNGI MAY BE BOTTOM–UP (RESOURCE) CONTROLLED

Considering that fungi are heterotrophs, populations of fungi may be highly dependent on organic carbon quality and quantity, unless they are generally more strongly controlled by consumers (i.e., fungal grazers or parasites). Soil fungi, for example, are generally regarded as bottom–up controlled by resources, as opposed to bacteria, which are hypothesized to be more strongly controlled by consumers (Wardle 2002). One of the mechanisms implicated in fungi being less strongly controlled by consumers is the ability of fungi to show compensatory growth (or overcompensation) in response to mycelium grazing. Additionally, and depending on their growth strategy, fungal hyphae can represent a food source of low palatability due to defensive compounds, high C:N ratios, or indigestible hyphal wall components. As a consequence of being often resource controlled, any ecosystem treatment that impacts on organic carbon quality or quantity could theoretically be expected to have significant effects on fungi. Drivers of climate change, such as elevated atmospheric CO_2, temperature increase, and precipitation changes, which all have potential to affect resources in ecosystems via different mechanisms, are no exception. Of course, in addition to organic carbon, fungi can also be concurrently limited by other parameters, such as nutrient concentrations, pollutants, substrate moisture, and temperature (Cooke and Whipps 1993).

9.1.3 MULTIPLE ROLES OF FUNGI IN ECOSYSTEMS; DIVERSITY OF MECHANISMS

In addition to having effects on fungal communities, factors of climate change can also induce responses from these altered populations and communities at the ecosystem or individual plant level, derived from the respective functions of fungi. Roles of fungi in terrestrial ecosystems are multifold (Dighton 2003) (Figure 9.1). Fungi can affect net primary productivity (NPP) by having either positive (e.g, mycorrhizae, *Neotyphodium* endophyte association) or negative (root or shoot pathogens) effects on plant growth and plant community composition (e.g., mycorrhizal fungi). Mostly mycorrhizal fungi can have strong effects on soil aggregation in hierarchically structured soils and, hence, can exert control on carbon storage. The role of fungi

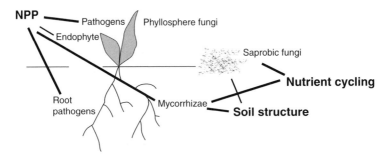

FIGURE 9.1 More than pathogens: Diverse roles of different groups of fungi in agroecosystems. Main ecosystem attributes affected by fungi are NPP (net primary productivity), which includes aspects of crop growth and yield; nutrient cycling (including nutrient uptake, immobilization, and decomposition of crop residues); and soil aggregation.

in soil aggregation is not completely understood mechanistically, but fungi may align primary particles along their hyphae, provide agents of stabilization of aggregates (e.g., glomalin produced by arbuscular mycorrhizal fungi), or may provide an enmeshing stabilization through their hyphal growth habit (*mesh bag* mechanism). A more traditionally widely acknowledged role of fungi in ecosystems relates to decomposition and mineralization carried out by saprobic fungi. Here, fungi are important because of their enzymatic versatility (including processing of recalcitrant polymers such as lignin), and their ability to exploit the interiors of large substrates owing to their invasive growth.

9.1.4 DIFFERENT MODES OF ACTION OF CLIMATE CHANGE FACTORS VS. ELEVATED CO_2

Due to the fact that fungi are heterotrophs, they are differentially affected by elevated CO_2 concentrations and factors of climate change (temperature, water) (Figure 9.2). Fungi can be affected by very high concentrations of CO_2, but the magnitude of changes considered in global change (e.g., double-ambient concentration) is generally not significant in this context. In addition, as far as soil-borne fungi are concerned, the soil air contains much higher concentrations of CO_2 than aboveground air, making any direct effects of elevated CO_2 on soil organisms highly unlikely. Hence, the effects of CO_2 on free-living fungi should be inconsequential. However, water and temperature changes can obviously affect free-living fungi directly.

Fungi commonly form symbioses with plants (and animals) in terrestrial ecosystems (see Figure 9.1) and, hence, it is important to distinguish direct effects on fungi from indirect effects mediated by the host plant (either the host of a pathogenic or mutualistic fungus). As shown in Figure 9.2, elevated CO_2 affects symbiotic fungi indirectly via effects on the host plant. By contrast, warming and precipitation changes can affect symbiotic fungi both directly and indirectly; this is because temperature and water availability are often important controllers of life history stages of fungi. These distinctions of modes of action are very important for the design of appropriate research studies aimed at elucidating mechanisms of fungal

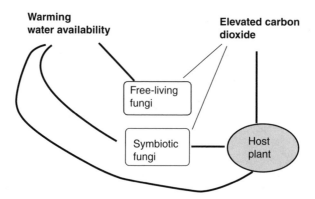

FIGURE 9.2 Factors of climate change (warming, water availability changes) and elevated CO_2 concentrations have different modes of action on fungi. Additionally, symbiotic fungi are affected by elevated CO_2 almost exclusively indirectly, whereas increased temperature and alter precipitation can have direct effects on fungal symbionts. This has important consequences for the design of experiments aimed at mechanisms of fungal responses.

responses (Rillig et al. 2002a). In addition, it must be highlighted that warming and water availability changes give rise to a multitude of other changes in soil (altered mineralization rates, nutrient availability) that can have important effects on soil microbiota, including fungi. It is also not represented in the simplified conceptual model that plants can have indirect effects on fungal litter decomposers, namely via changes in litter quality or quantity.

9.1.5 SCOPE OF CHAPTER

The goal of this chapter is to review key aspects of the effects of global change factors (precipitation change, warming, and CO_2) on various groups of fungi. These include saprobes, mycorrhizae, endophytes, and pathogens; hence, covering the range of functions in an agroecosytem, as much is possible from a mechanistic perspective.

9.2 BELOWGROUND RESPONSES: MYCORRHIZAE AND SAPROBES

In agroecosystems, just like in their natural counterparts, fungi are found in the above and belowground compartments of ecosystems; these merit a separate treatment because different functions and relationships are involved. In this section we will discuss symbiotic associations with plants (mycorrhizae), as well as free-living fungi (decomposers). In the former case, owing to the tight association of fungus and host plant, effects are much more directly mediated by the host. At the same time, the symbiotic associations also have a much greater potential to directly influence host plant growth and NPP.

9.2.1 MYCORRHIZAE

Mycorrhizae are root–fungus symbioses common in natural and managed ecosystems. There are several types of mycorrhizal associations, the two most common of which are ectomycorrhizae and arbuscular mycorrhizae (AM) (Smith and Read 1997). The different types of associations differ in their function, structure, and the phylogeny of the host and fungal partners (Smith and Read 1997). We will focus our discussion on the latter, since AM are common in important production crops and pasture plant species. AM fungi (all in the phylum Glomeromycota; formerly part of the phylum Zygomycota) inhabit the root cortical cells of host plants, forming intraradical structures such as hyphae, arbuscules (the main sites of carbon-nutrient exchange between the symbionts), and vesicles (lipid storage structures of the fungus). But importantly, a major part of the fungal mycelium extends from the root into the soil; this extraradical mycelium (which can represent 60 m g^1 soil in grasslands or more) is responsible for nutrient uptake, colonization of new roots, production of asexual spores (dispersal propagules), and is also involved in soil aggregation.

9.2.1.1 Elevated CO$_2$

Effects of various global change factors on AM have received relatively intense research attention, mostly owing to their widely acknowledged role in terrestrial ecosystems. Among the major drivers of climate change, CO$_2$ has in turn been the factor most studied. A useful framework for the study of AM responses to CO$_2$ exposure is the resource balance model, which predicts that with removal of limitation of a resource (i.e., carbon), allocation of resources to the acquisition of the next most limiting resource(s) should be stimulated. These limiting resources are frequently belowground (i.e., mineral nutrients), and with AM involved in their acquisition, carbon allocation to AM fungi is hypothesized to increase. For example, in the future plants may experience increased demand for P (Ghannoum et al. 2006, Chapter 3); this may in turn enhance demand for transport functions carried out by AM fungi. In part by virtue of their vast surface area in the soil (hyphae are generally only a few micrometers in diameter), AM fungi are of paramount importance in soil P uptake.

Elevated CO$_2$ responses of AM have been extensively reviewed previously, albeit with a more general focus on natural ecosystems (Staddon and Fitter 1998; Rillig and Allen 1999; Treseder and Allen 2000; Fitter et al. 2000; Rillig et al. 2002a; Staddon et al. 2002). Studies that have examined the responses to CO$_2$ under very controlled conditions in the greenhouse have tended to find no significant effects (e.g., Gavito et al. 2002), especially when controlling for effects of plant growth (Staddon and Fitter 1998), but others have found large responses that cannot be explained by greater plant biomass (e.g., Sanders et al. 1998; Rillig et al. 1999a). Studies in greenhouses, while important to mechanistic understanding, have limited relevance to actual responses in the field. Field evidence suggests an overall stimulation of AM, although the database is comparatively small (e.g., Rillig et al. 1999a,b). Addressing another important limitation of CO$_2$ studies, that of long-term vs. short-term responses, Rillig et al. (2000) showed (using a CO_2 spring in New Zealand) that after several decades, positive effects of increased CO_2 on AM can persist.

There are relatively few field studies carried out on AM responses to CO_2 in crop production agroecosystems, perhaps reflecting the distribution of experimental sites, such as free air CO_2 enrichment (FACE). Runion et al. (1994) assessed mycorrhizal colonization for cotton (*Gossypium hirsutum* L.) grown under ambient (370 µmol mol[1]) and elevated (550 µmol mol[1]) CO_2 concentrations; they found nonsignificant effects for both sampling dates. Rillig et al. (2001) documented increases in AM fungal extraradical hyphal length in a sorghum field in Arizona.

Shifts in community composition of AM fungi in response to CO_2 exposure are not clearly understood. It could be hypothesized that increased carbon supply from the host would favor more competitive AM fungal species at the expense of others, leading to community composition changes. There are differences in the response of individual AM fungal species, associated with the same host plant, to elevated CO_2 exposure (Klironomos et al. 1998), and there is also field evidence for shifts in coarse groups of AM fungi (coarse vs. fine endophyte AM fungal colonization; Rillig et al. 1999b; Rillig and Field 2003). Wolf et al. (2003), quantifying spore abundances in the field, found overall limited community-level shifts within the AM fungi, with minimal changes under plant polyculture conditions (16 plant species). However, one AM fungal species (*Glomus clarum*) responded significantly to CO_2 in plant monocultures. Hence, the potential for community shifts is clear, although clearly more work is necessary.

The discussion so far has focused on responses of AM fungi to enrichment with CO_2, but what about feedbacks? This question has not yet been adequately addressed. However, there are strong indications that there may be important feedbacks of altered AM fungal communities to plants, plant communities, and ecosystems (Figure 9.3). For example, Johnson et al. (2003) recently showed an important interaction of AM fungi presence and CO_2 concentration on plant communities. In their mesocosms, at ambient CO_2, AM fungi actually reduced plant species richness (through increased mortality of forbs), but at elevated CO_2, AM fungi had the opposite effect. The authors suggest that CO_2 enrichment may have ameliorated the carbon cost of AM symbioses for some hosts. We also found strong interactive effects of CO_2 concentration and mycorrhizal colonization on plant biomass (Figure 9.4). These studies examined the presence and absence of mycorrhizae, but it will be important to test effects of AM fungal communities altered by CO_2 exposure, compared to control communities.

As noted previously, AMF (arbuscular mycorrhizal fungi) also have effects on soil aggregation, partly mediated by the AMF-produced protein glomalin, which is highly correlated with aggregate water stability. With AMF growth generally stimulated with plants exposed to elevated CO_2, it has been hypothesized that this function of AMF could also be enhanced (Rillig and Allen 1999). Indeed, Rillig et al. (1999c) found that glomalin concentrations in California grasslands and a shrubland were increased, and concomitant increases in soil aggregate water stability were found. Rillig et al. (2001) found a similar response in an agroecosystem (sorghum in Arizona). Six et al. (2001) found increased aggregation (54% increase in mean

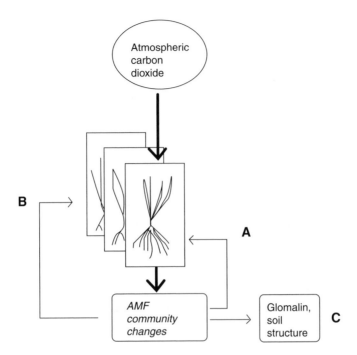

FIGURE 9.3 Conceptual overview of the potential different effects of altered AMF (arbuscular mycorrhizal fungi) community composition (or community-level physiology), as brought about by elevated CO_2 exposure. AMF can have feedbacks on individual host plants (A), such as crop species; can affect competitive interaction among different species in a community (B); and can have effects at the ecosystem scale (C), for example, via influencing soil aggregation (and, hence, a variety of biogeochemical processes) through production of the protein glomalin.

weight diameter) in *Lolium perenne* monocultures in a Swiss FACE experiment, whereas there was no significant difference between ambient and elevated CO_2 with *Trifolium repens* in the same experiment. Other investigators have documented decreases in soil aggregation in other ecosystems (Swiss alpine grassland; Niklaus et al. 2001, 2003). It is unclear how these results relate to AMF or glomalin production (because these parameters were not measured); other factors may have had overriding importance. For example, in the latter studies, soils were exposed to freezing-drying cycles (which are largely absent in the California or Arizona systems), and the Swiss FACE results by Six et al. (2001) suggest that residue quality may also be an important modifier of potential AMF effects.

9.2.1.2 Temperature and Precipitation

Compared to the relative clarity of hypotheses for elevated CO_2 responses of the AM symbiosis, there is a lack of a conceptual framework for responses to temperature and precipitation. Clearly, changes in temperature (Shaver et al. 2000) and precipitation (Weltzin et al. 2003) are challenging to study at the ecosystems scale,

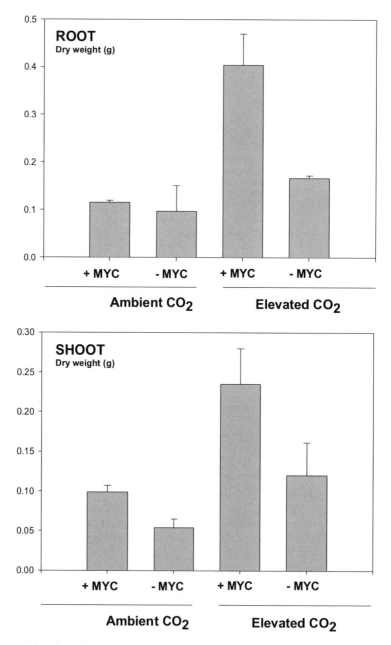

FIGURE 9.4 Arbuscular mycorrhizal fungi (AMF) can influence plant responses to elevated CO_2. *Bromus hordeaceus* plants were grown in a factorial experiment in which both ambient and elevated CO_2 concentrations were crossed with mycorrhizal inoculum (present and absent washed field AMF spore inoculum). The interaction term was significant ($P_{AMFxCO2} < 0.05$ for both root and shoot biomass). Note that AMF had no significant effects on root growth at ambient CO_2 concentrations, but had large effects under elevated CO_2; this suggests that mycorrhizal responsiveness can change under elevated CO_2. (From Rillig, M.C., unpublished).

due to the complexity of ensuing interactions with nutrient cycles. Importantly, for the study of symbioses, there is also the additional problem that temperature and precipitation can have direct *and* indirect effects on AM fungi (Rillig et al. 2002a), whereas CO_2 is generally assumed to only have indirect effects on the mycobionts, mediated by the plant (because soil air is generally already very enriched in CO_2). Due to these complexities, it is perhaps not surprising that the few available studies present apparently conflicting evidence (see Rillig et al. 2002a for review).

9.2.1.2.1 Temperature

Numerous points of the life cycle of AMF respond positively to increased incubation temperature, for example, germinating hyphae growth (reviewed in Fitter et al. 2000). However, few field experiments have been carried out to date on warming effects on AM, and apparently none in production agroecosystems. Rillig et al. (2002b) report an increase in mycorrhizal root colonization and soil AM fungal hyphal length in an annual grassland in California. In this study, however, soil concentrations of the AM-produced protein glomalin were decreased, concurrent with a small decrease in soil aggregate water stability. It appears that at least this function of AM in soil aggregation was impaired in the context of an overall stimulation of AM fungi. Rillig et al. (2002b) hypothesized that this result was due to enhanced decomposition of organic matter (and, hence, glomalin). After 7 years of treatment, Staddon et al. (2003) found mycorrhizal abundance in roots and in the soil significantly affected by warming and water additions. Winter warming increased the proportion of root length colonized, confirming positive effects of increased temperature on AMF. However, winter warming also decreased soil hyphal lengths. In fact, over the whole study there was a negative relationship between root length colonized at extraradical hyphal length, a response perhaps partly explained by vegetation changes. Heinemeyer et al. (2004), in a 1-year field soil warming (+3 °C) experiment, found no effects on AMF colonized root length (no data on soil hyphae were reported).

9.2.1.2.2 Precipitation

There is a rich literature on AMF and plant–water relations, summarized in an extensive review by Augé (2001). AMF can influence plant–water relations by a multitude of mechanisms, and AMF can be hypothesized to provide important feedbacks to plants under altered precipitation regimes. The mechanisms include hormonal involvement (nonhydraulic signaling), more extensive exploration of soil water by mycorrhizal roots (e.g., due to changes in root architecture), stimulation of gas exchange via AMF constituting a strong C sink, contributions of AMF soil hyphae themselves to water absorption, and fungal modifications of soil moisture release characteristics (*mycorrhizal soil*, e.g., changes in soil aggregation), in addition to, or in combination with, AMF-induced changes in plant growth and mineral nutrition (Augé 2001).

There are numerous studies that address the effects of mycorrhizal responses to drought, which are summarized in Augé (2001). Of the large number of studies reviewed, about half reported a change in the percentage AMF root colonization, with many studies finding an increase in the parameter; but several found a decreased percentage of root colonized. Mechanisms of fungal responses are not completely clear, especially given the multitude of effects on plant–water relations (see above). However, there may be direct negative effects of decreased soil water potential on soil hyphae, and there may be increased demand for AMF-mediated phosphate uptake because P availability to the plant may decrease with increased drought (leading to increased root colonization). The relative strength of these effects may in part help explain the wide range of responses to drought, in addition to plant host–fungal species identity and other soil parameters. In longer-term studies in the field, there seems to be an overall trend for increases in root colonization (Augé 2001). Similarly, spore germination has been observed to be increased, decreased, or not changed in response to soil drying. Of the studies reviewed in Augé (2001), only very few reported on changes in the extraradical mycelium of AMF: Seven reported a decrease in aspects of extraradical hyphal growth, and three reported an increase. Under extreme conditions (flooding or extreme drought), AMF seemed to respond negatively, for example with reduced sporulation. There is also evidence for increasing drought to reduce AMF species richness (Stahl and Christensen 1982); however, this may be confounded by decreases in AMF culturability under more arid conditions.

It is evident that responses to precipitation changes may be idiosyncratic depending on many ecosystem characteristics; there is also a need for further mechanistically dissecting responses of AMF to changes in precipitation. Given the importance of drought in crop production (or livestock grazing), mycorrhizal responses to this aspect of climate change are likely of great significance. However, the direction and magnitude of responses is highly uncertain (compounded with the uncertainty of global circulation models (GCM) predictions of regional changes in precipitation in the first place).

9.2.2 SAPROBES

Soil saprobic fungi form highly diverse communities (e.g., Christensen 1989), which are not close to being fully described, as evidenced by the recent discovery of a new clade of soil fungi (Schadt et al. 2003). The relatively recent development and application of eumycotan primer pairs (e.g., Vandenkoornhuyse et al. 2002) has led to a revival of ecological studies on this important group of fungi. In agroecosystems and nonmanaged ecosystems, saprobic fungi play important roles in residue decomposition (leading to nutrient mineralization), and exert control on soil organic carbon storage (Wardle 2002). Because soil saprobic fungal communities are so highly diverse and because decomposition is a broad process (i.e., a complex multistage process mediated by a large number of different organisms), there is a limited likelihood that changes in the decomposer fungal community (as potentially brought about by climate change and CO_2) would have large consequences on ecosystem process rates. Perhaps as a consequence, while numerous studies have been carried out concerning other fungal groups (such as mycorrhizal fungi), there is a comparative dearth of information on responses of saprobic fungi to aspects of climate change and CO_2.

In accordance with the hypothesis that fungi are bottom–up and bacteria are more top–down controlled (Wardle 2002), many but not all studies report an increase in the fungal energy channel in soil relative to the bacterial energy channel in response to elevated CO_2 (e.g., Klironomos et al. 1996; Rillig et al. 1999a; see also Niklaus et al. 2003). However, these responses also include nonsaprobic members of the soil fungal community. Many studies that have examined microbial community composition in more detail have used phospholipid fatty acid analysis (PLFA), which does not permit sufficient resolution within fungal groups to come to strong conclusions, and DNA-based methods have predominantly dealt with bacteria diversity (with Eumycota-specific primers only recently available).

Klironomos et al. (1997) have found increased total fungal spore abundances on decaying litter underneath *Populus tremuloides* grown under elevated CO_2. Interestingly, different genera of fungi were stimulated to vastly different degrees; this suggests that at least as far as spore production is concerned, different fungal species may become more abundant components of the decomposer community. If greater spore abundance is a common occurrence, then there may also be consequences for human health (Klironomos et al. 1997).

Although studies have apparently not yet been designed with the explicit goal of understanding climate change effects on soil fungal saprobes, there are numerous studies available that address temperature and water availability effects on saprobic soil fungi (Cooke and Whipps 1993; Dighton 2003). Also, classical seasonal studies can be utilized to understand climate change effects. However, many of these classical studies are based on culture-dependent methods, and it is currently not clear what proportion of fungal communities is captured.

9.3 ABOVEGROUND RESPONSES

9.3.1 FUNGAL ENDOPHYTES

Fungal endophytes have been defined in the broad sense as any fungi that live asymptomatically (e.g., not causing disease) within tissues of plants. In the narrower sense to be used in this discussion, this term refers to endophytes in the genus *Neotyphodium* (formerly *Acremonium*). These fungi are asexual, obligate seedborne symbionts that commonly form intercellular infections in leaves (also in inflorescences) in many cool-season grasses in the subfamily Pooideae (Schardl et al. 1997). The association has been mostly studied with introduced pasture grasses (chiefly using *Festuca* spp.), where it is considered to be mutualistic. Beneficial properties of *Neotyphodium* infection in exotic pasture grasses include increased plant resistance to herbivores and pathogens (by virtue of fungal alkaloids), and increased plant drought resistance (probably in part by increased leaf rolling). From the perspective of pasture grazing by livestock, increased herbivore resistance is viewed as undesirable. Interestingly, there is emerging evidence that in native grasses (e.g., *Festuca arizonica*) these fungi may actually be parasitic (Faeth and Sullivan 2003). Because these fungi are important in grazing systems, but probably not of widespread consequence in agricultural production systems, we will only briefly deal with climate change and CO_2 effects on this group of fungi.

Increasing evidence suggests that *Neotyphodium*–grass interactions depend on environmental conditions, specifically moisture levels (e.g., Morse et al. 2002). This points to potentially important responses of the symbiosis to factors of climate change, specifically precipitation changes. Furthermore, Marks and Clay (1996) demonstrated an endophyte (*Neotyphodium ceonophialum*) by temperature interaction. At temperatures above 35°C, endophyte-infected *Festuca arundinacea* plants had significantly higher rates of photosynthesis than endophyte-free plants (while at low leaf temperatures no differences were present between infected and uninfected plants). These types of interactions may be significant in warming scenarios.

Marks and Clay (1990) examined responses of endophyte-infected and uninfected plants to CO_2 ambient and elevated concentrations. They observed that under elevated CO_2 photosynthetic rates increased more in infected than uninfected plants, suggesting that endophytes could be rather important in determining plant responses to elevated CO_2.

Under elevated CO_2, increased plant carbon fixation could also lead to enhanced supply of C to the endophyte; this could be hypothesized to lead to increased production of secondary metabolites important in grazing tolerance (Newman, Chapter 10, this volume). Newman et al. (2003) found significant interactions of endophyte (*N. coenophialum*) presence and CO_2 concentration in *F. arundinacea* for the plant tissue response variables soluble crude protein (%) and acid detergent insoluble crude protein. For both these variables, increases with elevated CO_2 were more pronounced in the endophyte-free plants compared to the endophyte-infected plants. This suggests that nutritional value of plants under elevated CO_2 can depend on the presence of the endophyte, irrespective of any potential additional effects related to changes in alkaloid concentrations.

Overall it would appear that the endophyte–host interaction is quite sensitive in a number of physiological factors to elements of climate change and CO_2; clearly, endophyte responses and presence have to be considered in examining responses of these plant hosts (if a mechanistic understanding is the goal).

9.3.2 PHYLLOSPHERE FUNGI

The phyllosphere (surface of leaves) is a challenging microbial habitat due to a number of stressful factors, such as UV (ultraviolet) radiation, osmotic stress, low nutrient availability, rapid temperature fluctuations, and other adverse conditions (Lindow and Brandl 2003). The phyllosphere is inhabited by a diverse range of microbiota, with bacteria being the dominant (and most studied) component; but there are also filamentous fungi, yeast, and spores of fungi (Andrews and Harris 2000). Recently, Yang et al. (2001), the first to apply culture-independent microbial community analysis methods to this microbial habitat, uncovered a much greater than expected microbial diversity (based on culturable microbes). However, so far no study has yet exhaustively and specifically attempted to describe the fungal communities with nonculture-dependent methods and, hence, the true diversity of fungal phyllosphere communities is still unknown.

The significance of the phyllosphere stems from its large extent as a microbial habitat, its role as a reservoir of pathogenic organisms (but also its potential in plant protection), and possible contributions to foodborne illnesses (and economic importance). From a pure microbial ecology perspective, the phyllosphere also offers opportunities to study microbial interactions in an extreme environment.

There appears to be very little information on the consequences of factors of climate change or elevated CO_2 on the phyllosphere microbial, and specifically, fungal communities; this is clearly an area of potentially fruitful future research. However, parallel arguments to the discussion of foliar fungal pathogens (see Section 9.4) could be made for nonsymptomatic phyllosphere fungi. In particular, increased precipitation, and perhaps increased leaf wetness could be expected to have particularly strong effects on phyllosphere fungi. Unsurprisingly, certain factors of global change with a hypothesized strong direct effect on phyllosphere fungi have been studied. Clearly, radiation effects would be hypothesized to be of great importance in surficial populations of fungi. For example, Newsham et al. (1997) reported decreased phyllosphere yeast populations after leaves of *Quercus robur* were experimentally exposed to enhanced UV-B radiation. This study suggests that phyllosphere fungal communities are susceptible to factors of global change, and that potential responses to other global change factors in crop plants should be explored.

9.4 PATHOGENS ABOVEGROUND AND BELOWGROUND

9.4.1 GENERAL SIGNIFICANCE

Fungal pathogens are generally considered of minor importance in natural ecosystems, except for periodic outbreaks of disease, such as brought about by invasive pathogens or particularly favorable environmental conditions for the pathogen. The assumptions are that pathogen biomass is negligible, and that pathogens would only affect a subset of plant species present. Hence, fungal pathogens have been most often studied in agroecosystems. However, it is worth noting that a picture is emerging in the literature on natural ecosystems that pathogens do have important effects on plant communities and ecosystems outside of outbreak situations. Examples include the study by Packer and Clay (2000) involving soilborne pathogenic fungi that strongly influence the structure of forest tree communities; and an exclusion experiment (Mitchell 2003), focused on effects of foliar fungal pathogens on ecosystem NPP (here fungal pathogens had similar biomass to insect herbivores, and larger effects on plant biomass).

Fungal parasites and pathogens of roots and shoots are a major consideration in agroecosystems due to their potential to affect serious crop losses. More than 10,000 species of fungi can cause disease in plants (Agrios 1997), a number that includes above- and belowground pathogens. Some parasitic fungi are obligate biotrophs, that is, they can only grow in association with a living host plant (cell), whereas others are facultative parasites that can grow on dead organic matter (as saprobes) as well as on living hosts. Perhaps predictability of responses to climate change of facultatively parasitic fungi would be lower (compared to obligate biotrophs), since more factors than those mediated by the host plant directly affect their life histories.

9.4.2 Mechanisms

Climate change factors or CO_2 increases can affect fungal pathosystems via several fundamentally different mechanisms (Harvell et al. 2002): there can be direct or indirect effects on the growth and fecundity of the pathogenic fungus (including adaptations of the pathogen), there can be changes in the susceptibility of the host itself to the fungal-caused disease, and there can be climate-driven changes in the geographic ranges of the pathogen or host (bringing together new host-fungus combinations). There appear to be no examples for the latter case — clear cases in which a factor of climate change of elevated CO_2 has altered the host ranges of host and fungal pathogens, leading to novel encounters. However, there are numerous examples in the literature where the arrival of a new pathogen has had devastating effects on the resident hosts, such as Dutch elm disease or chestnut blight in the United States (Harvell et al. 2002).

We will attempt to distinguish between mainly host-side and fungal-side effects, although this is not always easy in phenomenological experiments (mostly focused on disease severity); a subset of studies is presented where this distinction is relatively clear. In addition, when the goal is to elucidate effects of elevated CO_2 or climate change over the longer term, it is also important to consider evolutionary changes in the pathosystem (Chakraborty, Special Example 4, this volume).

9.4.3 Elevated CO_2

Host susceptibility: Manning and Tiedemann (1995) and Coakley et al. (1999) reviewed potential mechanisms of CO_2 effects on plant pathosystems from the perspective of host plant susceptibility. Some of these hypotheses pertain to increased canopy growth with concomitant changes in shoot architecture; some of these architectural changes may be conducive to increased interception of the pathogenic fungus by the plant (also by altering microclimate in the canopy), thereby potentially increasing disease incidence. For example, Chakraborty et al. (2000) showed that an enlarged canopy of *Stylosanthes scabra* plants (a tropical pasture legume) under elevated CO_2 was more conducive to increased colonization by a pathogenic *Colleotrichum*, presumably through provision of enhanced canopy microclimates (particularly relative humidity). Conversely, reduced opening of stomata under elevated CO_2 may lead to decreased penetration ability of stomata-invading pathogens. Additionally, increased water use efficiency under elevated CO_2 could reduce water stress, enhancing disease resistance. For example, Jwa and Walling (2001) found that *Lycopersicon esculentum* (tomato) showed higher tolerance against *Phytophthora parasitica* (causing root rot) infection, a response that was likely caused by enhanced water relations of tomato under elevated CO_2 (leaf xylem pressure potential of infected elevated-CO_2-grown tomato plants was 17% higher than under ambient CO_2; no disease-related biochemical changes were found in this study). Increased carbohydrate content in shoots may make conditions more favorable for some sugar-dependent pathogens. However, Tiedemann and Firsching (1998) found only minor effects of a CO_2-increased carbohydrate content of wheat leaves on disease severity (*Puccinia* leaf rust). Belowground, increased root exudation may have either positive or negative consequences, depending on the relative stimulation of pathogenic and antagonistic soil fungi. Additionally, in parallel to aboveground canopy changes, root architecture may change in response to elevated CO_2, with more root biomass produced, and perhaps more fine roots (enhancing the probability of pathogen interception). Runion et al. (1994) assessed a rootborne *Rhizoctonia* spp. infestation in a cotton field exposed to elevated CO_2 with FACE; there was only a trend (P = 0.17) toward an increase in infestation, suggesting a potential for increased root disease. However, a higher disease incidence could not be confirmed with a bioassay for damping-off.

Effects on fungal fecundity and growth: There are direct effects of relatively high (5 to 10%) concentrations of CO_2 on fungal pathogens, extensively reviewed in Manning and Tiedemann (1995); these effects were largely stimulatory (in part probably through fungal CO_2 fixation). However, it is unlikely that CO_2 concentration increases relevant to global change scenarios would have strong effects. Additionally, soilborne fungal pathogens will likely not be directly affected by elevated CO_2 concentrations, as they are adapted to the already high CO_2 concentrations in soil air. Hence, direct effects may be of comparatively minor importance.

However, elevated CO_2 could still lead to increased fungal growth and sporulation as a result of interaction with the host. For example, Chakraborty et al. (2000) and Chakraborty and Datta (2003) documented increased fecundity (spore production) for pathogenic isolates of *Colleotrichum* grown on their host under elevated CO_2. Klironomos et al. (1997) measured aerial fungal spore concentrations of different fungal genera in response to *Populus tremuloides* grown under elevated CO_2, and found significant increases in total spore concentrations in the air. Importantly, genera that were increased in response to CO_2 included *Fusarium* and *Alternaria*, genera that are known to contain plant pathogens or weak parasites. This again suggests increased fecundity of fungal pathogens under elevated CO_2, confirmed also in other pathosystems (Hibberd et al. 1996; powdery mildew).

9.4.4 PRECIPITATION AND TEMPERATURE

Potential and documented responses of fungal pathosystems to climate change are highlighted in reviews by Coakley et al. (1999) and Harvell et al. (2002), and will not be discussed in great detail here. Even the classic disease triangle of plant pathology acknowledges the important role of physical environment in plant disease (the other "points" of the triangle being host and pathogen factor): No pathogen can induce disease on a highly susceptible host if climatic conditions are not favorable. There are clear direct effects of, for example, temperature on pathogen populations. Winter is a major bottleneck for many pathogens (causing high mortality), and greater overwintering success through milder winters could increase disease severity. This is supported by observational evidence on disease incidences following mild winters or during warmer temperatures, suggesting that climate warming would lead to increased disease severity (Coakley et al. 1999). Generation number of polycyclic pathogens or growth of others may increase through increased growing season length and accelerated pathogen development in a warmer climate.

Additionally, there are effects also on host susceptibility to disease and host growth. For example, higher plant population growth rates (due to warming effects) can lead to greater population densities, with the consequence of increased pathogen transmission success and disease severity (Harvell et al. 2002). Similar to the situation under elevated CO_2, increased canopy growth and structure can lead to more conducive conditions for pathogen interception and infection.

9.5 CONCLUSION

Mirroring the multifaceted roles of fungi in ecosystems, there are many fungal-supported functions that could potentially be impacted by factors of climate change and elevated CO_2. At the ecosystem level, fungi can influence NPP (via mycorrhizal fungi or pathogens), nutrient cycling (e.g., soil saprobic fungi, mycorrhizal fungi), and soil structure. At the plant community level, fungi can mediate competition and coexistence of plant species, for example in pastures or other multispecies agroecosystems. Fungi also have strong influences on plant health and growth (e.g., pathogens, endophytes, mycorrhizae).

In some cases the effects on fungi can be direct, such as with precipitation changes on soil fungi, while in others there are only indirect effects, such as with elevated CO_2 influences on mycorrhizal fungi (mediated by the host plant).

In many instances it is not known how diversity of fungi is impacted by factors of global change. For some groups of fungi (e.g., saprobes), diversity changes (if any) are unlikely to be limiting to process rates. Conversely, for mycorrhizal fungi, which are generally a (relatively) low diversity group, impacts on diversity mediated by climate change can be more significant; especially because many conventional agroecosystem management practices are already detrimental to this group of organisms.

Major gaps in our knowledge exist pertaining to the *relative importance* of impacts on various fungal functional groups on ecosystem processes or crop growth compared to other effects. For example, how important are potential climate warming-related changes in alkaloid production of endophyte fungi compared to effects on nutrient cycling in pastures? In some cases, for example, elevated CO_2 and soil structure, we do have strong indications that fungal responses are highly significant and are manifested at the ecosystem level. In many other cases fungal and plant responses and feedbacks are tightly intertwined, owing to the intimate symbiotic association of many fungi with their plant hosts (mycorrhizae, pathogens, endophytes, phyllosphere fungi), making it difficult to distinguish causes and effects.

REFERENCES

Agrios, G.N. (1997) *Plant Pathology*, 4th ed., Academic Press, San Diego, CA.
Andrews, J.H. and Harris, R.F. (2000) The ecology and biogeography of microorganisms on plant surfaces. *Annual Review of Phytopathology*, 38, 145–180.
Augé, R.M. (2001) Water relations, drought, and vesicular-arbuscular mycorrhizal symbiosis. *Mycorrhiza*, 11, 3–42.

Chakraborty, S. (2006) Evolution of pathogens under elevated CO_2. In *Agroecosystems in a Changing Climate*, Newton, P.C.D., Carran, R.A., Edwards, G.R., and Niklaus, P.A., Eds., CRC Press, Boca Raton, FL, 331–334.

Chakraborty, S. and Datta, S. (2003) How will plant pathogens adapt to host plant resistance at elevated CO_2 under a changing climate. *New Phytologist*, 159, 733–742.

Chakraborty, S., Pangga, I.B., Lupton, J., Hart, L., Room, P.M., and Yates, D. (2000) Production and dispersal of *Colletotrichum gloeosporioides* spores on *Stylosanthes scabra* under elevated CO_2. *Environmental Pollution*, 108, 381–387.

Christensen, M. (1989) A view of fungal ecology. *Mycologia*, 81, 1–19.

Coakley, S.M., Scherm, H., and Chakraborty, S. (1999) Climate change and plant disease management. *Annual Review of Phytopathology*, 37, 399–426.

Cooke, R.C. and Whipps, J.M. (1993) *Ecophysiology of Fungi*, Blackwell Science, Oxford, U.K.

Dighton, J. (2003) *Fungi in Ecosystem Processes*, Marcel Dekker, New York.

Faeth, S.H. and Sullivan, T.J. (2003) Mutualistic asexual endophytes in a native grass are usually parasitic. *American Naturalist*, 161, 310–325.

Fitter, A.H., Heinemeyer, A., and Staddon, P.L., (2000) The impact of elevated CO_2 and global climate change on arbuscular mycorrhizas: a mycocentric approach. *New Phytologist*, 147, 179–187.

Gavito, M.E., Bruhn, D., and Jakobsen, I. (2002) Phosphorus uptake by arbuscular mycorrhizal hyphae does not increase when the host plant grows under atmospheric CO_2 enrichment. *New Phytologist*, 154, 751–760.

Ghannoum, O., Searson, M.J., and Conroy, J.P. (2006) Nutrient and water demands of plants under global climate change. In *Agroecosystems in a Changing Climate*, Newton, P.C.D, Carran, R.A., Edwards, G.R., and Niklaus, P.A., Eds., CRC Press, Boca Raton, FL, 55–86.

Harvell, C.D., Mitchell, C.E., Ward, J.R., Altizer, S., Dobson, A.P., Ostfeld, R.S., and Samuel, M.D. (2002) Climate warming and disease risks for terrestrial and marine biota. *Science*, 296, 2158–2162.

Heinemeyer, A., Ridgway, K.P., Edwards, E.J., Benham, D.G., Young, J.P.W., and Fitter, A.H. (2004) Impact of soil warming and shading on colonization and community structure of arbuscular mycorrhizal fungi in roots of a native grassland community. *Global Change Biology*, 10, 52–64.

Hibberd, J.M., Whitbread, R., and Farrar, J.F. (1996) Effect of elevated concentrations of CO_2 on infection of barley by *Erysiphe graminis*. *Physiological and Molecular Plant Pathology*, 48, 37–53.

Johnson, N.C., Wolf, J., and Koch, G.W. (2003) Interactions among mycorrhizae, atmospheric CO_2 and soil N impact plant community composition. *Ecology Letters*, 6, 532–540.

Jwa, N.S. and Walling, L.L. (2001) Influence of elevated CO_2 concentration on disease development in tomato. *New Phytologist*, 149, 509–518.

Kendrick, B. (2000) *The Fifth Kingdom*, 3rd ed., Focus Publishing, Newburyport, MA.

Klironomos, J.N., Rillig, M.C., and Allen, M.F. (1996) Below-ground microbial and microfaunal responses to *Artemisia tridentata* grown under elevated atmospheric CO_2. *Functional Ecology*, 10, 527–534.

Klironomos, J.N., Rillig, M.C., Allen, M.F., Zak, D.R., Pregitzer, K.S., and Kubiske, M.E. (1997) Increased levels of airborne fungal spores in response to *Populus tremuloides* grown under elevated atmospheric CO_2. *Canadian Journal of Botany*, 75, 1670–1673.

Klironomos, J.N., Ursic, M., Rillig, M., and Allen, M.F. (1998) Interspecific differences in the response of arbuscular mycorrhizal fungi to *Artemisia tridentata* grown under elevated atmospheric CO_2. *New Phytologist*, 138, 599–605.

Lindow, S.E. and Brandl, M.T. (2003) Microbiology of the phyllosphere. *Applied and Environmental Microbiology*, 69, 1875–1883.

Manning, W.J. and Tiedemann, A.V. (1995) Climate change: potential effects of increased atmospheric carbon dioxide (CO_2), ozone (O3), and ultraviolet-B (UV-B) radiation on plant diseases. *Environmental Pollution*, 88, 219–245.

Marks, S. and Clay, K. (1990) Effects of CO_2 enrichment, nutrient addition, and fungal endophyte-infection on the growth of two grasses. *Oecologia*, 84, 207–214.

Marks, S. and Clay, K. (1996) Physiological responses of *Festuca arundinacea* to fungal endophyte infection. *New Phytologist*, 133, 727–733.

Mitchell, C.E. (2003) Trophic control of grassland production and biomass by pathogens. *Ecology Letters*, 6, 147–155.

Morse, L.J., Day, T.A., and Faeth, S.H. (2002) Effect of *Neotyphodium* endophyte infection on growth and leaf gas exchange of Arizona fescue under contrasting water availability regimes. *Environmental and Experimental Botany*, 48, 257–268.

Newman, J.A. (2006) Trophic interactions and climate change. In *Agroecosystems in a Changing Environment*, Newton, P.C.D., Carran, R.A., Edwards, G.R., and Niklaus, P.A., Eds., CRC Press, Boca Raton, FL, 231–260.

Newman, J.A., Abner, M.L., Dado, R.G., Gibson, D.J., Brookings, A., and Parsons, A.J. (2003) Effects of elevated CO_2, nitrogen and fungal endophyte-infection on tall fescue: Growth, photosynthesis, chemical composition, and digestibility. *Global Change Biology*, 9, 425–437.

Newsham, K.K., Low, M.N.R., McLeod, A.R., Roberts, J.D., Greenslade, P.D., and Emmett, B.A. (1997) Ultraviolet-B radiation influences the abundance and distribution of phylloplane fungi on pedunculate oak (*Quercus robur*). *New Phytologist*, 138, 287–297.

Niklaus, P.A., Alphei, J., Ebersberger, D., Kampichler, C., Kandeler, E., and Tscherko, D. (2003) Six years of *in situ* CO_2 enrichment evoke changes in soil structure and soil biota of nutrient-poor grassland. *Global Change Biology*, 9, 585–600.

Niklaus, P.A., Glöckler, E., Siegwolf, R., and Körner, C. (2001) Carbon allocation in calcareous grassland under elevated CO_2: a combined ^{13}C pulse-labelling/soil physical fractionation study. *Functional Ecology*, 15, 43–50.

Packer, A. and Clay, K. (2000) Soil pathogens and spatial patterns of seedling mortality in a temperate tree. *Nature*, 404, 278–281.

Rillig, M.C. and Allen, M.F. (1999) What is the role of arbuscular mycorrhizal fungi in plant-to-ecosystem responses to elevated atmospheric CO_2? *Mycorrhiza*, 9, 1–8.

Rillig, M.C., Allen, M.F., and Field, C.B. (1999a) Soil biota responses to long-term atmospheric CO_2 enrichment in two California annual grasslands. *Oecologia*, 119, 572–577.

Rillig, M.C., Field, C.B., and Allen, M.F. (1999b) Fungal root colonization responses in natural grasslands after long-term exposure to elevated CO_2. *Global Change Biology*, 5, 577–585.

Rillig, M.C., Wright, S.F., Allen, M.F., and Field, C.B. (1999c) Rise in carbon dioxide changes soil structure. *Nature*, 400, 628.

Rillig, M.C., Hernández, G.Y., and Newton, P.C.D. (2000) Arbuscular mycorrhizae respond to elevated atmospheric CO_2 after long-term exposure: evidence from a CO_2 spring in New Zealand supports the resource-balance model. *Ecology Letters*, 3, 475–478.

Rillig, M.C., Wright, S.F., Kimball, B.A., Pinter, P.J., Wall, G.W., Ottman, M.J., and Leavitt, S.W. (2001) Elevated carbon dioxide and irrigation effects on water stable aggregates in a Sorghum field: a possible role for arbuscular mycorrhizal fungi. *Global Change Biology*, 7, 333–337.

Rillig, M.C., Treseder, K.K., and Allen, M.F. (2002a) Mycorrhizal fungi and global change. In *Mycorrhizal Ecology*, Ecological Studies Series 157, Van der Heijden, M.G.A. and Sanders, I.R., Eds., Springer Verlag, Berlin, 135–160.

Rillig, M.C., Wright, S.F., Shaw, M.R., and Field, C.B. (2002b) Artificial ecosystem warming positively affects arbuscular mycorrhizae but decreases soil aggregation. *Oikos*, 97, 52–58.

Rillig, M.C. and Field, C.B. (2003) Arbuscular mycorrhizae respond to plants exposed to elevated atmospheric CO_2 as a function of soil depth. *Plant and Soil*, 254, 383–391.

Runion, G.B., Curl, E.A., Rogers, H.H., Backman, P.A., Rodríguez-Kábana, R., and Helms, B.E. (1994) Effects of free-air CO_2 enrichment on microbial populations in the rhizosphere and phyllosphere of cotton. *Agricultural and Forest Meteorology*, 70, 117–130.

Sanders, I.R., Streitwolf-Engel, R., van der Heijden, M.G.A., Boller, T., and Wiemken, A. (1998) Increased allocation to external hyphae of arbuscular mycorrhizal fungi under CO_2 enrichment. *Oecologia*, 117, 496–503.

Schadt, C.W., Martin, A.P., Lipson, D.A., and Schmidt, S.K. (2003) Seasonal dynamics of previously unknown fungal lineages in tundra soils. *Science*, 301, 1359–1361.

Schardl, C.L., Leuchtmann, A., Chung, K.R., Penny, D., and Siegel, M.R. (1997) Coevolution by common descent of fungal symbionts (*Epichloë* spp.) and grass hosts. *Molecular Biology and Evolution,* 14, 133–143.

Shaver, G.R., Canadell, J., Chapin, F.S., III, et al. (2000) Global warming and terrestrial ecosystems: a conceptual framework for analysis. *Bioscience*, 50, 871–882.

Six, J., Carpentier, A., van Kessel, C., Merckx, R., Harris, D., Horwath, W.R., and Lüscher, A. (2001) Impact of elevated CO_2 on soil organic matter dynamics as related to changes in aggregate turnover and residue quality. *Plant and Soil*, 234, 27–36.

Smith, S.E. and Read, D.J. (1997) *Mycorrhizal Symbiosis*, 2nd ed., Academic Press, San Diego, CA.

Staddon, P.L., Thompson, K., Jakobsen, I., Grime, J.P., Askew, A.P., and Fitter, A.H. (2003) Mycorrhizal fungal abundance is affected by long-term climatic manipulations in the field. *Global Change Biology*, 9, 186–194.

Stahl, P.D. and Christensen, M. (1982) Mycorrhizal fungi associated with *Bouteloua* and *Agropyron* in Wyoming sagebrush-grasslands. *Mycologia*, 74, 877–885.

Staddon, P.L. and Fitter, A.H. (1998) Does elevated atmospheric carbon dioxide affect arbuscular mycorrhizas? *Trends in Ecology and Evolution*, 13, 455–458.

Staddon, P.L., Heinemeyer, A., and Fitter, A.H. (2002) Mycorrhizas and global environmental change: research at different scales. *Plant and Soil*, 244, 253–261.

Tiedemann, A.V. and Firsching, K.H. (1998) Combined whole-season effects of elevated ozone and carbon dioxide concentrations on a simulated wheat leaf rust (*Puccinia recondita* f. sp. *tritici*) epidemic. *Zeitschrift fur Pflanzenkrankheiten und Pflanzenschutz*, 105, 555–566.

Treseder, K.K. and Allen, M.F. (2000) Mycorrhizal fungi have a potential role in soil carbon storage under elevated CO_2 and nitrogen deposition. *New Phytologist*, 147, 189–200.

Vandenkoornhuyse, P., Baldauf, S.L., Leyval, C., Straczek, J., and Young, J.P.W. (2002) Extensive fungal diversity in plant roots. *Science*, 295, 2051.

Wardle, D.A. (2002) *Communities and Ecosystems:Linking the Aboveground and Belowground Components*, Princeton University Press, Princeton, NJ.

Weltzin, J.F., Loik, M.E., Schwinning, S., et al. (2003) Assessing the response of terrestrial ecosystems to potential changes in precipitation. *BioScience*, 53, 941–952.

Wolf, J., Johnson, N.C., Rowland, D.L., and Reich, P.B. (2003) Elevated CO_2 and plant species richness impact arbuscular mycorrhizal fungal spore communities. *New Phytologist*, 157, 579–588.

Yang, C-H., Crowley, D.E., Borneman, J., and Keen, N.T. (2001) Microbial phyllosphere populations are more complex than previously realized. *Proceedings of the National Academy of Sciences, USA*, 98, 3889–3894.

10 Trophic Interactions and Climate Change

Jonathan A. Newman

CONTENTS

10.1 INTRODUCTION

Climate change biology is in some senses a young discipline, but in other senses it is nothing more than classical ecology and physiology aimed at producing predictions about the likely consequences of a set of abiotic changes. That is, we do not need new science but new applications of the science we have already developed, and that we are continuing to develop, to understand the effects of climate change.

Early work in climate change biology was conducted mainly by plant physiologists, who concentrated on the effects of CO_2 enrichment on changes in photosynthesis, plant growth, and chemical composition. Other aspects of climate change like temperature and drought stress had already been the subject of a great deal of physiological research before physiologists became interested in climate change per se. Later on, plant ecologists started to ask how the changes reported by the physiologists might translate into changes in the abundance and distribution of plant species, mainly via the mechanism of plant–plant competition. It was only later that animal ecologists really started to get involved in climate change research, asking whether the abundance of phytophagous insects would likely be higher, lower, or relatively unchanged in a high CO_2 world.

Species not only respond to global climate change as individuals, but also in the context of their relationships with competitors, predators, diseases, and symbionts. Climate change research moved from laboratory growth chambers to open-topped chambers and free air CO_2 enrichment (FACE) arenas because researchers recognized, even if only in a vague way, that the ecological context was as important as the plant physiology. We needed to know that the changes we observed in growth chambers would still be detectable against the backdrop of a multispecies community. In such communities, the plants respond to climate change as links in food chains — chains that are interwoven into more extensive food webs.

In a high-CO_2 world, a grossly generalized expectation of a *closed system* would be that plants would grow larger but would be less nutritious (e.g., lower N concentrations) to their primary consumers; and that primary consumers would attempt to compensate for the reduced nutritional quality by increasing their daily intake (see e.g., Emmerson et al., 2004, for further discussion). Two possible numerical responses could arise. First, primary consumers might compensate for the reduced nutritional quality leaving their numbers (and presumably those of their natural enemies) unchanged. The additional growth of the plant may or may not be offset by the increase in herbivore consumption. On the other hand, herbivores might be unable to compensate for the reduced nutritional quality by eating more, and their reduced growth rate and smaller size would leave them vulnerable to natural enemies for longer periods of time, which would result in reduced herbivore populations (and reduced natural enemy populations) but increased plant quantity.

Two points above warrant further comment. First, so far, I am only talking about increased CO_2 concentrations (I will broaden the context later in the chapter). How other climate variables combine with CO_2 to influence plant–herbivore relationships has, so far, resisted generalization. Nevertheless, CO_2 per se is the variable that has commonly been manipulated under the banner of "climate change research." This approach is at least initially justifiable because a complete understanding of plant–herbivore interactions, in the context of the *entire suite* of climate variables that we expect to change, will require an understanding of each of the main effects, which includes CO_2. Second, open-topped chambers and FACE experiments are, generally speaking, *open systems*. Herbivores and their natural enemies migrate freely between the experimental and current ambient conditions. For sedentary herbivores, like leaf miners, the fact that the system is open probably matters little.

The same cannot be said for the parsitoids that attack the leaf miners. It is not clear that theory derived for closed systems will apply to open-system experiments.

Despite this relatively simple theory, in making the move to study climate change in multispecies communities, we have observed the entire spectrum of responses at every trophic level. Experiments regarding the impact of climate change in this context yield a bewildering array of results. In summarizing a large body of experiments on plant–insect relationships, Coviella and Trumble (1999) concluded, "the effects of increased atmospheric CO_2 on herbivory will be not only be highly species-specific but also specific to each insect-plant interaction (p. 700)." This statement is less a surrender to complexity, than an acknowledgement that we are unlikely to observe a simple common phenomenon across all herbivores and natural enemies. It means only that we have to work harder to find the generality in the system responses.

In this chapter I will be concerned with interpretation and expectations of individual species' responses to climate change in the context of multispecies interactions. In such a context, we need to be cognizant of potential *indirect effects*. Indirect effects occur when one species affects another species through its interaction with a third species. We generally speak of two types of indirect effects (Wootton, 1994) — interaction chains and interaction modifications. Interaction chains occur when we alter the abundance of one species and this alters the abundance of another species, through its affects on the abundance of an intermediate species. Interaction chains are of the form: Species A affects species B, which in turn affects species C. Interaction modification occurs when altering the abundance of one species changes the way that two other species interact. For example, grass grown in a mixture with clover may be more attractive to herbivores if its nitrogen concentration increases. In this scenario, more clover might increase the level of herbivory on the grass.

Distinguishing between the two types of indirect effects is important as each implies something different about a system's response to a perturbation like climate change. In theory, prediction regarding interaction chain effects is fairly straightforward as long as we understand each of the pairwise interactions. Interaction modification, on the other hand, is much more difficult to predict. More than a decade ago, Tim Wootton (1994, p. 445) suggested, "At our present level of understanding, however, the quantitative consequences of interaction modification can only be determined by experimental manipulation within the context of the community of interest; they cannot be predicted ahead of time." While we now have better experimental evidence for interaction modification, we are probably little closer to being able to predict such effects. One might argue that we just want to know the likely *responses* we should expect from climate change and we are not so concerned about knowing the *mechanism* causing these responses. However, this logic is flawed. Without an understanding of the basic mechanisms of response, we will be unable to develop general theories or even to draw general conclusions about the likely responses. Without such general theories and conclusions, we will need to test every species and every combination of species under every combination of every aspect of climate change — surely both an unattainable and an unworthy goal. To provide the robust impact predictions society demands, we must develop general theories and to do so we need to understand the mechanisms that determine these impacts.

10.2 GENERAL THEORY OF MULTITROPHIC INTERACTIONS

In this section I develop the basic ecological theories about multitrophic interactions. Readers already familiar with these theories may wish to advance to the next section. Before developing the general theories, I will first discuss two classic problems in population ecology, the paradox of enrichment and the linear food chain. These simple examples both yield what may be counterintuitive conclusions and provide motivation for the general thinking required when considering population responses at multiple trophic levels.

10.2.1 THE PARADOX OF ENRICHMENT

Let us consider a classic, simple, heuristic model from population ecology — the Lotka–Volterra predator–prey model. The Lotka–Volterra models are really a family of similar models. The original predator–prey model was developed independently by Alfred Lotka, an actuarial scientist working for Metropolitan Life, and Vito Volterra, an Italian physicist who was helping his future son-in-law, Umberto D'Ancona, a marine biologist, to understand the dynamics of Adriatic fisheries following World War I (Kingsland, 1995).

The model is a classic of population biology; it can be found in every undergraduate textbook even now, 75 plus years after its first publication. I will first briefly introduce one version of the model and then use it to demonstrate a long-known result, the *paradox of enrichment*.

The model we will consider is sometimes referred to as the Rosenzweig–MacArthur system (Kot, 2001, pp. 132–137, Rosenzweig and MacArthur, 1963):

$$\underset{\substack{\text{rate of}\\\text{change of}\\\text{plants}}}{\frac{dV}{dT}} = \underset{\text{logistic growth}}{rV\left(1-\frac{V}{K}\right)} - \underset{\substack{\text{type II}\\\text{function}\\\text{response}}}{\frac{cVI}{a+V}},$$

$$\underset{\substack{\text{rate of}\\\text{change of}\\\text{herbivores}}}{\frac{dI}{dT}} = b\underset{\substack{\text{constant}\\\text{numerical}\\\text{response}}}{\left(\frac{cVI}{a+V}\right)} - \underset{\substack{\text{constant}\\\text{death}\\\text{rate}}}{mI}.$$

$$(10.1)$$

By introducing the change of variables $V = ax$, $I = r\frac{a}{c}y$, and $T = \frac{t}{r}$, we simplify Equation 10.1 to

$$\frac{dx}{dt} = x\left(1-\frac{x}{\gamma}\right) - \frac{xy}{1+x},$$

$$\frac{dy}{dt} = \beta\left(\frac{x}{1+x} - \alpha\right)y,$$

$$(10.2)$$

where

$$\alpha \equiv \frac{m}{bc}, \quad \beta \equiv \frac{bc}{r}, \quad \gamma = \frac{K}{a}. \tag{10.3}$$

We solve Equation 10.2 by setting $dx/dt = 0$ and $dy/dt = 0$. There are three possible equilibria to this system: (1) both species go extinct, (2) only the plants persist, or (3) some form of coexistence determined by:

$$x^* = \frac{\alpha}{1+\alpha}, \quad y^* = \left(1+x^*\right)\left(1 - \frac{x^*}{\gamma}\right). \tag{10.4}$$

The equilibrium (3) is shown in Figure 10.1. Now, without presenting the formal stability analysis, we can assess the stability of equilibrium (iii) visually (see e.g., Kot, 2001, for mathematical details). Equation 10.4 gives the zero-growth isoclines for plants and herbivores, respectively. If the plant zero-growth isocline intersects the herbivore zero-growth isocline to the right of the hump, the resulting equilibrium is a stable node. If the intersection occurs to the left of the hump, a limit cycle results.

The paradox of enrichment (Rosenzweig, 1971) is this: Enrichment of the plants is equivalent to increasing $\gamma = K/a$, and doing so can change the plant–herbivore interaction from a stable node into a cyclic dynamic (i.e., $\gamma_1 \rightarrow \gamma_2$ in Figure 10.1).

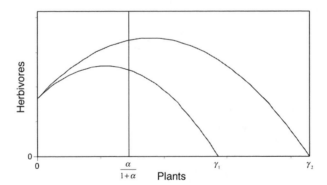

FIGURE 10.1 The paradox of enrichment. Plotted are the plant and herbivore zero-growth isoclines. An equilibrium occurs where these two isoclines cross. If the cross occurs to the right of the hump in the herbivore isocline, a stable node results. If the cross occurs to the left of the hump, a stable limit cycle occurs. The paradox of enrichment is seen by comparing the dynamics implied with $\gamma = \gamma_1$ to $\gamma = \gamma_2$. Enriching the system is equivalent to increasing the carrying capacity, which is equivalent to increasing $\gamma = \frac{K}{a}$. When $\gamma = \gamma_1$, a stable node results. By enriching the system ($\gamma = \gamma_2$), the equilibrium changes from a node to a cycle and the cycle is more prone to stochastic extinction. See Equations 10.1 through Equation 10.4 and the text for more detail.

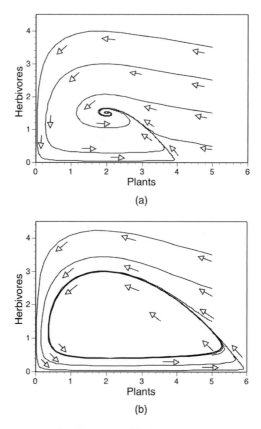

FIGURE 10.2 The paradox of enrichment. This figure depicts the phase plane diagram for the dynamics of the two equilibria shown in Figure 10.1. In Figure 10.2a the equilibrium is a stable node. In Figure 10.2b the equilibrium is a stable cycle.

These two dynamics are illustrated in Figures 10.2a and 10.2b. Notice that, in 10.2b the trajectories can get close to either axis, making stochastic extinction a real possibility. For this reason, Rosenzweig concluded that enriching ecosystems can be dangerous.

This overly simplistic model suggests, for example, that CO_2 enrichment will lead to a larger carrying capacity for the plants, and that this alone may destabilize plant–herbivore interactions via the mechanism of the paradox of enrichment. Of course, the situation with climate change is more complex because increasing the plant carrying capacity may also reduce plant quality and this might lead to increased herbivore consumption (i.e., increasing c in Equation 10.1) as insects attempt to compensate for this reduction in quality. If this is the case, increasing γ also has the effect of decreasing α and this always moves $\frac{\alpha}{1-\alpha}$ to the left in Figure 10.1, thereby further destabilizing the plant–herbivore interaction.

10.2.2 Simple Linear Food Chain

Let's consider a simple tritrophic system that exists as a linear food chain comprising a plant population, an insect population, and a predator population. The model we will consider is given by:

$$\underset{\substack{\text{rate of change} \\ \text{of the plants}}}{\frac{dV}{dt}} = \underset{\substack{\text{logistic growth of} \\ \text{plants in absence} \\ \text{of insects}}}{rV\left(1 - \frac{V}{K}\right)} - \underset{\substack{\text{Type I} \\ \text{functional} \\ \text{response of} \\ \text{insects}}}{(\varphi V)I} ,$$

$$\underset{\substack{\text{rate of change} \\ \text{of the insects}}}{\frac{dI}{dt}} = \underset{\substack{\text{constant numerical} \\ \text{response of insects}}}{\vartheta(\varphi V)I} - \underset{\substack{\text{Type I} \\ \text{functional} \\ \text{response of} \\ \text{predators}}}{(\delta I)P} , \qquad (10.5)$$

$$\underset{\substack{\text{rate of change} \\ \text{of the predators}}}{\frac{dP}{dt}} = \underset{\substack{\text{constant numerical} \\ \text{response of predators}}}{\varepsilon(\delta I)P} - \underset{\substack{\text{constant death} \\ \text{rate of the} \\ \text{predators}}}{\xi P} .$$

where V is the vegetation, I is the insect population, and P is the population size of the predator. This model assumes that vegetation grows logistically in the absence of herbivores $rV\left(1 - \frac{V}{K}\right)$, herbivores' and predators' per capita intake rates increase linearly with vegetation and herbivore density (φV and δI respectively), herbivores and predators convert intake into births at constant rates (ϑ and ε respectively), and predators die at a constant rate (ξ). Ignore, for the moment, that this model is overly simplistic; we use this model only to make a simple heuristic point.

We can solve this model by setting $\frac{dV}{dt} = \frac{dI}{dt} = \frac{dP}{dt} = 0$ (at equilibrium the rate of change of all populations will be zero) and solving for V, I, and P. This yields:

$$\hat{V} = K\left(1 - \frac{\varphi\xi}{\varepsilon\delta r}\right),$$

$$\hat{I} = \frac{\xi}{\varepsilon\delta}, \qquad (10.6)$$

$$\hat{P} = \frac{\vartheta\varphi K}{\delta}\left(1 - \frac{\varphi\xi}{\varepsilon\delta r}\right).$$

Although the solution may look impenetrable, there is one very straightforward conclusion that we can draw from inspection. Suppose that the effect of climate change on this system was really very simple: Plants photosynthesized more and so grow more. In Equation 10.5 that would have the affect of increasing the vegetation's carrying capacity (K). At equilibrium, Equation 10.6 tells us that increasing the vegetation's carrying capacity results in more predators, not more insects! Notice

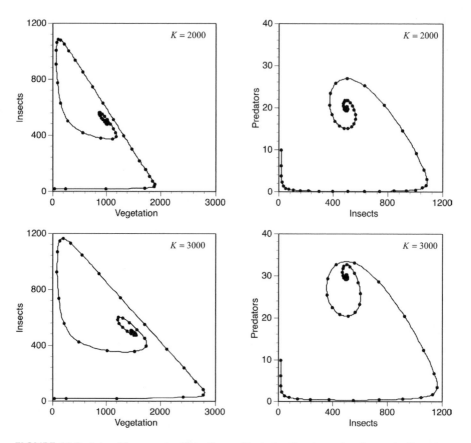

FIGURE 10.3 A trophic cascade. This figure illustrates the dynamics shown in Equation 10.5. Notice that increasing the carrying capacity of the vegetation yields increases in the vegetation and the predators, but the equilibrium value for the insects remains the same.

that K does not appear in the equilibrium solution for the insects. This result is shown graphically in Figure 10.3. Readers worried by the unrealistic type I functional responses in Equation 10.5 can be reassured that the same general conclusion regarding K follows if we substitute more realistic type II functional responses, although the algebra involved is more hideous (see e.g., Gurney and Nisbet, 1998, pp. 192–195).

The moral of this story is quite simple: It tells us to be careful about conclusions we draw at one trophic level if we are leaving out information from the next trophic level. If we did an experiment in which we examined the population growth of aphids feeding on wheat in high and low CO_2, and the results suggested more aphids under elevated CO_2, we should draw this conclusion hesitantly. It may be that the real result is not more aphids, but more aphid natural enemies! The reverse is also true. If our results suggest smaller aphid populations under elevated CO_2, it might be the case that aphid populations are relative unchanged, but parasitoid populations decline.

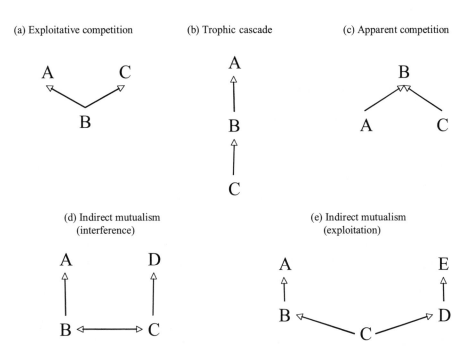

FIGURE 10.4 Wootton's (1994) five types of indirect effects. Horizontal arrows denote interference competition (arrows show impacted species); vertical arrows denote consumer–resource relationships. (Reprinted with permission, from the *Annual Review of Ecology and Systematics*, Volume 25 © 1994 by Annual Reviews www.annualreviews.org.)

10.2.3 INTERACTION CHAIN EFFECTS

Wootton (1994) enumerates five types of indirect effects. These are illustrated in Figure 10.4. I will briefly describe each of these and give examples. Note that many examples do not come from agroecosystems because the study of indirect effects has not been very popular in such systems, although there are good reasons for believing that such effects are both prevalent and strong in agroecosystems. I will return to this issue later.

10.2.3.1 Exploitative Competition

Some ecologists do not consider this to be a form of indirect effect, while others do. Often the distinction is linked to the identity of the limiting resource. Exploitative competition occurs when two species compete for a resource and the effects of competition are experienced through the drawing down of the limiting resource. For example, a crop and a weed might compete for limiting soil nitrogen. In this case, as the resource is not another species, we would not usually think of this as an indirect effect. On the other hand, two species of parasitoid used in the biocontrol of aphids in glasshouses might well fit this notion of exploitative competition as a form of indirect effect, one that has the interaction chain mechanism. Because such forms of competition have been a major focus for community ecologists for many decades, we consequently know quite a lot about this form of indirect interaction, both theoretically and empirically. See Golberg and Barton (1992) for a review of such interactions among plants.

10.2.3.2 Trophic Cascades

We sometimes say that the effect at one trophic level *cascades* to other trophic levels. This notion has led to a debate among ecologists as to whether the cascades tend to come from the top–down or the bottom–up. The top–down cascade model was initially proposed by Hairston et al. (1960, although the term *cascade* was apparently coined by Paine, 1980). The theory proposes that plant growth is resource limited, but herbivore populations are more strongly limited by their predators than by resource (plant) availability. Predators, on the other hand, are limited by the availability of prey (herbivores). In this scenario, predators regulate the herbivores and this, indirectly, benefits the plants. In bottom–up cascades, it is assumed that both herbivores and predators are resource limited, which leads to the conclusion presented above in the Section 10.2.2.

There are experimental examples of both top–down and bottom–up cascades. Top–down cascades should, in principle, be familiar to agroecologists as they are the theoretical foundation for the biological control of herbivorous pest insects. When glasshouse growers release parasitoid wasps to control aphid populations, they are relying on the principle of a top–down cascade. A particularly nice example of a top–down cascade in the field is a study by Marquis and Whelan (1994), who looked at biomass production of white oak (*Quercus alba*) saplings that were either caged to prevent birds from foraging on them, or left uncaged. In the caged treatment, *Lepidopteran* larvae were twice as abundant and leaf area loss was 25%. In contrast, uncaged saplings had half the larval population and leaf area loss was only 13%. A very clear example of a bottom–up cascade is work done by Wootton and Power (1993). They experimentally shaded portions of a stream thereby manipulating the carrying capacity of the algal species. They then monitored the biomass of algae, herbivores, and predators. Increasing the degree of shading decreased the algae and predator biomass, but resulted in no change to the herbivore biomass (this is another example of the theoretical system considered in Equation 10.5).

McCann et al. (1998) contrast the bottom–up cascade with something they call a "trophic trickle." They demonstrate that cascades such as that observed by Wootton and Power (1993) are theoretically likely when the per capita predation rate is a function of prey numbers alone. However, when the per capita predation rate depends on the ratio of predator to prey numbers, the addition of nutrients (for example) to the plants results in increased biomass at all three trophic levels. These so-called *ratio-dependent models* are likely to arise when predators interfere with each other in the sense that they directly influence each other's feeding rates. Simple top–down and bottom–up cascades are also probably more common in linear food chains rather than in more complex food webs, and Abrams (1993) suggests good reasons for this. Abrams conducted a thorough study of some simple food webs. He showed that webs in which: (1) there is competition among the predators, (2) there are competitive but invulnerable prey, or (3) there are competing prey species that share

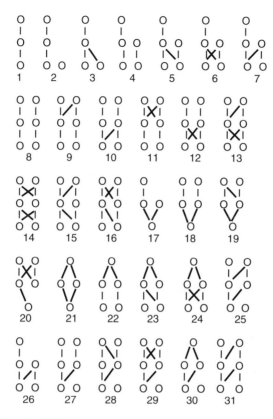

FIGURE 10.5 Abram's (1993) investigation of trophic interactions in simple food webs. The 31 possible food web configurations with three trophic levels, no omnivory (feeding within a trophic level), and a maximum of two species per level. The outcomes of increased nutrient input on the abundances of each level are denoted I for increases, D for decreases, U for unchanged, and A when all three results are possible. Results show changes from bottom (vegetation) to top (predators). IUI: 1–3, 8, 10, 12, 18, 27; IUA: 9, 11, 13–16, 19–20, 25, 28–29, 31; III: 4–5; IIU: 26; AIA: 6–7; UIU: 17; UAU: 21, IAI: 22, IAU: 30, AAA: 23–24.

a predator, such bottom–up effects are prevented (Figure 10.5). Finally, Strong (1992) has argued that trophic cascades are unimportant in terrestrial ecosystems due to the compensating effects of interspecific competition (see also Polis and Strong, 1996).

10.2.3.3 Apparent Competition

First used by Holt (1977), the term *apparent competition* might apply whenever two species share a common predator. The theory is that an increase in one prey species leads to an increase in the shared predator that leads to a decrease in the "apparent" competitor. Again, apparent competition may be familiar to agroecologists in the context of predatory biocontrol. Some biocontrol agents can be more effective if they are released together with an alternative prey species. A good example of this comes from the use of predatory mites to control herbivorous mites on grapevines (Karban et al., 1994). When *Metaselius occidentalis* (predator) was released together with *Eotetranychus willamettii* (alternative prey), *M. occidentalis* was better at controlling the pest mite species *Tetranychus pacificus*. A similar effect was seen when the variegated leafhopper (*Erythroneura variabilis*) was introduced in California (Settle and Wilson, 1990). The variegated leafhopper apparently increased the density of parasitic wasps, which in turn reduced the density of the grape leafhopper (*Erythroneura elegantula*). Another nice example comes from aphid–ladybird beetle interactions. Müller and Godfray (1997) demonstrate apparent competition between the nettle aphid (*Microlophium carnosum*) and the bird cherry-oat aphid (*Rhopalosiphum padi*) mediated by the seven-spotted ladybird beetle (*Coccinella septempunctata*). Such interactions are not confined to the herbivore trophic level. Morris et al. (2001) provide evidence that two species of aphid parasitoids each attacking different species of aphids exhibited apparent competition through the actions of secondary parasitoids that attack both primary parasitoids.

There has been some doubt as to whether such interactions are important or even whether they exist in more complex, *hyperdiverse* communities. Recently, such an interaction has been convincingly demonstrated from parasitoid wasps attacking leaf miner species in the forests of Belize. Morris et al. (2004) removed eight species of fly (*Calycomyza spp.*) and recorded lower parasitism and higher abundance in other species of dipteran leaf miners that shared parasitoids with the *Calycomyza* species, but which occurred on different plant species (and so could not be directly competing). This study is intriguing as it hints at both the strength and ubiquity of this indirect interaction in the structure of diverse communities. Obviously more studies of this nature are needed.

10.2.3.4 Indirect Mutualism

This indirect interaction can occur when we have two predator–prey relationships in which the prey species compete with each other (e.g., Vandermeer, 1980). In such cases, the two predators can appear to have mutualistic effects on each other: an increase in predator 1 decreases the population size of prey 1; this then results in an increase in the population of its competitor, prey 2, and this leads to an increase in predator 2. The predators will appear to be mutualistic because increasing one leads to an increase in the other. Sometimes this is referred to as *indirect commensalism* because, strictly speaking, the positive effects are not reciprocal between the two predators. A particularly clear example of indirect mutualism and commensalism is seen in the work of Thompson et al. (1991), where an increase in kangaroo rats (*Dipodomys spp.*) changed the balance between small-seeded and large-seeded plants (the rats preferentially feed on large seeds). The increase in the abundance of small-seeded plants then led to an increase in ant species that feed on small seeds.

10.2.4 INTERACTION MODIFICATION EFFECTS

Confusingly, interaction modification effects are sometimes simply referred to as *higher order interactions*. Because interaction modification effects are somewhat more amorphous than interaction chain effects, and because ecologists have a penchant for coining new jargon, many examples are known by a variety of other terms (as will become clear shortly).

There are many ways in which such effects can be present, but probably the best studied is the so-called *associational plant defense* (Hay, 1986). Somewhat counterintuitively, this interaction occurs where one competitor has a strong *positive* effect on another. The proposed mechanism is that an abundant but unpalatable plant can provide some protection for a rare but palatable plant. As farfetched as this idea may sound, there are several good experimental examples of the interaction, and it is the theoretical basis for some agricultural practices like intercropping with mustard plants. Hjalten et al. (1993) refer to this interaction as the *repellent-plant hypothesis*. They develop a model using foraging theory in patchy environments (the so-called *marginal value theorem*) to predict the circumstances under which such protection might be evident, and then test this theory using voles and mountain hares grazing in birch-alder mixtures. More recently Hjalten and Price (1997) demonstrated associational defense among clones of willow being attacked by stem gallers (*Euura lasiolepis*) and petiole gallers (*E. venusta Osten-Sacken* sp.). Holmes and Jepson-Innes (1989) demonstrate similar effects among perennial grasses (*Bouteloua gracilis* and *Aristida* spp.) being grazed by grasshoppers. These results all suggest that small-scale spatial heterogeneity can be an important determinant of herbivory.

While unpalatable plants providing protection for palatable plants has been the most commonly studied example, other similar interactions have also been demonstrated. For example, plants of lower palatability may receive protection from plants of high palatability — the so-called *attractant-decoy hypothesis* (see Hjalten et al., 1993 for more discussion). Rousset and Lepart (2002) provide an example of this interaction using sheep grazing *Buxus sempervirens* and a variety of highly palatable neighbors.

10.2.5 CLIMATE CHANGE AND INDIRECT EFFECTS IN AGROECOSYSTEMS

When are indirect effects likely to be important in agroecosystems and when are they not? I should preface this discussion by saying that it is difficult to generalize across agroecosystems, although I will try. Agroecosystems embrace vast monocultures, complex intercropping, rotation systems, multispecies pastures, agroforestry, orchards, vineyards, and so on. Any generalization is unlikely to apply to all of these systems all of the time. The important points to bear in mind are the principles rather than the generalizations.

So when would we expect to see indirect trophic effects in agroecosystems? Morin (1999) points out that strong direct effects are probably required in order for us to be able to observe indirect effects. Experiments that demonstrate large effects of, for example, CO_2 enrichment, are perhaps likely to be generating noticeable indirect effects as well if only we were to look for them (assuming we have not experimentally excluded the possibility). Trophic cascades seem more likely in simple, linear food chains than in more complex food webs (Polis and Strong, 1996). Strong (1992) forcefully argues that trophic cascades are *restricted* to systems of low diversity where one or a few species can exert great influence on the dynamics of others. It can be argued that agroecosystems tend to be such simple systems and are thus more likely to produce the conditions necessary for cascades (Duffy, 2002; Halaj and Wise, 2001). For this reason, Schmitz et al. (2000) specifically excluded agroecosystems from their meta-analysis on trophic cascades because agroecosystems "can be predisposed to show such top-down effects" (p. 142). Halaj and Wise (2001) point out that *only* in managed agricultural systems (mainly crops) do predators sometimes produce substantial indirect effects, despite a strong annual disturbance regime. They suggest that this results from the relatively simple habitat structure and the more homogeneous and vulnerable plant community in crop systems. Generally speaking, the more tightly connected the species in the community, the more important indirect trophic effects are likely to be. Two species are said to be *tightly connected* if they depend strongly on one another (positively or negatively). Are agroecosystems more likely to be such tightly connected systems than natural systems? My suspicion is "yes," but little data exists to support a generalization in either direction. On average, tighter connections should be produced in communities of low diversity that have a preponderance of specialist herbivores and predators. Agroecosystems are certainly less diverse than natural systems, and they often involve highly specialized pest species, species-specific mutualisms, and perhaps highly specialized biocontrol species; thus yielding a tight connection between species.

When are indirect effects unlikely to be important in agroecosystems? Wootton (1994) notes that even when strong direct effects are detectable, they may still be sufficiently weak that they are damped out when linked in long chains. This seems less likely in agroecosystems because of the general simplification of the community and, hence, shorter interaction chains. Wootton (1994) also notes that environmental variation is likely to make indirect effects less important if such variation keeps species numbers low so that interactions among species remain weak. In some agroecosystems this may well be the case. Repeated applications of pesticides and herbicides may function much like fire or flooding in more natural ecosystems, in that they keep populations low thereby reducing strong direct effects, the prerequisite for indirect effects. On the other hand, this tendency must be balanced by the applications of fertilizers, which encourage the strong direct effects. Lastly, indirect effects may not be evident if they are opposed by other strong (direct or indirect) interactions (Wootton, 1994). However, Halaj and Wise (2001) show that the net cancellation of effects generated by different tropic levels is less common in agroecosystems than in natural systems.

In the next section, I will look at some candidate systems for investigating indirect trophic effects in agroecosystems and consider how these systems might be altered by climate change.

10.3 SOME EXAMPLES

To say that multitrophic interactions are everywhere in agroecosystems is a trite observation. Obviously, I cannot review all of these interactions in this one chapter, nor will I necessarily review the most important of these, assuming we know which interactions these are. What I have tried to do in this section is to review three examples for which something (in some of these cases this is not much) is known about the response of the system to climate change. In general, multitrophic interactions under climate change remain an unanswered challenge for climate change biologists.

The first example illustrates a straightforward trophic cascade problem involving grass, aphids, and parasitoids. Although it is structurally a trophic cascade, it behaves more like a trophic trickle (see previous section). Other examples of such trophic cascades are beginning to be considered by climate change ecologists (Roth and Lindroth, 1995; Stacey and Fellowes, 2002; Stiling et al., 1999). The remaining two examples might fairly be described as examples of interaction modification. In both examples, the presence of a fungus modifies the interaction between herbivores and plants. The first of these considers the effects of endophytic fungi and the second example examines arbuscular mycorrhizal fungi.

10.3.1 TROPHIC CASCADES: GRASSES, CEREAL APHIDS, AND THEIR PARASITOIDS

There is a growing body of literature on the effects of climate change on aphid population dynamics. Lawton (2000) reviewed aphid responses to elevated CO_2 (see also Bezemer and Jones, 1998; Whittaker, 1999; Whittaker, 2001) and concluded that we cannot even predict the sign of likely change — increase, decrease, or none. Bezemer et al. (1999) presage this frustration in the title of their paper "How General are Aphid Responses to Elevated CO_2?" and they concluded that aphid responses are often species specific and idiosyncratic. My colleagues and I tried to find some mechanism or mechanisms by which we might see generality in aphid responses to CO_2. We used a model to suggest that all of the observed results could be explained by aphid species differences in their nitrogen requirements and density-dependent winged morph responses, as well as by differences in soil nitrogen input (Newman et al., 2003b). I then went on to consider the interaction between CO_2 and rising temperature on aphid responses and concluded that, at least for reasonably temperate climates like southern Britain, there is an interaction between these two factors of climate change. When both CO_2 and temperature are elevated, the resulting aphid populations may be more similar to current ambient conditions than the results of either factor on its own would suggest because in this temperate climate, the positive effects of increased temperature tend to cancel out the negative effects of increased CO_2 (see Newman, 2004, for further discussion).

Despite the continuing experimental and theoretical research programme aimed at resolving the ambiguity in aphid responses, a remaining nagging suspicion is that none of these predictions are likely to be correct because no one has previously considered whether the plant–aphid interaction under climate change is significantly altered by the presence of the aphids' natural enemies — the third trophic level. Bezemer et al. (1998) attempted to do just this, but an unfortunate contamination of their samples prevented this comparison. On the face of it, this tritrophic interaction seems a likely candidate for a trophic cascade. To examine this possibility, Julia Hoover and I developed a mechanistic model of the parasitoid–aphid–grass

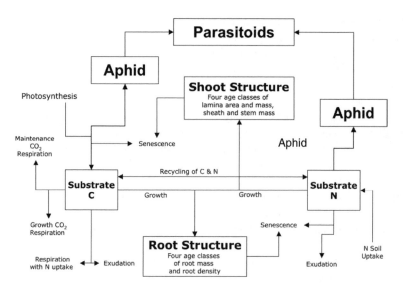

FIGURE 10.6 Schematic diagram of the Hoover and Newman plant–aphid–parasitoid model. (Hoover, J.K. and Newman, J.A. [2004] *Global Change Biology*, 10, 1197–1208. With permission.)

interaction (Hoover and Newman, 2004). The model was capable of responding mechanistically to changes in atmospheric CO_2 concentrations and temperature. Sparing the reader the mathematical details, which are published elsewhere (Newman et al., 2003b; Hoover and Newman, 2004) the model is shown schematically in Figure 10.6. Figure 10.7a shows the predicted relative total aphid populations in the presence and absence of parasitoids in the four combinations of temperature and CO_2. All of the points lie above the one-to-one line indicating that parasitoids are effective at reducing aphid populations (although not very if they do not hit the aphid colony when it is just starting). However, strikingly, the pattern of response to both aspects of climate change in the aphid is the same regardless of the presence or absence of parasitoids. Figure 10.7b plots the total parasitoids against the total unparasitised aphids. Here we see that the effects of these two aspects of climate change produce similar predictions for the parasitoids as for the aphids, suggesting that this response is more of a trophic trickle than a trophic cascade (McCann et al., 1998). In this model, the trophic trickle occurs *locally* because the parasitoids are introduced too late in the aphid colony development for the parasitoids to be able to *strongly* control the aphid populations. Thus, parasitoid populations strongly reflect aphid populations but the reverse relationship is much weaker. If parasitoids were introduced early on in the aphid colony's development, the parasitoids invariably wipe out the aphid population before the exponential phase of population growth really begins. These responses are due to differences in the generation times of the aphids and their parasitoids (cf. Kindlmann and Dixon, 1999). On a *regional* basis (i.e., many aphid colonies arranged spatially) it is not clear how either aphid or parasitoid numbers respond to climate change. No experimental or modelling work has yet addressed this question.

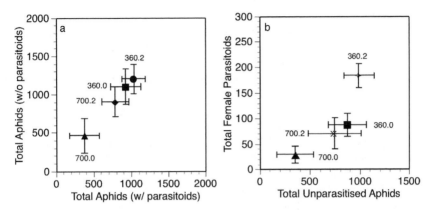

FIGURE 10.7 Grass–Aphid–Parasitoid model results. (a) Shows total aphid population (× $10^4 m^2$, parasitized and unparasitised aphids summed across days) with and without parasitoids. Plotted are the mean responses of six hypothetical aphid species (±sem) for each of the four environmental conditions: 360.0 = current ambient; 360.2 = elevated temperature; 700.0 = elevated CO_2; 700.2 = elevated CO_2 *and* temperature. (b) Shows total female parasitoid population (× $10^4 m^2$, summed across days) and the total unparasitized aphids (× $10^4 m^2$, in the presence of parasitoids). Plotted again are the means for six hypothetical aphids in each of the four environmental conditions. The relationship between the environmental conditions in (a) remains the same whether parasitoids are present or not, suggesting that parasitoids do not fundamentally alter the aphid responses to the climate change variables. The relationship between the environmental variables in (b) are similar between the parasitoids and the aphids suggesting that both trophic levels respond to changes in these climate variables in a similar manner, i.e., this seems to be more of a trophic trickle than a trophic cascade. (Hoover, J.K. and Newman, J.A. [2004] *Global Change Biology*, 10, 1197–1208. With permission.)

Previous work by Roth and Lindroth (1995) and Stacey and Fellows (2002) suggested that the effects of elevated CO_2, mediated by the plant, would be damped at the third trophic level. However, it should be noted that these studies did not examine the complete *dynamics* of the tritrophic system. On the other hand, our model suggests that such damping does not necessarily happen and generally confirms the results of Bezemer et al. (1998) for the main effects of CO_2 and temperature. Importantly, we concluded that parasitoids did not significantly change the aphid responses to these aspects of climate change. That is, the effect is more a trickle than a cascade. The effects of climate change led to changes at all three trophic levels. Perhaps importantly, in the context of this book, this may be a case in which we can tentatively study the herbivore–plant interaction without worrying about multitrophic effects.

10.3.2 Interaction, Modification I: Grasses, Endophytic Fungi, and Herbivores

Many species of grass are sometimes found in apparent mutualistic associations with endophytic fungi. The interaction is particularly well known to agricultural scientists as endophytes are often present in two of the most important cool-season agricultural grasses worldwide, tall fescue (*Lolium arundinaceum*) and perennial ryegrass

(*Lolium perenne*). The endophytes that sometimes infect these two plants are closely related, *Neotyphodium coenophialum* and *N. lolii*. The endophytes are reported to be beneficial to their grass hosts by conferring protection for the mature plant, seedlings, and seeds against herbivory; conferring drought resistance; resistance to other abiotic factors (e.g., fire, cold); and increasing intra- and interspecific competitiveness (see Faeth and Bultman, 2002, for review and further references). It is not my intention to review the entire body of literature on endophytes. Interested readers should see Roberts et al. (2005).

Herbivore resistance is achieved through the production, *in planta*, of alkaloid compounds. Several classes of alkaloids can be produced via this fungal–plant interaction. Ergot alkaloids (ergovaline in tall fescue and lolitrem B in perennial ryegrass) are known to cause several health problems for grazing livestock. In tall fescue these problems are known collectively as fescue toxicosis. They involve such problems as fescue foot (necrosis of the flesh on the feet of cattle and sheep), thermoregulation problems, and problems during pregnancy. In perennial ryegrass, one often reported problem is known as ryegrass staggers, a neurological condition in sheep caused by lolitrem B. Alkaloids are also involved in invertebrate herbivore deterrence. In this case, a different class of alkaloids, mainly peramine and pyrrolizidine alkaloids (mainly lolines; see Hunt and Newman, 2005 for further references and discussion).

Although considerable work is still needed in the basic biochemistry of the endophyte–plant interaction, the effects of the endophytes' presence on insects is often remarkable. For example, aphid responses to the endophyte presence are sometimes so strong that it has been suggested that they might be used to *detect* endophyte infection (Latch et al., 1985). Table 10.1 provides a brief overview of documented effects on herbivores of the two major pasture grasses we have been considering. The table is not complete, but gives a good idea of the likely importance of endophyte infection for modifying plant–animal interactions. Note also that there is at least some evidence that this interaction produces effects further up the food chain (Bultman et al., 2003; Goldson et al., 2000).

So how does the presence of the endophyte mediate the plant–animal interaction in the context of climate change? Endophytes might alter the interaction between plants and invertebrates through three different mechanisms — changes in plant preference, changes in feeding pressure, and changes in insect mortality. Many insects are more strongly affected by plant quality than by plant quantity, so before considering animal responses directly, let us first examine how endophytes modify plant quality changes in response to aspects of climate change.

We know that endophytes can alter plant photosynthetic responses to temperature, drought, and soil fertility. Marks and Clay (1996) have shown that at temperatures above 35°C, endophyte-infected tall fescue plants photosynthesize faster than endophyte-free plants. Amalric et al. (1999) showed that perennial ryegrass infected with its endophyte photosynthesized at a higher rate under drought stress than endophyte-free plants. My colleagues and I (Newman et al., 2003a) have shown endophyte-infected plants may photosynthesize faster than endophyte-free plants under high soil N conditions. Endophyte-infected plants wilt less and recover more

TABLE 10.1
Insect Herbivores of Tall Fescue (TF) or Perennial Ryegrass (PRG) Infected with *Neotyphodium* spp. Endophytes

Insect	Feeding Deterrence		Reduced Survival		Lower Densities	
	TF	PRG	TF	PRG	TF	PRG
Coleoptera						
Argentine stem weevil (*Listronotus bonariensis*)	√	√	√	√	√	√
Black Beetle (*Heteronychus arator*)		√				√
Bluegrass billbug (*Sphenophorus parvulus*)			√	√		√
S. venatus			√	√		
S. inaequalis			√	√		
S. minimus			√	√		
Corn flea beetle (*Chaetocnema pulicaria*)					√	
Flour beetle (*Tribolium* sp.)					√	
Red flour beetle (*T. castaneum*)			√	√		
Japanese beetle (*Popillia japonica*)	√				√	
Grass grub (*Costelytra zealandica*)			u			
Hemiptera						
Hairy cinch bug (*Blissus leucopterus hirtus*)		√		√		
Sharpshooter leafhopper (*Draculacephala* spp.)					√	
Painted leafhopper (*Endria inimical*)					√	
Leafhopper (*Agallia constricta*)					√	
Graylawn leafhopper (*Exitianus exitiosus*)					√	
Back-faced leafhopper (*Graminella nigrifrons*)					√	
Two-lined spittlebug (*Prosapia bicincta*)					√	
Large milkweed bug (*Oncopeltus fasciatus*)					u	
Homoptera						
Pasture mealy bug (*Balanococcus poae*)						√
Bird cherry-oat aphid (*Rhopalosiphum padi*)	√		√			
Corn leaf aphid (*R. maidis*)		√				
Greenbug aphid (*Schizaphis graminum*)	√	√	√	√		
Russian wheat aphid (*Diuraphis noxia*)			√	√		
Yellow sugarcane aphid (*Sipha flava*)		√				
Lepidoptera						
Sod webworms (*Crambus* spp.)	√	√	√			√
Bluegrass webworm (*Parapediasia teterrella*)		√		√		
Fall armyworm (*Spodoptera frugiperda*)	√	√	√	√		
Southern armyworm (*S. eridania*)				√		
Common armyworm (*Mythimna convecta*)		√				
Cutworm (*Agrostis infusa*)		√				
Cutworm (*Graphania mutans*)		√		√		

Source: Adapted from Clement et al. (1994) and Prestidge and Ball (1997).

rapidly from severe drought conditions (Ravel et al., 1997, see also Cheplick et al., 2000), and these effects are likely to interact with temperature.

One commonly observed grass response to elevated CO_2 is a reduction in crude protein. This can be extremely important in livestock production, particularly among young animals and lactating females where protein demand is high. My colleagues and I have found that in tall fescue, endophyte-infected plants grown under low N conditions and high CO_2 do not suffer the reduction in crude protein seen in endophyte-free plants (Newman et al., 2003a). Idso and Idso (2003) have actually suggested that our result is a reason that livestock farmers should welcome climate change. This may be the case, but only if endophyte-infected plants also produce lower alkaloid concentrations in elevated CO_2 than in ambient CO_2. Others have suggested that this may be the case (Coley, 1998). Alkaloids are nitrogen-based compounds and, as nitrogen is more limiting in high CO_2, more may be devoted to plant growth leaving less available for alkaloid production. Surprisingly, there is little published data directly testing this hypothesis. Rufty et al. (1989) have shown that the primary pyridine alkaloids in tobacco decline with CO_2 enrichment. The question has yet to be addressed in tall fescue, but in perennial ryegrass the opposite seems to be true. Hunt et al. (2005) present data that suggests alkaloid concentrations decline with increasing N fertiliser at ambient CO_2, but stay constant across N fertiliser at elevated CO_2. Obviously it is still early to be answering this question, but it is a question that needs answering.

For many of the insects in Table 10.1, changes in crude protein and alkaloid concentrations are probably the most important aspects of plant quality, but for sap-feeding insects like aphids, the real indicator of quality is the amino acid composition in the phloem sap. Unfortunately, little is known about changes in phloem sap composition with rising CO_2, and nothing is know for grass–endophyte complexes. This remains an important open question.

Even if we do not know the mechanism, we can still examine the phenomenon of insect responses to plant–endophyte systems and climate change. There have been a few studies in this regard. Marks and Lincoln (1996) investigated the interacting effects of elevated CO_2 and endophyte infection on the relationship between tall fescue and the fall armyworm (*Spodoptera frugiperda*). They found that fall armyworms had the lowest efficiency of feed conversion and relative growth rates when feeding on endophyte-infected grass grown in elevated CO_2. Hunt (2003) has investigated the interaction between the bird cherry–oat aphid (*R. padi*) and both tall fescue-*N. coenophialum* and perennial ryegrass-*N. lolii* complexes in response to elevated CO_2. Because these experiments were conducted in the field using FACE arenas, they are statistically quite weak tests. Nevertheless, Hunt did show that for perennial ryegrass-*N. lolii*, *R. padi*'s preference for endophyte-free plants was stronger under elevated CO_2 and low to moderate nitrogen, but that this preference disappeared (or even reversed) under elevated CO_2 and high nitrogen.

The phenomenon of endophyte infection modifying the interaction between herbivores and plants has been well documented in current ambient conditions. While we have some information on the mechanisms behind these changes, our knowledge of the biochemistry of this interaction is far from complete. However, given the very strong phenomenological effects in current ambient conditions, and the effects of endophyte infection on plant vigour and chemical composition under conditions of drought and elevated CO_2, it seems likely that there will be further interaction modification, at least for some plant–herbivore interactions. Despite the logical potential for such alterations with climate change, there are as yet few studies investigating this possibility. This remains a fruitful area of research.

10.3.3 INTERACTION MODIFICATION II: ARBUSCULAR MYCORRHIZAL FUNGI, PLANTS, AND INSECTS

More than 90% of plants form associations with mycorrhizal fungi, and about two-thirds of these are arbuscular mycorrhizal fungi (AMF); Fitter and Moyersoen, 1996; Smith and Read, 1997). This includes many species of agricultural interest. The fungi are obligate symbionts, using carbon fixed by the plant while supplying phosphorus to its host plant. The host plant is not an obligate partner, and depending on the environmental conditions, the AMF–plant relationship can range from parasitic to symbiotic (Gange and Ayres, 1999). In addition to nutrient uptake, AMF may increase drought resistance (Davies et al., 1992), competitive abilities (West, 1997a), and resistance to fungal pathogens (West, 1997b).

So what is the likely impact of elevated CO_2 and global climate change on AMF–plant–insect interactions? The short answer is that we do not know; no one has yet looked at these interactions in this environmental context. We can, however, talk about the likely changes in the AMF–plant interaction and speculate a bit about what these changes might mean for the herbivorous insects. In a major review of the issue, Staddon et al. (2002) concluded that the effects of elevated CO_2 on AMF are dependent on the effects of CO_2 on the host plant. Many previous studies have argued that AMF colonisation increases with increased belowground carbon, which often results from elevated CO_2. However, Staddon and Fitter (1998, see also Fitter et al., 2000) have argued that this is probably not the case as previous studies have not controlled for the differential plant growth, and when such plant growth-independent measures are used, there is no effect on AMF colonisation (see Hartwig et al., 2002). These conclusions are based on single plant species colonised by single AMF species. In multispecies assemblies, the situation may be rather different; there may be competition among AMF species for increased belowground carbon or competition between AMF and other mycorrhizal species (notably ectomycorrhizas, e.g,. Godbold et al., 1997). Rilig et al. (1998) have shown that host plant responses to elevated CO_2 can alter the composition of mycorrhizal communities. Regardless of the ambiguous effect of elevated CO_2 on AMF colonisation, AMF colonisation has been shown to affect the plant's response to elevated CO_2. Hartwig et al. (2002) showed that the response of perennial ryegrass to elevated CO_2 was greater for plants colonised by AMF than for plants that remained uncolonised (see Gavito et al., 2000).

Brown and Gange (2002) point out, given the plethora of studies on the effects of AMF colonisation, it is perhaps surprising that little work has been done on the herbivore trophic level. Since AMF fundamentally alter critical carbon source–sink relationships, they have the potential to affect plant growth and plant quality in many different ways and have been shown to affect leaf nitrogen and some chemical defences (see above). For example, AMF can increase the photosynthetic rates of their host plant and that may alter the plant size and overall condition, potentially making it a more acceptable host to phytophagous insects. Alternatively, increased C might negatively affect herbivores. AMF-colonised plants have been shown to reduce growth rates in the Lepidopteran larvae: *Arctia caja* (Gange and West, 1994), *Spodoptera frugiperda*, and *Heliothis zea* (Rabin and Pacovsky, 1985). The mechanism to explain these decreased growth rates is unclear. AMF-colonised plants sometimes show increased C:N ratios in the leaves and this can lead to declines in larval growth rates. Gange and West (1994) found increased carbon-based allelochemicals (iridoid glycosides, aucubin, and catapol) in AMF-colonised *Plantago lanceolata*, but Rabin and Pacovsky (1985) found no chemical changes in plant quality of soybean (*Glycine max*). Negative effects of AMF colonisation on insects have also been found in the root-feeding black vine weevil (*Otiorhynchus sulcatus*, Gange et al., 1994). This negative interaction seems to be mediated by the particular combinations of AMF species (Gange, 2001). On the other hand, the Mexican bean beetle larvae (*Epilachna varivestis*) showed *increased* survival and pupal mass on soybean, but this effect was only evident at low levels of phosphorous (Borowicz, 1997). Positive effects of AMF have also been shown for the common blue butterfly larvae (*Polyommatus icarus*) feeding on *Lotus corniculatus* that had been colonized by single AMF species and by groups of AMF species (Goverde et al., 2000).

So, as far as folivorous insects are concerned, there is evidence for both positive and negative bottom–up effects of AMF on insect life histories and population dynamics. However, there is also evidence that top–down dynamics play a role in this interaction. Gange et al. (2002) found that AMF colonization of *P. lanceolata* was reduced in response to insect herbivory. Klironomos et al. (2004) demonstrate that simulated herbivory (clipping) can have positive or negative effects of AMF colonization, depending on the AMF species and the micorrhizal structures being assessed.

It is not clear how the reasonably well-studied changes in foliage quality, which result from AMF colonisation, translate into changes in sap quality (if at all). Borowicz (1997) suggested that sap feeders would be more affected by AMF as we could expect bigger chemical changes in the plant's sap than in the structural mass. Gange and West (1994) found that amino acid, soluble sugar, and total nitrogen were lower in AMF-colonised plants than in AMF-free plants. Sap-feeding insects seem to be either unaffected or positively affected by AMF colonisation. The greenbug aphid (*Schizaphis graminum*) feeding on sorghum was unaffected by AMF presence, but aphid weight, embryo content, and fecundity increased in the peach aphid (*Myzus persicae*) feeding on *P. lanceolata* (Gange and West, 1994; Gange et al., 1999). Whether these positive effects of AMF colonisation on aphid populations are caused by changes in the phloem sap quality, or by changes in the physical structure of the plant are unclear. Gange et al. (1999) suggest the possibility that aphids are better at finding the phloem in AMF-colonised plants.

So what is the scope for tritrophic interactions in this system? As we have seen, AMF can produce both positive and negative effects at the herbivore trophic level, and herbivores can produce both positive and negative effects at the AMF trophic level. That the literature seems to lack a common phenomenon in terms of the sign and direction (top–down vs. bottom–up) of this interaction is perhaps not surprising. The literature comprises studies that use different AMF species, different plant species, and different edaphic conditions. Until we develop a more complete mechanistic understanding of how these differences combine to produce changes in both plant growth and plant quality, we will be unable to generalize about subsequent tritrophic effects. Nevertheless, given that we *sometimes* observe strong indirect effects, both positive and negative, at the AMF and insect trophic levels, this system seems to be a good model for considering the *importance* of tritrophic dynamics under climate change. To clarify the interpretation of previous and future work on the AMF–plant responses to climate change, particularly work conducted in the field in either open-topped chambers or FACE arenas, where herbivory is often uncontrolled, it seems that we will need to develop an understanding of the multitrophic dynamics of these systems.

10.4 FUTURE RESEARCH DIRECTIONS

Many, if not all, FACE experiments are conducted on parts of, or whole, extant communities. Results from FACE experiments can be either a mixture of direct and indirect effects or just direct effects, and as such the interpretation of results from these experiments requires careful thought. If we are to develop the robust predictions of climate change impacts that society requires, we have to do more than 'kick it and see' experiments. That is, it is not sufficient to just raise the CO_2 concentration for a period of time and see what happens to the community. Such experiments are useful as early forms of exploratory research, but to make powerful long-term predictions, we must identify and understand the mechanisms that structure the particular agroecosystem of interest. We cannot hope to investigate all agroecosystems under all conditions. Our only hope is to identify the important mechanisms so that we may develop general theory.

Whittaker (1997) points out that we need two kinds of experiments: Those in which multitrophic interactions are a feature of the experimental design, and those that manipulate or eliminate the interaction in natural communities. In the former, carefully controlled factorial experimental designs are used to understand the detailed mechanisms of such interactions. In the latter, we can test whether mechanisms identified in the highly controlled manipulative experiments are either detectable or ecologically relevant in the field. Whittaker (1997) rightly stresses this last point, noting that just because we can demonstrate such interactions under controlled experimental conditions does not mean that they are either relevant or general.

If 'kick it and see' experiments are not sufficient, then what kinds of experiments are necessary? Wootton (1994) suggests that we must examine a variety of variables, preferably throughout the community or ecosystem. These experiments will be more valuable if they test specific mechanistic models of community structure and dynamics. They will also be more valuable if they estimate the interaction strength between species in the field. Interaction strength is not a topic I have discussed in this chapter, but it is a key component of many models in community ecology and refers to the per capita rate of change in the population of one species as a direct consequence of the per capita effects of another species. Two common experimental approaches in community ecology experiments on indirect effects have been termed *press* and *pulse* experiments (Bender et al., 1984). In press experiments, the density of one or more species is permanently changed while in pulse experiments the density of one or more species is altered and then the system is allowed to relax back to its premanipulation equilibrium. Bender et al. (1984) argue that press experiments include both direct and indirect effects, but that pulse experiments shed light only on direct effects. Although the usual discussion of these experimental designs is in the context of species additions and removals, we might also think of them in terms of environmental change that alters some species directly but not others. The length of time for which the environmental manipulation is maintained (relative to data collection and community dynamics) corresponds to press and pulse experiments. It seems to me that long-term manipulations of communities via FACE experiments are roughly equivalent to a press experiment (hence, difficult to interpret the results). Pulse experiments would correspond to shorter-term FACE experiments in which the CO_2 is stopped and the community is monitored as it returns to its ambient CO_2 conditions. To my knowledge, this type of experiment has not been done, even when FACE arenas have been taken out of operation. Such experiments might prove particularly valuable for comparing the strength of direct and indirect effects in agroecosystems.

10.5 CONCLUSIONS

We are still a long way from understanding whether knowledge of indirect multitrophic effects is important or not in predicting the impacts of climate change on our essential agroecosystems. There are many candidates for such indirect interactions in agroecosystems, from biocontrol to fungal symbionts. I have discussed just three of many possible such interactions. Even in these cases we know very little except that perhaps we should be looking for multitrophic indirect effects in these systems.

10.6 FURTHER READING

For more discussion of the very basics of multitrophic interactions, I recommend Peter Morin's (1999) excellent text on community ecology. For more depth and a plethora of further examples, there are two relatively recent edited volumes on multitrophic interactions: Tscharntke and Hawkins (2002) and Gange and Brown (1997). There is also a very good, but somewhat older, review in *Annual Reviews of Ecology and Systematics* by Tim Wootton (1994). In particular, this review has some good advice on methodology for conducting experiments to seek out indirect effects. For more on the historical debate between top–down and bottom–up dynamics, there was a special feature in volume 73 (number 3, 1992) of the journal *Ecology*.

ACKNOWLEDGMENTS

I am grateful to Ann Rypstra and Neo Martinez for valuable discussion of indirect effects in agroecosystems, John Klironomos for discussion of AMF interactions, Neil Rooney who kindly reviewed my revised version of this chapter, Paul Newton and Grant Edwards for editorial comments; and to an anonymous referee for his or her thorough review.

REFERENCES

Abrams, P.A. (1993) Effect of increased productivity on the abundances of trophic levels. *The American Naturalist,* 141, 351–371.

Amalric, C., Sallanon, H., Monnet, F., Hitmi, A., and Coudret, A. (1999) Gas exchange and chlorophyll fluorescence in symbiotic and non-symbiotic ryegrass under water stress. *Photosynthetica,* 37, 107–112.

Bender, E.A., Case, T.J., and Gilpin, M.E. (1984) Perturbation experiments in community ecology: theory and practice. *Ecology,* 65, 1–13.

Bezemer, T.M. and Jones, T.H. (1998) Plant-insect herbivore interactions in elevated atmospheric CO_2: Quantitative analyses and guild effects. *Oikos,* 82, 212–222.

Bezemer, T.M., Jones, T.H., and Knight, K.J. (1998) Long-term effects of elevated CO_2 and temperature on populations of the peach potato aphid *Myzus persicae* and its parasitoid *Aphidius matricariae. Oecologia,* 116, 128–135.

Bezemer, T.M., Knight, K.J., Newington, J.E., and Jones, T.H. (1999) How general are aphid responses to elevated CO_2? *Annals of the Entomological Society of America,* 92, 724–730.

Borowicz, V.A. (1997) A fungal root symbiont modifies plant resistance to an insect herbivore. *Oecologia,* 112, 534–542.

Brown, V.K. and Gange, A.C. (2002) Tritrophic below- and above-ground interactions in succession. In *Multitrophic Level Interactions,* Tscharntke, T. and Hawkins, B.A., Eds., Cambridge University Press, Cambridge, U.K., 197–222.

Bultman, T.L., McNeill, M.R., and Goldson, S.L. (2003) Isolate-dependent impacts of fungal endophytes in a multitrophic interaction. *Oikos,* 102, 491–496.

Cheplick, G.P., Perera, A., and Koulouris, K. (2000) Effect of drought on the growth of *Lolium perenne* genotypes with and without fungal endophytes. *Functional Ecology,* 14, 657–667.

Clement, S.L., Kaiser, W.J., and Eichenseer, H. (1994) Acremonium endophytes in germplasms of major grasses and their utilisation for insect resistance. In *Biotechnology of Endophytic Fungi of Grasses,* Bacon, C.W. and White, J.F., Eds., CRC Press, Boca Raton, FL, 185–199.

Coley, P.D. (1998) Possible effects of climate change on plant-herbivore interactions in moist tropical forests. *Climate Change,* 39, 455–472.

Coviella, C.E. and Trumble, J.T. (1999) Effects of elevated atmospheric carbon dioxide on insect-plant interactions. *Conservation Biology,* 13, 700–712.

Davies, F.T., Potter, J.R., and Linderman, R.G. (1992) Mychorriza and repeated drought exposure affect drought resistance and extra radical hyphae development on pepper plants independent of plant size and nutrient content. *Journal of Plant Physiology,* 139, 289–294.

Duffy, J.E. (2002) Biodiversity and ecosystem function: the consumer connection. *Oikos*, 99, 201–219.

Emmerson, M., Bezemer, T.M., Hunter, M.D., Jones, T.H., Masters, G.J., and van Dam, M.M. (2004) How does global change affect the strength of trophic interactions? *Basic and Applied Ecology*, 5, 505–514.

Faeth, S.H. and Bultman, T.L. (2002) Endophytic fungi and interactions among host plants, herbivores, and natural enemies. In *Multitrophic Level Interactions*, Tscharntke, T. and Hawkins, B.A., Eds., Cambridge University Press, Cambridge, U.K., 89–123.

Fitter, A.H., Heinemeyer, A., and Staddon, P.L. (2000) The impact of elevated CO_2 and global climate change on arbuscular mycorrhizas: a mycocentric approach. *New Phytologist*, 147, 179–187.

Fitter, A.H. and Moyersoen, B. (1996) Evolutionary trends in root-microbe symbioses. *Philosophical Transactions of the Royal Society of London B*, 351, 1367–1375.

Gange, A.C. (2001) Species-specific responses of a root- and shoot-feeding insect to arbuscular colonization of its host plant. *New Phytologist*, 150, 615–618.

Gange, A.C. and Ayres, R.L. (1999) On the relationship between mycorrhizal colonization and plant "benefit." *Oikos*, 87, 615–621.

Gange, A.C., Bower, E., and Brown, V.K. (1999) Positive effects of an arbuscular mycorrhizal fungus on aphid life history. *Oecologia*, 120, 123–131.

Gange, A.C., Bower, E., and Brown, V.K. (2002) Differential effects of insect herbivory on arbuscular mycorrhizal colonization. *Oecologia*, 131, 103–112.

Gange, A.C. and Brown, V.K., Eds. (1997) *Multitrophic Interactions in Terrestrial Systems*, Blackwell Science, London.

Gange, A.C., Brown, V.K., and Sinclair, G.S. (1994) Reduction of black vine weevil larval growth by vesicular-arbuscular mycorrhizal infection. *Entomologia Experimentalis et Applicata*, 70, 115–119.

Gange, A.C. and West, H.M. (1994) Interactions between arbuscular-mycorrhizal fungi and foliar-feeding insects in *Plantago lancelolata* L. *New Phytologist*, 128, 79–87.

Gavito, M.E., Curtis, P.S., Mikklelsen, T.N., and Jakobsen, I. (2000) Atmospheric CO_2 and mycorrhiza effects on biomass allocation and nutrient uptake of nodulated pea (*Pisum sativum* L.) plants. *Journal of Experimental Botany*, 51, 1931–1938.

Godbold, D.L., Berntson, G.M., and Bazzaz, F.A. (1997) Growth and mycorrhizal colonization of three North American tree species under elevated CO_2. *New Phytologist*, 137, 433–440.

Goldberg, D.E. and Barton, A.M. (1992) Patterns and consequences of interspecific competition in natural communities: a review of field experiments with plants. *The American Naturalist*, 139, 771–801.

Goldson, S.L., Proffitt, J.R., Fletcher, L.R., and Baird, D.B. (2000) Multitrophic interaction between the ryegrass *Lolium perenne*, its endophyte *Neotyphodium lolii*, the weevil pest *Listronotus bonariensis*, and its parasitoid *Microctonus hyperodae*. *New Zealand Journal of Agricultural Research*, 43, 227–233.

Goverde, M., van der Heijden, M.G.A., Wiemken, A., Sanders, I.R., and Erhardt, A. (2000) Arbuscular mychorrhizal fungi influence life history traits of a lepidopteran herbivore. *Oecologia*, 125, 362–369.

Gurney, W.S.C. and Nisbet, R.M. (1998) *Ecological Dynamics*, Oxford University Press, Oxford, U.K.

Hairston, N.G., Smith, F.E., and Slobodkin, L.B. (1960) Community structure, population control, and competition. *American Naturalist*, 94, 421–425.

Halaj, J. and Wise, D.H. (2001) Terrestrial trophic cascades: how much do they trickle? *American Naturalist*, 157, 262–281.

Hartwig, U.A., Wittman, P., Braun, R., et al. (2002) Arbuscular mycorrhiza infection enhances the growth response of *Lolium perenne* to elevated atmospheric pCO_2. *Journal of Experimental Botany,* 53, 1207–1213.

Hay, M.E. (1986) Associational plant defenses and the maintenance of species diversity: turning competitors into accomplices. *American Naturalist,* 128, 617–641.

Hjalten, J., Kjell, D., and Lundberg, P. (1993) Herbivore avoidance by association: vole and hare utilization of woody plants. *Oikos,* 68, 125–131.

Hjalten, J. and Price, P.W. (1997) Can plants gain protection from herbivory by association with unpalatable neighbours? A field experiment in a willow-sawfly system. *Oikos,* 78, 317–322.

Holmes, R.D. and Jepson-Innes, K. (1989) A neighborhood analysis of herbivory in Bouteloua gracilis. *Ecology,* 70, 971–976.

Holt, R.D. (1977) Predation, apparent competition, and the structure of prey communities. *Theoretical Population Biology,* 12, 197–229.

Hoover, J.K. and Newman, J.A. (2004) Tritrophic interactions in the context of climate change: a model of grasses, cereal aphids and their parasitoids. *Global Change Biology,* 10, 1197–1208.

Hunt, M.G. and Newman, J.A. (2005) Reduced herbivore resistance from a novel grass-endophyte association. *Journal of Applied Ecology,* 42, 762–769.

Hunt, M.G. (2003) *Effects of Environmental Change on Endophyte-Plant-Insect Relationships*. Dissertation. Department of Zoology. University of Oxford, Oxford, U.K.

Hunt, M.G., Rasmussen, S., Newton, P.C.D., Parsons, A.J., and Newman, J.A. (2005) Near-term impacts of elevated CO_2, nitrogen and fungal endophyte-infection on perennial ryegrass: growth, chemical composition and alkaloid production. *Plant, Cell and Environment,* 28, 1345–1354.

Idso, S. and Idso, C. (2003) Effects of CO_2 on forage quality. *CO_2 Science Magazine.* Center for the Study of Carbon Dioxide and Global Change http://www.co2science.org/scripts/CO2ScienceB2C/Index.jsp, accessed November 11, 2005.

Karban, R., Hougen-Eitzmann, D., and English-Loeb, G. (1994) Predator-mediated apparent competition between two herbivores that feed on grapevines. *Oecologia,* 97, 508–511.

Kindlmann, P. and Dixon, A.F.G. (1999) Generation time ratios–determinants of prey abundance in insect predator-prey interactions.*Biological Control,* 16, 133–138.

Kingland, S.E. (1995) *Modeling Nature: Episodes in the History of Population Ecology,* University of Chicago Press, Chicago.

Klironomos, J.N., McCune, J., and Moutoglis, P. (2004) Species of arbuscular mycorrhizal fungi affect mycorrhizal response to simulated herbivory. *Applied Soil Ecology,* 26, 133–141.

Kot, M. (2001) *Elements of Mathematical Ecology,* Cambridge University Press, Cambridge, U.K.

Latch, G.C.M., Christensen, M.J., and Gaynor, D.L. (1985) Aphid detection of endophyte infection in tall fescue. *New Zealand Journal of Agricultural Research,* 28, 129–132.

Lawton, J.H. (2000) *Community Ecology in a Changing World,* Ecology Institute, Luhe, Germany.

Marks, S. and Clay, K. (1996) Physiological responses of *Festuca arundinacea* to fungal endophyte infection. *New Phytologist,* 133, 727–733.

Marks, S. and Lincoln, D.E. (1996) Antiherbivore defense mutualism under elevated carbon dioxide levels: a fungal endophyte and grass. *Environmental Entomology,* 25, 618–623.

Marquis, R.J. and Whelan, C.J. (1994) Insectivorous birds increase growth of white oak through consumption of leaf-chewing insects. *Ecology,* 75, 2007–2014.

McCann, K.S., Hastings, A., and Strong, D.R. (1998) Trophic cascades and trophic trickles in pelagic food webs. *Proceedings of the Royal Society, London, B,* 265, 205–209.

Morin, P.J. (1999) *Community Ecology,* Blackwell Science, Oxford, U.K.

Morris, R.J., Lewis, O.T., and Godfray, H.C.J. (2004) Experimental evidence for apparent competition in a tropical forest food web. *Nature,* 428, 310–313.

Morris, R.J., Müller, C.B., and Godfray, H.C.J. (2001) Field experiments testing for apparent competition between primary parasitoids mediated by secondary parasitoids. *Journal of Animal Ecology,* 70, 301–309.

Müller, C.B. and Godfray, H.C.J. (1997) Apparent competition between two aphid species. *Journal of Animal Ecology,* 66, 57–64.

Newman, J.A. (2004) Climate change and cereal aphids: the relative effects of increasing CO_2 and temperature on aphid population dynamics. *Global Change Biology,* 10, 5–15.

Newman, J.A., Abner, M.L., Dado, R.G., Gibson, D.J., Brookings, A., and Parsons, A.J. (2003a) Effects of elevated CO_2, nitrogen and fungal endophyte-infection on tall fescue: growth, photosynthesis, chemical composition and digestibility. *Global Change Biology,* 9, 425–437.

Newman, J.A., Gibson, D.J., Parsons, A.J., and Thornley, J.H.M. (2003b) How predictable are aphid population responses to elevated CO_2? *Journal of Animal Ecology,* 72, 556–566.

Paine, R.T. (1980) Food webs: linkage, interaction strength and community infrastructure. *Journal of Animal Ecology,* 49, 667–685.

Polis, G.A. and Strong, D.R. (1996) Food web complexity and community dynamics. *The American Naturalist,* 147, 813–846.

Prestidge, R.A. and Ball, O.J.P. (1997) A Catch 22: the utilization of endophytic fungi for pest management. In *Multitrophic Interactions in Terrestrial Systems,* Gange, N. and Brown, V.K., Eds., Blackwell Scientific Press, Oxford, U.K.

Rabin, L.B. and Pacovsky, R.S. (1985) Reduced larva growth of two Lepidotera (Noctuidae) on excised leaves of soybean infected with a mycorrhizal fungus. *Journal of Economic Entomology,* 78, 1358–1363.

Ravel, C., Courty, C., Coudret, A., and Charmet, G. (1997) Beneficial effects of *Neotyphodium lolii* on the growth and the water status in perennial ryegrass cultivated under nitrogen deficiency or drought stress. *Agronomie Paris,* 17, 173–181.

Rillig, M.C., Allen, M.F., Klironomos, J.N., Chiariello, N.R., and Field, C.B. (1998) Plant species-specific changes in root-inhabiting fungi in a California annual grassland: responses to elevated CO_2 and nutrients. *Oecologia,* 113, 252–259.

Roberts, C.A., West, C.P., and Spiers, D.E. (2005) *Neotyphodium in Cool-Season Grasses.* Blackwell Scientific Press, Oxford, U.K.

Rosenzweig, M.L. (1971) Paradox of enrichment: destabilization of exploitation ecosystems in ecological time. *Science,* 171, 385–387.

Rosenzweig, M.L. and McArthur, R.H. (1963) Graphical representation and stability conditions of predator-prey interactions. *American Naturalist.* 97, 209–223.

Roth, S.K. and Lindroth, R.L. (1995) Elevated atmospheric CO_2 effects on phytochemistry, insect performance and insect parasitoid interactions. *Global Change Biology,* 1, 173–182.

Rousset, O. and Lepart, J. (2002) Neighbourhood effects on the risk of an unpalatable plant being grazed. *Plant Ecology,* 165, 197–206.

Rufty, T.W., Jackson, M.D., Severson, R.F., Lamm, J.J., and Snook, M.E. (1989) Alterations in growth and chemical constituents of tobacco in response to CO_2 enrichment. *Journal of Agricultural Food Chemicals,* 37, 552–555.

Schmitz, O.J., Hamback, P.A., and Beckerman, A.P. (2000) Trophic cascades in terrestrial systems: a review of the effects of carnivore removals on plants. *American Naturalist,* 155, 141–153.

Settle, W.H. and Wilson, L.T. (1990) Invasion by the variegated leafhopper and biotic interactions: parasitism, competition, and apparent competition. *Ecology,* 71, 1461–1470.

Smith, S.E. and Read, D.J. (1997) *Mycorrhizal Symbiosis,* Academic Press, London, U.K.

Stacey, D.A. and Fellowes, M.D.E. (2002) Influence of elevated CO_2 on interspecific interactions at higher trophic levels. *Global Change Biology,* 8, 668–678.

Staddon, P.L. and Fitter, A.H. (1998) Does elevated atmospheric carbon dioxide affect arbuscular mycorrhizas? *Trends in Ecology & Evolution,* 13, 455–458.

Staddon, P.L., Heinmeyer, A., and Fitter, A.H. (2002) Mycorrhizas and global environment change: research at different scales. *Plant and Soil,* 244, 253–261.

Stiling, P., Rossi, A.M., Hungate, B., Dijkstra, P., and Hinkle, C.R. (1999) Decreased leafminer abundance in elevated CO_2: reduced leaf quality and increased parasitoid attack. *Ecological Applications,* 9, 240–244.

Strong, D.R. (1992) Are trophic cascades all wet? Differentiation and donar-control in speciose ecosystems. *Ecology,* 73, 747–754.

Thompson, D.B., Brown, H.H., and Spencer, W.D. (1991) Indirect facilitation of granivorous birds by desert rodents: experimental evidence from foraging patterns. *Ecology,* 72, 852–863.

Tscharntke, T. and Hawkins, B.A. (2002) *Multitrophic Level Interactions,* Cambridge University Press, Cambridge, U.K.

Vandermeer, J. (1980) Indirect mutualism: variations on a theme by Stephen Levine. *American Naturalist,* 116, 441–448.

Watt, A.D. and McFarlane, A.M. (2002) Will climate change have a different impact on different trophic levels? Phenological development of winter moth *Operophtera brumata* and its host plants. *Ecological Entomology,* 27, 254–256.

West, H.M. (1997a) Influence of arbuscular mychorrizal infection on competition between *Holcus lanatus* and *Dactylis glomerata. Journal of Ecology,* 84, 429–438.

West, H.M. (1997b) Interactions between arbuscular mycorrhizal fungi and foliar pathogens: consequences for host and pathogen. In *Multitrophic Interactions in Terrestrial Systems,* Gange, A.C. and Brown V.K., Eds., Blackwell Science, Oxford, U.K.

Whittaker, J.B. (1997) Concluding remarks. In *Multitrophic Interactions in Terrestrial Systems,* Gange, A.C. and Brown, V.K., Eds., Blackwell Science, Oxford, U.K.

Whittaker, J.B. (1999) Impacts and responses at population level of herbivorous insects to elevated CO_2. *European Journal of Entomology,* 96, 149–156.

Whittaker, J.B. (2001) Presidential address: insects and plants in a changing atmosphere. *Journal of Ecology,* 89, 507–518.

Wootton, J.T. (1994) The nature and consequences of indirect effects in ecological communities. *Annual Review of Ecology and Systematics,* 25, 443–466.

Wootton, J.T. and Power, M.E. (1993) Productivity, consumers, and the structure of a river food chain. *Proceedings of the National Academy of Science, USA,* 90, 1384–1387.

11 Future Weed, Pest, and Disease Problems for Plants

Lewis H. Ziska and G. Brett Runion

CONTENTS

11.1 INTRODUCTION

Recently, the global human population surpassed 6 billion. Increasing populations require increasing resources, particularly with respect to energy and food. As the global demand for power and agricultural land intensifies, fossil fuel burning and deforestation will continue to be human-derived sources of atmospheric carbon dioxide. Since the mid-1950s, records of carbon dioxide concentration $[CO_2]$ obtained from the Mauna Loa observatory in Hawaii have shown an increase of about 20% from 311 to 375 parts per million (ppm) (Keeling and Whorf 2001). The current rate of $[CO_2]$ increase (~0.5%) is expected to continue with concentrations exceeding 600 ppm by the end of the 21st century (Schimel et al. 1996). Interestingly, because the observatory at Mauna Loa and other global monitoring sites (cdiac.esd.ornal.gov/home.html) sample air at high elevations, away from anthropogenic sources, actual ground-level $[CO_2]$ may be significantly higher. For example, urban areas in Phoenix and Baltimore show $[CO_2]$ values exceeding 500 ppm, and suburban values near Washington, D.C. and Sydney, Australia, report $[CO_2]$ above 420 ppm (Idso et al. 1998, 2001; Ziska et al. 2000). This suggests that while the Mauna Loa data may reflect $[CO_2]$ for the globe as a whole, regional increases in $[CO_2]$ may already be occurring as a result of urbanization.

Overall, the documented increases in atmospheric $[CO_2]$ are likely to change the biology of agricultural weeds, insects, and diseases in two fundamental ways. The first is related to climate stability. The observed change in atmospheric $[CO_2]$ has been accompanied by documented increases in other radiation-trapping gases such as methane (CH_4) (0.9% increase per year), nitrous oxide (N_2O) (0.25% per year), and chlorofluorocarbons (CFCs) (4% per year). Recent evaluations by the Intergovernmental Panel on Climate Change (IPCC) based, in part, on an assessment by the U.S. National Academy of Sciences has indicated that the rise of $[CO_2]$ and associated *greenhouse gases* could lead to a 3 to 12°C increase in global surface temperatures, with subsequent consequences on precipitation frequency and amounts.

The second likely impact is the $[CO_2]$ *fertilization effect*. That is, plants evolved at a time when the atmospheric $[CO_2]$ appears to have been four or five times present values (Bowes 1996). Because CO_2 remains the sole source of carbon for plant photosynthesis (and, hence, 99% of all living terrestrial organisms), and because at present $[CO_2]$ is less than optimal, as atmospheric $[CO_2]$ increases, photosynthesis will be stimulated accordingly. Elevating $[CO_2]$ stimulates net photosynthesis in plants with the C_3 photosynthetic pathway by raising the CO_2 concentration gradient from air to leaf and by reducing the loss of CO_2 through photorespiration. Specifically, because oxygen competes with CO_2 for active sites of the enzyme, ribulose-bisphosphate carboxylase/oxygenase (Rubisco, the principle enzyme that incorporates carbon into the plant), elevating the concentration of CO_2 (relative to O_2) increases net carbon uptake by stimulating photosynthesis and reducing CO_2 lost via photorespiration. Because the competition between O_2 and CO_2 for active sites is temperature sensitive (favoring O_2 with increasing temperature), the stimulation of net photosynthesis by elevated CO_2 should increase as the temperature increases. While some studies have suggested that the photosynthetic response to $[CO_2]$ is limited by nutrients (Diaz et al. 1993), this seems less likely in managed agroeco-systems where nitrogen, phosphorus, and so forth are likely to be optimal for crop growth. Thus, for plants that rely solely on the C_3 photosynthetic pathway, (~95% of all plant species), increasing $[CO_2]$ and temperatures associated with climate change should be favorable for increased growth.

Alternatively, plants with the C_4 photosynthetic pathway (~4% of all known plant species) have an internal mechanism for concentrating CO_2 around Rubisco; therefore, increases in external CO_2 concentration should have little effect on net photosynthesis in C_4 plants (for reviews, see Bowes 1996; Ghannoum et al. 2000, Ghannoum et al., Chapter 3, this volume). However, one of the most consistent responses of both C_3 and C_4 species to elevated $[CO_2]$ is a decrease in stomatal conductance (Morison 1985; Eamus 1991). The decrease in stomatal conductance can result in significant increases in leaf transpiration efficiency (CO_2 assimi-lated/H_2O transpired) or water use efficiency (dry matter obtained/H_2O transpired). Hence, under water-limiting conditions, elevated CO_2 should result in significant increases in photosynthesis and biomass for both C_3 and C_4 plant species.

At the whole plant level, the $[CO_2]$-induced stimulation of photosynthesis is associated with an decrease in Rubisco investment and an increase in the ratio of C:N (Bowes 1996). Enhanced $[CO_2]$ also can result in increased growth, leaf area, tillering, and total biomass as well as allometric shifts between plant organs (Kimball et al. 1993, Poorter 1993). In addition, enhanced $[CO_2]$ can alter germination (Heichel and Jaynes 1974; Ziska and Bunce 1993), flowering times (Reekie et al. 1997), pollen output (Ziska and Caulfield 2000), seed yield (Allen et al. 1987), and the onset of senescence (Sicher 1998).

For weeds, insects, and diseases, what are the consequences of a direct CO_2 fertilization effect, and how are these consequences likely to be altered by concurrent changes in temperature or precipitation? Are these changes likely to increase or decrease crop production in agroecosystems? Surprisingly, while the detrimental effects of the above pests are well recognized, most research to date has focused on how individual crop species will respond to $[CO_2]$ and climate (for reviews, see Kimball et al. 1993; Poorter 1993; Curtis and Wang 1998). This may be due, in part, to the complex nature of ecological systems (even managed ones), and the challenge of implementing ecologically relevant experiments that address spatial and temporal interactions between organisms. Yet, not understanding these complex interactions and deriving suitable responses to avoid or mitigate their resulting impacts will certainly have critical and potentially damaging consequences with respect to crop productivity and global food security.

To that end, we are beginning to address the probable impacts of climate change and atmospheric $[CO_2]$ on pest biology. These impacts are multifaceted and include the direct effects of CO_2 and concomitant changes in climate on weed growth and weed–crop competition; secondary CO_2-induced effects on crop hosts (digestibility, chemical defenses, canopy microclimate), which may affect insect fecundity and pathogen success, and temperature and precipitation changes that directly alter where and when pathogen or insect outbreaks occur. While providing a tentative mecha-nistic basis for both the direct and indirect consequences of $[CO_2]$ and climate, we also hope to determine, in part, how current pest management efforts might fare in a future climate. We recognize that given the scarcity of available data, any review is likely to be problematic; however, our overall goal is to begin a synthesis of what is known and to derive a preliminary set of key climatic questions to address in the context of pest biology and agroecosystems.

11.2 RISING [CO_2] AND WEED BIOLOGY

11.2.1 CO_2 FERTILIZATION

Weeds are an anthropogenic classification given to plants that are generally recognized as objectionable or undesirable to human activities. However, there are biological similarities among such plants, including colonization of disturbed environments, vigorous growth, prodigious seed production, and seed longevity (see Baker 1974). Historically, one of the earliest accepted classifications of weeds was recognition of those plant species that interfered with crop production. Human selection of agronomically desirable species has led to inadvertent selection for other undesirable plant species that mimic the biology of the crop (e.g., commercial and wild oat, sorghum, and Johnson grass). Therefore, while numerous studies have shown that crop species will demonstrate enhanced photosynthesis, growth, and yield with increasing atmospheric [CO_2] (Kimball 1983; Kimball et al. 1993; Poorter 1993; Curtis and Wang 1998), similar benefits are likely for weedy competitors as well. David Patterson, a weed specialist at North Carolina State University has classified the relative responses of a range of crops and weeds to a projected increase of ~300 ppm [CO_2] for plants with the C_3 and C_4 type of photosynthesis. He found that C_3 and C_4 crops showed a range of responses from 1.10 to 2.34x and 0.98 to 1.24x, while C_3 and C_4 weeds showed a range of response from 0.95 to 2.72 and 0.7 to 1.61x, respectively (Patterson and Flint 1990; Patterson 1993; Patterson et al. 1999). Recent data on the specific response of noxious weeds to recent increases in atmospheric [CO_2] during the 20th century (284 to 380 ppm) shows a much stronger response: 1.77 to 2.78x relative to 1.15 to 1.55 in other plants (Sage 1995; Ziska 2003a). Overall, the greater range of responses observed for weeds with increasing atmospheric [CO_2] is consistent with the suggestion of Treharne (1989), that weeds have a greater genetic diversity and, hence, physiological plasticity, relative to crop species.

11.2.2 CO_2 FERTILIZATION AND CLIMATIC INTERACTIONS

Because increasing temperatures result in greater photorespiratory carbon loss, the impact of increasing [CO_2] on net carbon uptake should increase with increasing temperature (see Long 1991 and earlier discussion); however, this is not always observed. For example, increasing day and night temperatures with a doubling of [CO_2] can either decrease leaf area and biomass (Ackerly et al. 1992, Coleman and Bazzaz 1992) or have no effect (Tremmel and Patterson 1993) in velvetleaf (*Abutilon theophrasti*), a common agronomic weed. Similarly, CO_2 enrichment and temperature did not interact for two C_4 weed species (*Echinochloa crus-galli* and *Eleusine indica*) (Potvin and Strain 1985). Alternatively, spurred anoda (*Anoda cristata*) biomass increased at 700 ppm when day and night temperatures increased from 26/17 to 32/23°C (Patterson et al. 1988). At present, there is little unequivocal evidence for significant differences in response to [CO_2] with increasing temperatures. Overall, theoretical limitations based on biochemical models have generally assumed no growth temperature effects on carboxylation kinetics (V_{Cmax}) and no limitation on the potential rate of electron transport (J_{max}) (Long 1991); however, more recent studies indicate that long-term adaptation to growth temperature may adjust both parameters, lowering the temperature sensitivity of CO_2-induced photosynthetic stimulation (Bunce 2000; Ziska 2001a).

Potential increases in global temperature may be accompanied by changes in the pattern and amount of precipitation. However, because of the indirect effect of CO_2 on stomatal aperture, elevated CO_2 can still stimulate plant photosynthesis and growth even if water is limiting (Patterson 1986; Chaudhury et al. 1990). For some C_4 weeds, increased photosynthesis and growth at elevated [CO_2] may only occur under dry conditions, presumably due to increased water use efficiency (WUE, the ratio of carbon gained to water lost) and reduced water stress. Although water shortages should not limit the response to elevated [CO_2], no assessment on CO_2 response under flooded conditions is available for weedy species.

11.3 RISING [CO_2] AND WEED–CROP COMPETITION

11.3.1 CO_2 FERTILIZATION

Any resource that affects the growth of an individual alters its capacity to compete with individuals of the same or different species (Patterson and Flint 1990). Consequently, induced changes in competition can be associated not only with limited resources, but with resource enhancement as well. For example, in weed–crop competition it was thought that the addition of nitrogen would reduce crop losses due to weeds by increasing the availability of a resource (Vengris et al. 1955). However, because weeds utilize nitrogen more efficiently than crops, weed competition was actually favored when additional nitrogen was applied (Appleby et al. 1976; Carlson and Hill 1985). Analogous to the nitrogen response, differential responses of weeds relative to crops are likely as atmospheric [CO_2] increases.

If differential responses to increasing [CO_2] occur between crops and weeds, will crop losses due to weedy competition increase or decrease? Early, subjective classification of "worst" weeds by Holm et al. (1977) indicated that a majority (14 out of 18) of the world's worst weeds are C_4, whereas of the 86 crop species that make up 95% of the world's food supply, only five are C_4 (Patterson 1995a). As a consequence of this observation, many initial experiments analyzed C_3 crop–C_4 weed competition (Alberto et al. 1996; Bunce 1995; Carter and Peterson 1983; Patterson et al. 1984). These studies uniformly reported that increasing [CO_2] resulted in a greater ratio of crop-to-weed vegetative biomass (C_3:C_4), which is consistent with the known carboxylation kinetics of the C_3 and C_4 pathways. Hence, some global models have suggested less crop loss due to weedy competition as atmospheric [CO_2] increases (e.g., Rosenzweig and Hillel 1998).

However, the general perspective that weeds are C_4 and crops C_3 is somewhat misleading. For example, it can also be stated that 4 of the top 10 producing crops globally are C_4 (Maize, *Zea mays*; Millet, *Sorghum proviso*; Sorghum, *Sorghum bicolor*; and Sugarcane, *Saccharum officinarum*), and that of the 33 most invasive weeds globally (which can certainly be considered among the worst weeds categorically), only two are C_4 (*Spartina anglica* and *Imperata cylindrica* L. Beauv) (www.issg.org/database). In reality, crop–weed competition varies significantly by region; consequently, depending on temperature, precipitation, soil, and so forth, C_3 and C_4 crops will interact with C_3 and C_4 weeds (Bridges 1992). In addition, a C_3 crop vs. C_4 weed interpretation does not address weed–crop interactions where the photosynthetic pathway is the same (e.g., Bunce 1995; Potvin and Vasseur 1997). Yet many of the worst, most troublesome weeds for a given crop are genetically similar and frequently possess the same photosynthetic pathway (e.g., sorghum and Johnson grass [*Sorghum halapense*], both C_4; oat and wild oat, both C_3).

Even within crops and weeds of the same photosynthetic pathway, it is unclear how CO_2-induced variations in reproduction could alter competitive outcomes or persistence within the seed bank. Reproduction is often increased in response to rising CO_2 as additional carbon is allocated both to flowers and to increased nodes and branches (see Ward and Strain 1999 for a review). In common ragweed (*Ambrosia artemisiifolia*), time to reproduction was shortened, in part, by faster growth rates (Ziska et al. 2003), although for other species, elevated CO_2 may alter the size at which plants initiate reproduction (Reekie and Bazzaz 1991). However, no clear distinction between flowering times between weeds and crops is evident in response to [CO_2]. Furthermore, while seed yield is easy to determine in cultivated crops, it is difficult to assess seed production, particularly *in situ*, for weeds given that seed shattering (e.g., lambsquarters, *Chenopodium album*) is endemic to such species. Consequently, meta-analyses comparing the reproductive output of crop and "wild" species may only include a handful of common agronomic weeds (e.g., velvetleaf, sicklepod [*Cassia obtusifolia*], barnyard grass [*Echinochloa glabrescens*], and ladysthumb [*Polygonum persicaria*], see Appendix 1, Jablonski et al. 2002). In addition, comparisons of weed–crop reproductive success do not consider many invasive weeds, such as Canada thistle (*Cirsium arvense*), which may reproduce asexually from belowground structures, which, in turn, may be particularly sensitive to rising [CO_2] (Ziska 2003a).

Overall, data regarding vegetative or reproductive competition between crops and weeds as a function of increasing [CO_2] remain scarce. The studies that are available fall into two general categories, one where crops and weeds have the same photosynthetic pathway, and another where the pathway differs. The majority of studies involving different photosynthetic pathways have focused on a C_3 crop in competition with a C_4 weed (Table 11.1). In this comparison, increasing CO_2 increased the crop-to-weed biomass ratio, consistent with the known biochemical response. However, it is interesting to point out that biomass and yield of grain sorghum (C_4 crop) was reduced when grown in the presence of either velvetleaf or cocklebur (*Xanthium strumarium*), both C_3 weeds (Ziska 2001b, 2003b). Most comparisons with the same photosynthetic pathway for crops and weeds resulted in significant increases in weed-to-crop biomass when weed and crop emerged simultaneously (Table 11.1). In a study comparing

TABLE 11.1
Summary of Studies Examining Whether Weed or Crops Were Favored as a Function of Elevated [CO_2]

Crop	Weed	Increasing [CO_2] Favors?	Environment	Reference
A. C_3 Crops/C_3 Weeds				
Soybean	*Chenopodium album*	Weed	Field	Ziska 2000
Lucerne	*Taraxacum officinale*	Weed	Field	Bunce 1995
Pasture	*Taraxacum and Plantago*	Weed	Field	Potvin & Vasseur 1997
Pasture	*Plantago lanceolata*	Weed	Chamber	Woodward 1988
Sugarbeet	*Chenopodium album*	Crop*	Chamber	Houghton & Thomas 1996
B. C_4 Crops/C_4 Weeds				
Sorghum	*Amaranthus retroflexus*	Weed	Field	Ziska 2003b
C. C_3 Crops/C_4 Weeds				
Fescue	*Sorghum halapense*	Crop	Glasshouse	Carter & Peterson 1983
Soybean	*Sorghum halapense*	Crop	Chamber	Patterson et al. 1984
Rice	*Echinochloa glabrescens*	Crop	Glasshouse	Alberto et al. 1996
Pasture	*Paspalum dilatatum*	Crop	Chamber	Newton et al. 1996
Lucerne	Various grasses	Crop	Field	Bunce 1995
Soybean	*A. retroflexus*	Crop	Field	Ziska 2000
D. C_4 Crops/C_3 Weeds				
Sorghum	*Xanthium strumarium*	Weed	Glasshouse	Ziska 2001b
Sorghum	*Albutilon theophrasti*	Weed	Field	Ziska 2003b

Notes: Favored indicates whether elevated [CO_2] produced significantly more crop or weed biomass. *Pasture* refers to a mix of C_3 grass species. The asterisk (*) refers to earlier emergence of the crop relative to weeds at elevated [CO_2].

lambsquarters to sugarbeet (*Beta vulgaris*), the competitive advantage of sugarbeet at elevated [CO_2] was attributed to late emergence of the weed species within the experiment (Houghton and Thomas 1996).

Although these studies have reported changes in the ratio of crop and weed biomass, only two studies have actually quantified changes in crop yield with weedy competition as a function of rising [CO_2] (Ziska 2000b, 2003). In these studies, two crop species, one C_3 (soybean), and one C_4 (dwarf sorghum) were grown with lambsquarters (C_3) and redroot pigweed (*Amaranthus retroflexus*, C_4) and velvetleaf (C_3) and redroot pigweed, respectively, at a density of two weeds per meter of row. Although, soybean yield losses were less from pigweed, all other crop–weed interactions resulted in increased yield loss at elevated [CO_2]. Interestingly, the presence of any weed species negated the ability of the crop to respond either vegetatively or reproductively to additional [CO_2]. This is significant because CO_2 enhancement studies and crop modeling efforts rarely consider crop–weed competition. However, additional field-based studies are needed to confirm and amplify the results presented here.

11.3.2 CO_2, Environmental Interactions, and Competition

As with field evaluations of crop loss, almost no data has examined how crop weed competition will be altered by simultaneous increases in $[CO_2]$ and climate. Only a single study has evaluated the interactions among temperature, $[CO_2]$, and crop–weed competition (Alberto et al. 1996). In this experiment, competition between rice (*Oryza sativa*, C_3) and a weedy competitor, barnyard-grass (*Echinocholoa glabrescens*, C_4) was assessed at two different $[CO_2]$ (ambient and ambient +200 ppm) and two different temperatures (day and night of 27/21 and 37/29°C). This study confirmed that at 27/21°C, increased $[CO_2]$ favored the crop (the C_3 species); however, with concomitant changes in both $[CO_2]$ and temperature, the C_4 weed was favored, primarily because higher temperatures resulted in increased seed yield loss for rice (Alberto et al. 1996).

It is, of course, difficult to generalize based on a single experiment. Hypothetically, there are a number of additional potential interactive effects related to temperature, $[CO_2]$, and weed–crop competition. Weeds of the tropics show a large stimulation to small air temperature changes (Flint et al. 1984; Flint and Patterson 1983), but it is unknown if a greater synergy with rising $[CO_2]$ would be anticipated for these weeds relative to tropical crops. It is likely that such potential changes in competition will be species specific.

No studies are available on the interactions among drought, rising CO_2, and weed–crop competition. Empirically, crops and weeds have similar responses to drought; consequently, the overall impact of weeds may be reduced because of decreased growth of both crops and weeds in response to water availability (Patterson 1995b). Although competition was not determined directly, the proportion of weed biomass increased with $[CO_2]$ to a similar extent in wet and dry treatments in a pasture mixture (Newton et al. 1996).

11.4   CLIMATIC EFFECTS ON WEED BIOLOGY AND COMPETITION

In addition to the direct CO_2 fertilization effect, climatic change, particularly precipitation and temperature, will have significant effects on weed biology. Temperature and precipitation remain primary abiotic variables that control vegetative distribution (Woodward and Williams 1987), and as such will impact the geographical distribution of weeds with subsequent effects on their growth, reproduction, and competitive abilities.

Increasing temperature may mean expansion of weeds into higher latitudes or higher altitudes. High-latitude temperature limits of tropical species are set by accumulated degree days (Patterson et al. 1999), while low-latitude limits are determined in part by competitive ability at low temperatures (Woodward 1988). Many of the weeds associated with warm-season crops originated in tropical or warm temperature areas; consequently, northward expansion of these weeds may accelerate with warming (Patterson 1993; Rahman and Wardle 1990). For example, detailed studies of itchgrass (*Rottboelliia cochinchinensis*) indicate that a warming of 3°C (day night temperature increase from 26/20 to 29/23°C) increased biomass and leaf area by 88 and 68%, respectively (Patterson et al. 1979). Empirically, based on its temperature response, itchgrass could effectively increase its percent of maximum growth from 50 to 75% in the Middle Atlantic states to 75 to 100% (Patterson et al. 1999). Northward expansion of other weeds, such as cogongrass (*Imperata cylindrica*) and witchweed (*Striga asiatica*), is also anticipated (Patterson 1995b), although warming may restrict the southern expansion of some exotic weeds such as wild proso millet (*Panicum miliaceum*) due to increased competition (Patterson et al. 1986).

One of the more intriguing examples of potential northward expansion is that of kudzu (*Pueraria lobata*), a ubiquitous invasive weed. Approximately 15 years ago, a seminal work by Tom Sasek and Boyd Strain at Duke University noted that the current latitudinal distribution of kudzu was limited to areas south of the Ohio Valley and the Mason-Dixon line by low winter temperatures (see Figure 7, Sasek and Strain 1990). Interestingly, recent observations have noted kudzu populations near the Chicago area (www.chicagobotanic.org) and in northwestern Massachusetts (www.cyberonic.com). How much of this distribution is due to increasing winter temperatures is unclear, but the northward spread is consistent with the Sasek and Strain predictions.

Changes in weed distribution and the resultant changes in weed–crop competition remain unclear. If temperature changes the ranges of both agronomic and noxious weeds, such changes could indirectly alter weed–crop competition by changing the ratio of C_4 weeds to C_3 crops. For example, estimated crop losses due to weeds without the use of herbicides are substantially larger in the south than in the north in both corn (22 vs. 35%) and soybeans (22 vs. 64%) (Bridges 1992). This may be associated with the occurrence in the South of some very aggressive weeds whose presence is limited in the northern states by low temperatures (see Table 2, Bunce and Ziska 2000). Alternatively, greater increases in nighttime relative to daytime temperatures projected with global warming (McCarthy et al. 2001) could decrease seed production to a greater extent in crop relative to weed species (cowpea, *Vigna unguiculata*, Ahmed et al. 1993) with subsequent competitive effects. Differential responses of seed emergence to temperature could influence species establishment and subsequent weed–crop competition (Houghton and Thomas 1996).

Response to drought in agronomic conditions is dependent on species and cultural conditions. In general, decreased water availability may favor the crop by reducing the competitive impact of the weed (see Table 1 in Patterson 1995b). That is, when potential yield is already limited by water, weed competition for other resources has less impact. Water availability may also affect the duration of weed-free periods during crop development. Coble et al. (1981) demonstrated that in competition with common ragweed, a critical period to avoid competitive effects was 2 weeks in a dry year and 4 weeks in a wet year. However, the duration of the critical period varied by weed and crop (Harrison et al. 1985; Jackson et al. 1985).

11.5 RISING [CO_2] AND INSECTS

Although increasing [CO_2] can result in narcoleptic and behavioral changes in insects, projected concentrations of atmospheric [CO_2] up to 1000 ppm are unlikely to affect insects directly (Nicolas and Sillans 1989). Rather, it is more probable that insect biology will be impacted by the direct physiological effects of [CO_2] on host plant metabolism. Specific [CO_2]-induced changes in plant metabolism include increased C:N ratios, altered concentrations of defensive (allelopathic) compounds, increased starch and fiber content (Coble et al. 1981), and increased water content. Overall, these metabolic changes are likely to impact insect–crop interactions in two principal ways: First by altering feeding behavior, and second, by altering plant defenses (Newman, Chapter 10, this volume).

11.5.1 FEEDING TRAITS

Because of qualitative changes at the leaf level, insect feeding behavior and mortality can be affected both positively and negatively by elevated [CO_2] (Lincoln et al. 1993, Bezermer and Jones 1998). For example, increased growth and development were observed for larvae of *Polyommatus icarus* (Lepitdoptera) feeding on *Lotus corniculatus* (Goverde et al. 1999). This was due in part because increased [CO_2] resulted in both increased leaf digestibility and carbohydrate concentration. In contrast, larvae of Colorado beetle (*Leptinotarsa decemlineata*) feeding on potato (*Solanum tuberosum*), southern army worm (*Spodoptera eridania*) feeding on peppermint (*Mentha piperita*), cabbage looper (*Trichoplusia ni Hubner*) feeding on lima bean (*Phaseolus lunata*), and buckeye butterfly (*Junona coena*) feeding on plantain *(Plantago lanceolate)* demonstrated increased feeding rates but lower growth and increased mortality associated with [CO_2]-induced changes in leaf quality, specifically N concentrations (Fajer et al. 1989, Lincoln and Couvet 1989, Miglietta et al. 2000, Osbrink et al. 1987) . Overall, higher C:N ratios associated with increasing [CO_2] may result in compensatory increases in foliar consumption rates by insects. These increased consumption rates are often accompanied (but not always, see Watt et al. 1996) by a decrease in the efficiency of plant tissue conversion to body mass, reduced larval growth rate, and reduced pupal mass.

Leaf-sucking insects would also be affected by qualitative leaf changes associated with enhanced [CO_2]. For mites, increased epidermal or leaf thickness could reduce infestation (Joutei et al. 2000). However, positive effects of [CO_2] on mite infestation have been observed and were associated with an increase in the concentration of nonstructural carbohydrates (Heagle et al. 2002). Ostensibly, phloem feeders such as aphids should be less responsive to [CO_2]-induced changes in leaf quality as they avoid the majority of plant-derived secondary metabolites. However, increases in population density were observed for aphids (*Myzus persicae*) on groundsel *(Senecio vularis)* and annual blue grass *(Poa annua)* (Bezemer et al. 1998), and increases in the daily rate of nymph production were observed for the aphid *Aulacorthum solani* on bean (*Vicia faba*) (Awmack et al. 1997). The basis for the increased performance of these aphids at elevated-[CO_2]-grown plants is unclear; presumably, if all other limiting factors on aphid populations remain unchanged, then aphid damage and diseases carried by aphids could be more severe as atmospheric [CO_2] increases.

Whether the response observed at the leaf or plant level is consistent with the community response is undetermined. Certainly there are compensatory changes, particularly in leaf production, that could overcome insect related damage (Hughes and Bazzaz 1997), but whether the production of new tissue could stimulate additional feeding is unknown. For scrub oak and marsh ecosystems, less infestation of leaf eaters was observed at the higher $[CO_2]$ concentration (Thompson and Drake 1994; Stiling et al. 2002). Whether this is related to leaf qualitative changes (e.g., N), or more complex community level processes is unknown. Certainly, preferential herbivore feeding on one species may have a positive benefit on another plant species less affected by $[CO_2]$. Overall, however, most data to date have focused on single insect–host crop interactions, making a more complex assessment of insect risks to agroecosystems with increasing $[CO_2]$ tentative.

11.5.2 $[CO_2]$ AND PLANT DEFENSES

Because increasing atmospheric $[CO_2]$ alters C:N ratios, nutritional quality, and photosynthate supply, the production of secondary compounds will be affected. It has been widely observed that herbivore feeding is strongly influenced by leaf allelochemicals as well as by leaf nutritional quality (Lincoln and Couvet 1989). The carbon to nutrient balance hypothesis (Bryant et al. 1983) predicts that the increase in internal C availability will activate the synthesis of C-based secondary defense chemicals, with subsequent reductions in leaf quality for leaf-feeding insects. A number of studies have shown that the level of secondary (carbon-based) products tends to increase with enhanced $[CO_2]$ (Lavola and Julkunen-Titto 1994; Lindroth 1996; Lindroth and Kinney 1998; Lindroth et al. 1993, 1997), although this response is not universal (Kerslake et al. 1998). However, even if no increase in secondary compounds was observed, a decline in leaf protein levels under CO_2 enrichment would result in an increase in the ratio of allelochemicals per unit of protein, with subsequent negative effects on insects due to higher consumption rates and increased exposure.

11.6 CLIMATE AND INSECTS

11.6.1 WARMING

Although high temperature stress could increase crop vulnerability to insects directly, temperature is widely recognized as the principle abiotic factor controlling insect growth and development. Porter et al. (1991) hypothesized that climate warming could alter geographical distribution, increase overwintering, and lead to a subsequent reduction in generation time and increased generational number. Patterson et al. (1999) provides an extensive list of climatic thresholds, associated phenological responses, and potential shifts in the expansion of insect ranges. They suggest that for temperate regions, warming may lead to increased winter survival, while at northern latitudes, warming would speed up growth and increase fecundity. However, range expansion of insects will be dependent on plant host expansion as well, and it is likely that such expansion is species specific. Gutierrez (Figure 5, 2000) has suggested that predator and insect herbivores are likely to respond differently to increasing temperature, with possible reductions in insect predation. The synchronization between crop and insect may also be affected if increased temperature alters photoperiod sensitivity. Increased invasion by migratory exotic pests has also been hypothesized because pests could colonize crops present at distant locations. This was suggested by Cannon (1998), who examined the spread of nonindigenous species in northwestern Europe, and the implications for insect invasion in the United Kingdom. Overall, most of these projected changes are likely to be detrimental to crop production. However, given species-specific adaptation (Bale et al. 2002) and the complexity of ecosystem interactions and human mitigation responses, long-term predictions regarding temperature remain conditional.

11.6.2 WATER AVAILABILITY

Precipitation extremes such as droughts or floods are associated with changes in insect herbivory and projected shifts in availability will have significant impacts on agricultural ecosystems (Fuhrer 2003). Intense precipitation has been shown to act as a deterrent to the occurrence and success of oviposition by insects (e.g., European corn-borer, Davidson and Lyon 1987). Flooding may also have a negative impact on soil-dwelling insects (Watt and Leather 1986) or indirect effects on pathogens, predators, and parasites, as has been shown for *Helicoverpa zea* pupae in a corn system (Raulston et al. 1992). Conversely, drought, which concentrates carbohydrates or sugars at the leaf level, may make the host plant more attractive to insect pests, particularly phytophagous insects (Mattson and Haack 1987). As with temperature, projected changes in extreme precipitation events are likely to shift the occurrence and frequency of insect outbreaks. For example, Drake (1994) demonstrated that increased variability in precipitation was one factor in determining the size and quality of insect populations in Australia. Increased precipitation from the El Niño event of 1997 and 1998 was hypothesized as one factor in the spread of the little fire ant (*Wasmannia auropunctata*), an alien species in the Galapagos Islands (Roque-Albelo and Causton 1999).

11.7 CO$_2$ AND PLANT PATHOGENS

Overall, plant pathogens are recognized as a significant limitation on agronomic productivity. As with insects, while plant pathogens can be directly affected by high levels of CO$_2$ (e.g., > 5%), current and anticipated atmospheric concentrations (0.03 to 0.07%) are likely to have little direct effect on these microorganisms; this is particularly true for soilborne pathogens, which are exposed to much higher concentrations of CO$_2$ in the edaphic environment than exist in the atmosphere (Lamborg et al. 1983). However, one factor affecting pathogenesis is the condition of the crop host. In general, any condition that promotes plant health will better enable plants to either resist or tolerate infection by pathogenic microbes. A basic concept of plant pathology is the *disease triangle* (Figure 11.1), which simplistically demonstrates that susceptible host, pathogenic microorganism, and environment interact in a variety of ways to affect infection and disease development. For example, even with a susceptible host, many fungal pathogens cannot achieve infection unless sufficient moisture is present (environment). It is not difficult to perceive that changes in

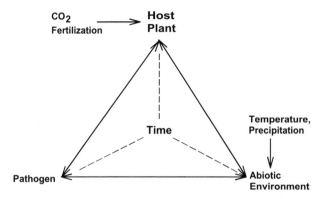

FIGURE 11.1 Interrelationships among time, pathological organisms, abiotic environment, host plant, and potential alterations as a function of climate and CO_2 fertilization.

atmospheric composition might influence disease incidence and severity; in fact, there are a number of recognized [CO_2]-induced changes that could, potentially, alter the susceptibility of crops to disease.

At the leaf level, reductions in stomatal aperture could reduce infection by stomatal-inhibiting pathogens such as rust and downy mildew fungi (Rudolph 1993). Similarly, increased epicuticular waxes, papillae, leaf surface thickness, and silicon accumulation at the sites of penetration could decrease disease incidence by pathogens that infect via direct penetration (Thomas and Harvey 1983; Hibberd et al. 1996). Improved water use efficiency and leaf water content could promote sporulation by foliar fungi (Thompson and Drake 1994; Woolacott and Ayres 1984). Increases in leaf carbohydrate content could also promote growth and reproduction of pathogens once infection occurs (Hibberd et al. 1996). Alternatively, reductions in leaf nitrogen content could reduce pathogen load and disease severity (Thompson and Drake 1994; Thompson et al. 1993). Delays in leaf senescence could increase disease severity by increasing the exposure time to the pathogen (Tiedemann and Firsching 1998) or pathogen load (Malmstrom and Field 1997), while accelerated ripening and senescence would reduce the infection period. Increased fungal fecundity (spores produced/lesion area) has also been reported to occur under elevated CO_2 (Chakraborty et al. 2000). An increase in spore numbers implies increased inoculum pressure for subsequent infection cycles and, generally, an increase in the spread and severity of disease. This could also have important implications for the functional duration of disease resistance in crop plants.

For whole plants, stimulation of plant biomass (leaves, shoots, tiller, leaves flowers, and so forth) by higher levels of atmospheric CO_2 increases the mass of host tissue for infection and use by pathogens. However, larger plants may tolerate more severe levels of infection without subsequent reductions in yield. Elevated [CO_2] can increase the production of defensive compounds (e.g., prunasin, phenolics, tannins), which could effect subsequent changes in pathogenesis (Gleadow et al. 1998, Runion et al. 1999). Although this has been studied in regard to insect performance and herbivory (Fajer et al. 1989; Lincoln and Couvet 1989), no work has addressed the impact of CO_2-altered allelochemical production on plant–microbe interactions. This could be of critical importance because infection by pathogenic microbes often elicits phytoalexin production (Mansfield 1982), while this may or may not be the case for insect herbivory (Zangerl and Bazzaz 1992). Below ground, increased root biomass and root length may increase the proportion of host tissue available for mycorrhizal fungi, nitrogen-fixing bacteria, or pathogens; similarly, increased root exudation could stimulate both beneficial and pathogenic microflora (Manning and Tiedemann 1995).

At the community level, increased density and height could increase humidity within the crop canopy, promoting growth and sporulation of most leaf-infecting fungi (e.g., rusts, powdery mildew, anthracnose; Chakraborty et al. 2000). Increased canopy residues at the end of the growing season could also potentially improve conditions for pathogenic overwintering (Manning and Tiedemann 1995). Essentially, any condition that improves production or survival of pathogen inocula on a host in one season has crucial implications for development of disease in subsequent cropping cycles.

Overall, while there are numerous CO_2-induced mechanisms that could alter disease susceptibility in host plants, field-based observations regarding the interaction of elevated CO_2 on pathogen biology have remained scarce, although exceptions exist (Chakraborty and Data 2003; Mitchell 2003; Montealegre et al. 2000). There is little doubt that increasing atmospheric $[CO_2]$ will elicit complex changes in plant–microbe interactions; the current challenge is to determine predictable ways in which these effects will vary (e.g., biotrophic vs. necrotrophic pathogens; Runion et al. 1994)). To date, the extremely limited attention given this field of study precludes any ability to make generalized predictions with confidence; diseases may increase, decrease, or show no change (Coakley 1995). The single consistent conclusion provided from the literature is that more research is needed to fill this important and fundamental void.

11.8   CLIMATE AND PLANT PATHOGENS

The relationship between climate variation, particularly temperature and water availability, and the incidence and severity of plant disease has long been recognized (Colhoun 1973); these are principle factors in the abiotic environment portion of the disease triangle (see Figure 11.1). The effects of temperature and moisture variation on pathogenic microbes and on disease development have been researched for over a century with varying effects depending on the exact environmental conditions and the microorganism or pathosystem of interest (e.g., Lonsdale and Gibbs 1996). While these results are too detailed to discuss here, it should be noted that disease development is generally favored by warm, moist conditions. Mild winters and warmer weather have been associated with increased outbreaks of powdery mildew, leaf spot disease, leaf rust, and rizomania disease (see Patterson et al. 1999), presumably in part because overwintering results in an increase in the amount of initial inoculum present in the spring. Warm, humid conditions may result in earlier and stronger incidence of late potato blight (*Phytophthora infestans*), a devastating disease of significant historical importance (Parry et al. 1990). Warmer temperature would also be likely to shift the occurrence of disease into cooler regions (Treharne 1989).

Increased precipitation per se is likely to increase the spread of diseases since rain and splash water both spread spores (Royle et al. 1986) and wet plant surfaces are necessary for spore germination and infection to proceed. Conversely, increasing aridity could lessen disease problems, although some diseases such as powdery mildew are promoted by hot, dry daytime conditions if nighttime temperatures result in dew formation (Gouk and Hill 1990). While extreme climatic conditions (drought, flood) will undoubtedly impact microorganisms directly, their effects on host plant stress are also of concern. Stressed plants are often more susceptible to pathogen attack and abiotic stresses, such as those caused by drought, and are often cited as primary contributors to disease complexes such as diebacks and declines (Manion and LaChance 1992). Drought can also effect production and concentration of plant defensive compounds; thus, these secondary plant metabolites may increase or decrease depending upon the duration and severity of the stress (Gershenzon 1984).

To make matters even more complex, it is known that elevated atmospheric $[CO_2]$ interacts with micrometeorological parameters to affect growth and health of plants. The primary example of this is the effect of increased CO_2 on water use efficiency. Increasing CO_2 generally increases photosynthesis and reduces stomatal conductance, thus increasing leaf-level water use efficiency (Rogers and Dahlman 1993). Therefore, under drought conditions, elevated CO_2 may provide a mechanism for moderation of stress. However, CO_2-induced increases in plant growth may offset increased leaf-level water use efficiency and, thus, the ability of elevated CO_2 to ameliorate the effects of drought (Runion et al. 1999). There is little doubt that these interactions will not only impact plant growth but also interactions with pathogenic microorganisms.

Although the importance of these aspects of plant–pest interactions are recognized and beginning to be addressed (Coakley et al. 1999; Rosenzweig et al. 2000), our ability to predict the impacts of climatic changes on natural and managed ecosystems and their interactions with pathogenic microorganisms is severely hampered by a lack of rigorous scientific knowledge. Hence, our ability to determine the impact on food security is tenuous at best. It is crucial, therefore, that we further our understanding of these interactions and of the mechanisms that drive them, if we are to continue to provide adequate supplies of food and fiber to a world having a future, altered environment.

11.9 IMPLICATIONS FOR THE MANAGEMENT OF WEEDS, INSECTS, AND DISEASES

It is difficult to derive the exact cost of managing pests. The cost of herbicide alone for the United States (~US$3 billion) is approximately equal to the value of lost crop production due to weedy competition (~US$4 billion) for a minimum estimate of US$7 billion (Bunce and Ziska 2000). However, the technology and associate costs of pest control vary by location and do not always include indirect environmental costs such as soil erosion or pollution.

11.9.1 CHEMICAL MANAGEMENT, CLIMATIC EFFECTS

Clearly, any direct or indirect impacts from a changing climate will have a significant effect on chemical management. Changes in temperature, wind speed, soil moisture, and atmospheric humidity can influence the effectiveness of applications (reviewed in Muzik 1976). For example, drought can result in thicker cuticle development or increased leaf pubescence, with subsequent reductions in pesticide entry into the leaf. These same variables can also interfere with crop growth and recovery following pesticide application. Overall, pesticides are most effective when applied to plants that are rapidly growing and metabolizing — those free from environmental stress.

11.9.2 CHEMICAL MANAGEMENT, DIRECT CO_2 EFFECTS

Although the effects of climate on pesticide efficacy have been well studied, can $[CO_2]$ directly affect chemical control? There are an increasing number of studies (Ziska and Teasdale 2000; Ziska et al. 1999) that demonstrate a decline in efficacy with rising $[CO_2]$ (Table 11.2). In theory, rising $[CO_2]$ could reduce foliar absorption of pesticides by reducing stomatal aperture or number, or by altering leaf or cuticular thickness. In addition, $[CO_2]$-induced changes in transpiration could limit uptake of soil-applied pesticides. For weed control, timing of application could also be affected if elevated $[CO_2]$ decreases the time the weed spends in the seedling stage (the time of greatest chemical susceptibility). For perennial weeds, increasing $[CO_2]$ could stimulate greater belowground growth (rhizomes, tubers, roots), diluting the active ingredient and making chemical control more difficult and costly. At the biochemical level, $[CO_2]$ could alter herbicide-specific chemistry in such a way as to directly reduce the efficacy of the active ingredient. For example, glyphosate inhibits aromatic amino acid production through the shikimic acid pathway; if $[CO_2]$ reduces the protein content per gram of tissue (Bowes 1996), this would result in less demand for aromatic amino acids.

At present, little is known mechanistically regarding how $[CO_2]$ would directly alter pesticide efficacy. For herbicides, short-term switching of quackgrass to the elevated $[CO_2]$ condition just prior to spraying did not increase tolerance, suggesting that stomatal closure did not play a factor in efficacy (Ziska et al. 1999). Recent unpublished data for Canada thistle indicated that significant increases in belowground relative to shoot biomass (Figure 11.2) could be associated with increased herbicide tolerance (Table 11.2). This is in agreement with the idea that tolerance may be simply a dilution effect of systemic herbicides associated with the large stimulation of root relative to shoot biomass at elevated $[CO_2]$.

It can be argued that $[CO_2]$-induced changes in efficacy are irrelevant given the rate of atmospheric $[CO_2]$ increase (other pesticides will be developed in the future). However, pesticide use can persist over decades (e.g., 2-4D has been in continuous use since 1950), coinciding with significant increases in atmospheric $[CO_2]$ (310 to 375 ppm from 1950 to 2003 for 2-4D). At present, many commercial ventures are investing in genetically modified crops specific for a given herbicide; consequently, it is likely that the use of these associated herbicides (e.g., glyphosate) would persist for decades. Furthermore, as mentioned previously, atmospheric $[CO_2]$ is not uniform

TABLE 11.2
Changes in Efficacy Determined as Changes in Growth (g day[1]) Following Herbicide Application for Weeds Grown at Either Current or Projected Future Levels of Carbon Dioxide

Species (Common name)	CO_2 p.p.m.	Environment	Herbicide	Growth g day[-1]
lambsquarters	365	GH	glyphosate	0.09 (death)
	723	GH	glyphosate	1.37
red-root pigweed	365	GH	glyphosate	0.04 (death)
	723	GH	glyphosate	0.18
quackgrass	388	GH	glyphosate	−0.05 (death)
	721	GH	glyphosate	1.14
Canada thistle	421	OTC	glyphosate	0.55
	771	OTC	glyphosate	1.37
Canada thistle	421	OTC	glufosinate	0.52
	771	OTC	glufosinate	1.14

Notes: Plants were followed for a 2- to 4-week period. Data for Canada Thistle (unpublished) are based on top (shoot) growth only. All weeds were sprayed with manufacturer recommended levels of the herbicide. Data for lambsquarters, red-root pigweed, and quackgrass are from Ziska and Teesdale, 2000. GH and OTC are greenhouse and open-top chamber, respectively.

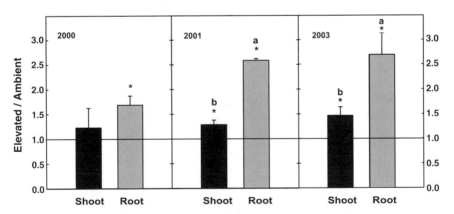

Canada Thistle

FIGURE 11.2 Shoot and root biomass at elevated (765 ppm) relative to ambient (414 ppm) CO_2 concentration for the invasive weed, Canada thistle, in 2000, 2001, and 2003 at the time of herbicide application from the unsprayed plots. Shoot biomass refers to all aboveground herbaceous material. Root biomass was obtained from soil subsamples (2.43 l in volume, to a depth of 30 cm). Bars indicate ±SE. The asterisk (*) indicates a significant difference (P < 0.05) relative to unity (no [CO_2] treatment effect); different letters indicate differences in the relative stimulation of shoots and roots.

and can differ as a function of urbanization. For example, effective chemical control of ragweed in an urban setting with high [CO_2] and temperature is likely to be different than in a farm setting for the same herbicide concentration (see Ziska et al. 2003). Overall, chemical control will still be possible with climatic changes or rising [CO_2], but additional spraying or increasing concentrations may be necessary. How these potential changes will alter economic and environmental costs is unclear.

11.9.3 Biological Control

Biological control of pests by natural or manipulated means is likely to be affected by increasing atmospheric [CO_2] and climatic change (Norris 1982; Froud-Williams 1996). Climate as well as [CO_2] could alter the efficacy of the biocontrol agent by potentially altering the development, morphology, and reproduction of the target pest. Direct effects of [CO_2] would also be related to changes in the ratio of C:N and alterations in the feeding habits and growth rate of herbivores. As pointed out by Patterson (1995a), warming could also result in increased overwintering of insect populations and changes in their potential range. Although this could increase both the biological control of some weeds, it could also increase the incidence of specific crop pests, with subsequent indirect effects on crop–weed competition. Overall, synchrony between development and reproduction of biocontrol agents and their selected targets is unlikely to be maintained in periods of rapid climatic change or climatic extremes. Whether this will result in a positive or negative result remains unclear.

11.9.4 Mechanical Control

A principal means of controlling weed populations, and the one most widely used in developing countries, is mechanical removal of the undesired plant. Tillage (by animal or mechanical means) is regarded as a global method of weed control in agronomic systems. Elevated [CO_2] could lead to further belowground carbon storage with subsequent increases in the growth of roots or rhizomes, particularly in perennial weeds (see Rogers et al. 1994, for a review). Consequently, mechanical tillage may lead to additional plant propagation in a higher [CO_2] environment, with increased asexual reproduction from belowground structures and negative effects on weed control (e.g., Canada thistle).

11.10 CONCLUSIONS

It is remarkable given the importance of weeds, insects, and diseases to crop production and food security, that so few experimental data are available assessing the impact of rising atmospheric [CO_2] or rapid climatic change on their biology. Furthermore, most of the data that are available are based on studies in controlled environment chambers or glasshouses, usually with a single host–pest focus. This represents a significant limitation because extrapolation of such studies to field environments may differ due to light or edaphic factors (see Ghannoum et al. 1997).

How can our current knowledge base be improved? At the whole plant level, quantification of $[CO_2]$- and climate-induced changes in host plant performance, including a temporal and spatial analysis of secondary compounds and allelopathic effects, anatomical and qualitative changes that alter pest susceptibility, and an integrated assessment of weed and crop germination, phenology, and allometry are necessary. If possible, evaluation of these parameters should be conducted at projected extremes of temperature and water availability. At the community level, a mechanistic understanding of how $[CO_2]$ and climate alter weed populations and associated seed banks, canopy microclimate, and pathogen load is also essential. For pathogens in particular, information on stages and rates of development and changes in the physiology of host–pathogen interactions is critical (Coakley et al. 1999). These suggestions are by no means inclusive, and given the paucity of data, there will be a number of fertile areas of inquiry. It should be emphasized, however, that hypotheses that consider multifactor responses, particularly at the ecosystem level, are highly preferable.

As has been illustrated in this review, much of what is currently known regarding the consequences of climatic change on pest biology and crop productivity is based on single-factor experiments. While useful, such experiments are limited in their ability to predict consequences for agroecosystems. We could, of course, continue to use existing knowledge to project how cropping systems might respond, but few data are available to validate whether current paradigms will accurately predict crop–pest relationships in a changing climate. There are likely to be unforseen differences that will influence our understanding of potential impacts (e.g., temperature induced shifts in insect ranges), and subsequent mitigation and management efforts (e.g., $[CO_2]$-induced changes in herbicide efficacy). Overall, any accurate assessment of future threats will be dependent on modeling efforts that consider variable combinations of limiting factors while recognizing that interactions at the agroecosystem level are governed by a complex set of feedbacks among soil, plant, atmosphere, and pest populations.

ACKNOWLEDGMENTS

Our thanks to Danielle Reed for editorial assistance, and to Dr. Stella Coakley for her valuable insight and comments.

REFERENCES

Ackerly, D.D., Coleman, J.S., Morse, S.R., and Bazzaz, F.A. (1992) CO_2 and temperature effects on leaf area production in two annual plant species. *Ecology* 73, 1260–1269.

Ahmed, F.E., Hall, A.E., and Madore, M.A. (1993) Interactive effect of high temperature and elevated carbon dioxide concentration on cowpea [*Vigna unguiculata* (L.) Walp.]. *Plant, Cell and Environment*, 16, 835–842.

Alberto, A.M.P., Ziska, L,H,, Cervancia, C.R., and Manalo, P.R. (1996) The influence of increasing carbon dioxide and temperature on competitive interactions between a C_3 crop, rice (*Oryza sativa*) and a C_4 weed (*Echinochloa glabrescens*). *Australian Journal of Plant Physiology*, 23, 795–802.

Allen, L.H., Jr. et al. (1987) Response of vegetation to rising carbon dioxide: photosynthesis, biomass and seed yield of soybean. *Global Biogeochemical Cycles* 1, 1–14.

Appleby, A.P., Olson, P.D., and Colbert, D.R. (1976) Winter wheat yield reduction from interference by Italian ryegrass. *Agronomy Journal* , 68, 463–466.

Awmack, C.S., Harrington, R., Leather, S.R. (1997) Host plant effects on the performance of the aphid *Aulacorthum solani* (Kalt.) (Homoptera: Aphididae) at ambient and elevated CO_2. *Global Change Biology*, 3, 545–549.

Baker, H.G. (1974) The evolution of weeds. *Annual Review of Ecology and Systematics*. 5, 1–24.

Bale, J.S. et al. (2002) Herbivory in global climate change research: direct effects of rising temperature on insect herbivores. *Global Change Biology*, 8, 1–16.

Bezemer, T.M. and Jones, T.H. (1998) Plant-insect herbivore interactions in elevated atmospheric CO_2: quantitative analyses and guild effects. *Oikos*, 82, 212–217.

Bezemer, T.M., Jones, T.H., and Knight, K.J. (1998) Long-term effects f elevated CO_2 and temperature on populations of the peach potato aphid (*Myzus persicae*) and its parasitoid, *Aphidius matricariae*. *Oecologia*, 116, 128–135.

Bowes, G. (1996) Photosynthetic responses to changing atmospheric carbon dioxide concentration. In *Photosynthesis and the Environment*, Baker, N.R., Eds., Kluwer Publishing, Dordrecht, The Netherlands, 387–407.

Bridges, D.C. (1992) *Crop Losses Due to Weeds in the United States*. Weed Science Society of America, Champaign, IL.

Bryant, J.P., Chapin, F.S., and Klein, D.R. (1983) Carbon nutrient balance of boreal plants in relation to vertebrate herbivory, *Oikos*, 40, 357–368.

Bunce, J.A. (1993) Growth, survival competition and canopy carbon dioxide and water vapor exchange of first year alfalfa at an elevated CO_2 concentration. *Photosynthetica*, 29, 557–565.

Bunce, J.A. (1995) Long-term growth of alfalfa and orchard grass plots at elevated carbon dioxide. *Journal of Biogeography*, 22, 341–348

Bunce, J.A. (2000) Acclimation to temperature of the response of photosynthesis to increased carbon dioxide concentration in *Taraxacum officinale*. *Photosynthesis Research*, 64, 89–94.

Bunce, J.A. and Ziska, L.H. (2000) Crop ecosystem responses to climatic change: crop/weed interactions. In *Climate Change and Global Crop Productivity*, Reddy, K.R. and Hodges, H.F., Eds., CABI Publishing, New York, 333–348.

Cannon, R.J.C. (1998) The implications of predicted climate change for insect pests in the U.K., with emphasis on non-indigenous species. *Global Change Biology*, 4, 785–796.

Carlson, H.L. and Hill, J.E. (1985) Wild oat (*Avena fatua*) competition with spring wheat: effects of nitrogen fertilization. *Weed Science*, 34, 29–33.

Carter, D.R. and Peterson, K.M. (1983) Effects of a CO_2 enriched atmosphere on the growth and competitive interaction of a C_3 and C_4 grass. *Oecologia*, 58, 188–193.

Chakraborty, S. and Data, S. (2003) How will plant pathogens adapt to host plant resistance at elevated CO_2 under a changing climate? *New Phytologist*, 159, 733–742.

Chakraborty, S., Pangga, I.B., Lupton, J., Hart, L., Room, P.M., and Yates, D. (2000) Production and dispersal of *Colletotrichum gloeosporioides* spores on *Stylosanthes scabra* under elevated CO_2. *Environ Pollution*, 108, 381–387.

Chaudhury, U.N., Kirkham, M.B., and Kanemasu, E.T. (1990) Carbon dioxide and water level effects on yield and water use of winter wheat. *Agronomy Journal*, 82, 637–641.

Coakley, S.M. (1995) Biospheric change: will it matter in plant pathology? *Canadian Journal of Plant Pathology*, 17, 147–153.

Coakley, S.M., Scherm, H., and Chakraborty, S. (1999) Climate change and plant disease management. *Annual Review of Phytopathology*, 37, 399–426.

Coble, H.D., Williams, F.M., and Ritter, R.L. (1981) Common ragweed (*Ambrosia artemisiifolia*) interference in soybeans (*Glycine max*). *Weed Science*, 29, 339–342.

Coleman, J.S. and Bazzaz, F.A. (1992) Effects of CO_2 and temperature on growth and resource use on co-occuring C_3 and C_4 annuals. *Ecology*, 73, 1244–1259.

Colhoun, J. (1973) Effects of environmental factors on plant disease. *Annual Review of Phytopathology*, 11, 343–364.

Curtis, P.S. and Wang, X. (1998) A meta-analysis of elevated CO_2 effects on woody plant mass, form, and physiology, *Oecologia*, 113, 299–313.

Davidson, R.H. and Lyon, W.F. (1987) *Insect Pests of Farm, Garden and Orchard*. John Wiley & Sons, New York.

Diaz, S., Grime, J.P., Harris, J., and McPherson, E. (1993) Evidence of a feedback system limiting plant response to elevated carbon dioxide. *Nature*, 364, 616–617.

Drake, V.A. (1994) The influence of weather and climate on agriculturally important insects— an Australian perspective. *Australian Journal of Agricultural Research*, 45, 487–509.

Eamus, D. (1991) The interaction of rising CO_2 and temperatures with water-use-efficiency. *Plant, Cell and Environment* 14, 843–852.

Fajer, E.D., Bowers, M.D., and Bazzaz, F.A. (1989) The effects of enriched carbon dioxide atmospheres on plant-insect herbivore interactions. *Science*, 243, 1198–1200.

Flint, E.P. and Patterson, D.T. (1983) Interference and temperature effects on growth in soybean (*Glycine max*) and associated C_3 and C_4 weeds. *Weed Science*, 31, 193–199.

Flint, E.P., Patterson, D.T., Mortensen, G.H., and Beyers, J.L. (1984) Temperature effects on growth and leaf production in three weed species. *Weed Science*, 32, 655–663.

Froud-Williams, R.J. (1996) Weeds and climate change: implications for their ecology and control. *Aspects of Applied Biology*, 45, 187–196

Fuhrer, J. (2003) Agro-ecosystem responses to combinations of elevated CO_2, ozone and global climate change. *Agriculture, Ecosystems and Environment* 97, 1–20.

Gershenzon, J. (1984) Changes in the levels of plant secondary metabolites under water stress and nutrient stress. *Recent Advances in Phytochemistry*, 18, 273–320.

Ghannoum, O., von Caemmerer, S., Barlow, E.W.R., and Conroy, J.P. (1997) The effect of CO_2 enrichment and irradiance on the growth, morphology and gas exchange of a C_3 (*Panicum laxum*) and a C_4 (*Panicum antidotale*) grass. *Australian Journal of Plant Physiology*, 24, 227–237.

Ghannoum, O., von Caemmerer, S., Ziska, L.H., and Conroy, J.P. (2000) The growth response of C_4 plants to rising atmospheric CO_2 partial pressure: a reassessment. *Plant, Cell and Environment*, 23, 931–942.

Ghannoum, O., Searson, M.J., and Conroy, J.P. (2006) Nutrient and water demands of plants under global climate change. In *Agroecosystems in a Changing Environment*, Newton, P.C.D., Carran, R.A., Edwards, G.R., and Niklaus, P.A., Eds., CRC Press, Boca Raton, FL, 55–86.

Gleadow, R.M., Foley, W.J., and Woodrow, I.E. (1998) Enhanced CO_2 alters the relationship between photosynthesis and defense in cyanogenic *Eucalyptus cladocalyx* F. Muell. *Plant, Cell and Environment* 21, 12–22.

Gouk, S.C. and Hill, R.A. (1990) Effect of climate change on diseases of fruit crops and woody plants. In *The Impact of Climate Change on Pests, Diseases, Weeds and Beneficial Organisms Present in New Zealand Agricultural and Horticultural Systems*, New Zealand Ministry for the Environment, Wellington, New Zealand, 147–158.

Goverde, M., Bazin, A., Shykoff, J.A., and Erhardt, A. (1999) Influence of leaf chemistry of *Lotus corniculatus* (Fabacea) on larval development of *Polyommatus icarus* (Lepidopter, Lycaenidae): effects of elevated CO_2 and plant genotype. *Functional Ecology*, 13, 801–810.

Gutierrez, A.P. (2000) Crop ecosystem responses to climatic change: pests and population dynamics. In *Climate Change and Global Crop Productivity*, Reddy, K.R. and Hodges, H.F., Eds., CABI Publishing, New York, 353–370.

Harrison, S.K., Williams, C.S., and Wax, L.D. (1985) Interference and control of giant foxtail (*Setaria faberi*) in soybeans (*Glycine max*). *Weed Science*, 33, 203–208.

Heagle, A.S., Burns, J.C., Fisher, D.E., and Miller, J.E. (2002) Effects of carbon dioxide enrichment on leaf chemistry and reproduction by two-spotted spider mites (Acari: Tetrachynidae) on white clover. *Environmental Entomology* 31, 594–601.

Heichel, G.H. and Jaynes, R.A. (1974) Stimulating emergence and growth of Kalmia genotypes with carbon dioxide. *HortScience*, 9, 61–63.

Hibberd, J.M., Whitbread, R., and Farrar, J.F. (1996) Effect of elevated concentrations of CO_2 on infection of barley by *Erysiphe graminis*. *Physiological and Molecular Plant Pathology*, 48, 37–53.

Holm, L.G., Plucknett, D.L., Pancho, J.V., and Herberger, J.P. (1977) *The Worlds Worst Weeds. Distribution and Biology*, University of Hawaii Press, Honolulu, HI.

Houghton, S.K. and Thomas, T.H. (1996) Effects of elevated carbon-dioxide concentration and temperature on the growth and competition between sugar beet (*Beta vulgaris*) and fat-hen (*Chenopodium album*). *Aspects of Applied Biology*, 45, 197–204.

Hughes, L. and Bazzaz, F.A. (1997) Effect of elevated CO_2 on interactions between the western flower thrips, *Frankliniella occidentalis* (Thysanoptera: Thripidae) and the common milkweed, *Ascelias syriaca*. *Oecologia*, 109, 286–290.

Idso, C.D., Idso, S.B., and Balling, R.C., Jr. (1998) The urban CO_2 dome of Phoenix, Arizona, *Physical Geography* 19, 95–108.

Idso, C.D., Idso, S.B., and Balling, R.C., Jr. (2001) An intensive two-week study of an urban CO_2 dome in Phoenix Arizona, USA. *Atmospheric Environment*, 35, 995–1000.

Jablonski, L.M., Wang, X., and Curtis, P.S. (2002) Plant reproduction under elevated CO_2 conditions: a meta-analysis of reports on 79 crop and wild species. *New Phytologist*, 156, 9–26.

Jackson, L.A., Kapusta, G., and Mason, D.J.S. (1985) Effect of duration and type of natural weed infestations on soybean yield. *Agronomy Journal*, 77, 725–729.

Joutei, A.B., Roy, J., Van Impe, G., and Lebrun, P. (2000) Effect of elevated CO_2 on the demography of a leaf-sucking mite feeding on bean. *Oecologia*, 123, 75–81.

Keeling, C.D. and Whorf, T.P. (2001) Atmospheric CO_2 records from sites in the SIO air sampling network. In *Trends: A Compendium of Data on Global Change*, Carbon Dioxide Information Analysis Center, Oak Ridge National Laboratory, U.S. Department of Energy, Oak Ridge, TN, 14–21.

Kerslake, J.E., Woodin, S.J., and Hartley, S.E. (1998) Effects of carbon dioxide and nitrogen enrichment on a plant-insect interaction: the quality of *Calluna vulgaris* as a host for *Operophtera brumata*. *New Phytologist*, 140, 43–53.

Kimball, B.A. (1983) Carbon dioxide and agricultural yield: an assemblage and analysis of 430 prior observations. *Agronomy Journal* 75, 779–788.

Kimball, B.A., Mauney, J.R., Nakayama, I.S., and Idso, S.B. (1993) Effects of increasing atmospheric CO_2 on vegetation, *Vegetatio*, 104/105, 65–75.

Lamborg, M.R., Hardy, R.W.F., and Paul, E.A. (1983) Microbial effects. In *CO_2 and Plants: The Response of Plants to Rising Levels of Atmospheric Carbon Dioxide*, Lemon, E.R., Ed., Westview Press, Boulder, CO, 131–176.

Lavola, A. and Julkunen-Titto, R. (1994) The effect of elevated carbon dioxide and fertilization on primary and secondary metabolites in birch, *Betula pendula* (Roth). *Oecologia*, 99, 315–321.

Lincoln, D.E. and Couvet, D. (1989) The effect of carbon supply on allocation to allelochemicals and caterpillar consumption of peppermint. *Oecologia*, 78, 112–114.

Lincoln, D.E., Fajer, E.D., and Johnson, R.H. (1993) Plant insect herbivore interactions in elevated CO_2 environments. *Trends in Ecology and Evolution,* 8, 64–68.

Lindroth, R.L. (1996) CO_2-mediated changes in tree chemistry and tree-Lepidoptera interactions. In *Carbon Dioxide and Terrestrial Ecosystems*, Koch, G.W. and Mooney ,H.A., Eds., Academic Press, San Diego, CA., 105–120.

Lindroth, R.L. and Kinney, K.K. (1998) Consequences of enriched atmospheric CO_2 and defoliation for foliar chemistry and gypsy moth performance. *Journal of Chemical Ecology*, 24, 1677–1695

Lindroth, R.L., Kinney, K.K., and Platz, C.L. (1993) Responses of deciduous trees to elevated atmospheric CO_2: productivity, phytochemistry, and insect performance. *Ecology*, 74, 763–777.

Lindroth, R.L., Roth, S., Kruger, E.L., Volin, J.C., and Koss, P.A. (1997) CO_2-mediated changes in aspen chemistry: effects on gypsy moth performance and susceptibility to virus. *Global Change Biology* 3, 279–289.

Long, S.P. (1991) Modification of the response of photosynthetic productivity to rising temperature by atmospheric CO_2 concentrations: has its importance been underestimated? *Plant, Cell and Environment.* 14, 729–739.

Lonsdale, D. and Gibbs, J.N. (1996) Effects of climate change on fungal diseases of trees. In *Fungi and Environmental Change*, Frankland, J.C., Magan, N., and Gadd, G.M., Eds., Cambridge University Press, Cambridge, U.K., 1–20.

Malmstrom, C.M. and Field, C.B. (1997) Virus-induced differences in the response of oat plants to elevated carbon dioxide. *Plant, Cell and Environment* 20, 178–188.

Manion, P.D. and LaChance, D. (1992) *Forest Decline Concepts.* American Phytopathological Society Press, St. Paul, MN.

Manning, W.J. and Tiedemann, A.V. (1995) Climate Change: potential effects of increased atmospheric carbon dioxide (CO_2), ozone (O_3) and ultraviolet-B (UV-B) radiation on plant diseases. *Environmental Pollution,* 88, 219–245.

Mansfield, J.W. (1982) *Phytoalexins*, John Wiley & Sons, New York.

Mattson, W.J. and Haack, R.A. (1987) The role of drought in outbreaks of plant-eating insects. *Bioscience*, 37, 110–118.

McCarthy, J.J. et al. (2001) *Climate Change 2001: Impacts, Adaptation, and Vulnerability.* Cambridge University Press, Cambridge, U.K.

Miglietta, F., Bindi, M., Vaccari, F.P., Schapendonk, A.H.C.M., Wolf, J., and Butterfield, R.E. (2000) Crop ecosystem responses to climatic change: root and tuberous crops. In *Climate Change and Global Crop Productivity*, Reddy, K.R. and Hodges, H.F., Eds., CABI Publishing, New York, 189–209.

Mitchell, C.E., Reich, P.B., Tilman, D., and Groth, J.V. (2003) Effects of elevated CO_2, nitrogen deposition, and decreased species diversity on foliar fungal plant disease. *Global Change Biology*, 9, 438–451.

Montealegre, C.M., van Kessel, C., Blumenthal, J.M., Hier, H.G., Hartwig, H.A., and Sadowsky, M.J. (2000) Elevated atmospheric CO_2 alters microbial population structure in a pasture ecosystem. *Global Change Biology*, 6, 475–482.

Morison, J.I.L. (1985) Sensitivity of stomata and water use efficiency to high CO_2. *Plant, Cell and Environment*, 8, 467–474.

Muzik, T.J. (1976) Influence of environmental factors on toxicity to plants. In *Herbicides: Physiology, Biochemistry, Ecology*, Audus, L.J., Ed., Academic Press, New York, 203–247.

Newman, J.A. (2006) Trophic interactions and climate change. In *Agroecosystems in a Changing Environment*, Newton, P.C.D., Carran, R.A., Edwards, G.R., and Niklaus, P.A., Eds., CRC Press, Boca Raton, FL, 231–260.

Newton, P.C.D., Clark, H., Bell, C.C., and Glasglow, E.M. (1996) Interaction of soil moisture and elevated CO_2 on the above-ground growth rate, root length density and gas exchange of turves from temperate pasture. *Journal of Experimental Botany*, 47, 771–779.

Nicolas, G. and Sillans, D. (1989) Immediate and latent effects of carbon dioxide on insects. *Annual Review of Entomology*, 34, 97–116.

Norris, R.F. (1982) Interactions between weeds and other pests in the agro-ecosystem. In *Biometeorology in Integrated Pest Management*, Hatfield, J.L. and Thomason, I.J., Eds., Academic Press, New York, 343–406.

Osbrink, W.L.A., Trumble, J.T., and Wagner, R.E. (1987) Host suitability of *Phaseolus lunata* for *Trichoplusia ni* (Lepitoptera: Noctuidae) in controlled carbon dioxide atmospheres. *Environmental Entomology*, 16, 639–644.

Parry, M.L., Porter, J.H., and Carter, T.R. (1990) Agriculture: climatic change and its implications. *Trends in Ecology and Evolution*, 5, 318–322.

Patterson, D.T. (1986) Responses of soybean (*Glycine max*) and three C_4 grass weeds to CO_2 enrichment during drought. *Weed Science* 34, 203–210.

Patterson, D.T. (1993) Implications of global climate change for impact of weeds, insects and plant diseases. *International Crop Science*, 1, 273–280.

Patterson, D.T. (1995a) Weeds in a changing climate. *Weed Science,* 43, 685–701.

Patterson, D.T. (1995b) Effects of environmental stress on weed/crop interactions. *Weed Science,* 43, 483–490.

Patterson, D.T. and Flint, E.P. (1990) Implications of increasing carbon dioxide and climate change for plant communities and competition in natural and managed ecosystems. In *Impact of Carbon Dioxide, Trace Gases and Climate Change on Global Agriculture*, Kimball, B.A., Rosenburg, N.J., and Allen, L.H., Jr., Eds., ASA Special Publication No. 53, American Society of Agronomy, Madison, WI, 83–110.

Patterson, D.T., Flint, E.P., and Beyers, J.L. (1984) Effects of CO_2 enrichment on competition between a C_4 weed and a C_3 crop. *Weed Science,* 32, 101–105.

Patterson, D.T., Highsmith, M.T., and Flint, E.P. (1988) Effects of temperature and CO_2 concentration on the growth of cotton (*Gossypium hirsutum*), spurred anoda (*Anoda cristata*), and velvetleaf (*Abutilon theophrasti*). *Weed Science* 36, 751–757.

Patterson, D.T., Meyer, C.R., Flint, E.P., and Quimby, P.C., Jr. (1979) Temperature responses and potential distribution of itchgrass (*Rottboellia exaltata*) in the United States. *Weed Science,* 27, 77–82.

Patterson, D.T., Westbrook, J.K., Joyce, R.J.C., Lingren, P.D., and Rogasik, J. (1999) Weeds, insects and diseases. *Climatic Change*, 43, 711–727.

Patterson, D.T. et al. (1986) Effects of temperature and photoperiod on Texas panicum (*Panicum texanum*) and wild proso millet (*Panicum miliaceum*). *Weed Science*, 34, 876–882.

Poorter, H. (1993) Interspecific variation in the growth response of plants to an elevated ambient CO_2 concentration. *Vegetatio*, 104/105, 77–97.

Porter, J.H., Parry, M.L., and Carter, T.R. (1991) The potential effects of climatic change on agricultural insect pests. *Agriculture and Forest Meteorology* 57, 221–240.

Potvin, C. and Strain, B.R. (1985) Effects of CO_2 enrichment and temperature on growth in two C_4 weeds, *Echinochloa crus-galli*, and *Eleusine indica*. *Canadian Journal of Botany*, 63, 1495–1499.

Potvin, C. and Vasseur, L. (1997) Long-term CO_2 enrichment of a pasture community. *Ecology*, 78, 666–677.

Rahman, A. and Wardle, D.A. (1990) Effects of climate change on cropping weeds in New Zealand. In *The Impact of Climate Change on Pests, Diseases, Weeds and Beneficial Organisms Present in New Zealand Agricultural and Horticultural Systems*, Prestridge, R.A. and Pottinger, P.P., Eds., New Zealand Ministry for the Environment, Wellington, NZ, 107–112.

Raulston, J.R., Pair, S.D., Loera, J., and Cabanillas, H.E. (1992) Prepupal and pupal parasitism of *Helicoverpa zea* and *Spodoptera frugiperda* (Lepidopters: Noctuidae) by *Steinernema* sp. in corn fields in the lower Rio Grande Valley. *Journal of Economic Entomology* 85, 1666–1670.

Reekie, E.G. and Bazzaz, F.A. (1991) Phenology and growth in four annual species grown in ambient and elevated CO_2. *Canadian Journal of Botany*, 69, 2475–2481.

Reekie, J.Y.C., Hicklenton, P.R., and Reekie, E.G. (1997) The interactive effects of carbon dioxide enrichment and daylength on growth and development in Petunia hybrida, *Annals of Botany*, 80, 57–64

Rogers, H.H. and Dahlman, R.C. (1993) Crop responses to CO_2 enrichment. *Vegetation*, 104/105, 117–131.

Rogers, H.H., Runion, G.B., and Krupa, S.V. (1994) Plant responses to atmospheric CO_2 enrichment, with emphasis on roots and the rhizosphere. *Environmental Pollution*, 83, 155–189.

Roque-Albelo, L. and Causton, C. (1999) El Nino and introduced insects in the Galapagos Islands: different dispersal strategies, similar effects. *Noticias de Galapagos*, 60, 1–9.

Rosenzweig, C. and Hillel, D. (1998) Effects on weeds, insects and diseases. In *Climate Change and the Global Harvest*, Rosenzweig, C. and Hillel, D., Eds., Oxford University Press, Cambridge, U.K., 101–122.

Rosenzweig, C., Iglesias, A., Yang, X.B., Epstein, P.R., and Chivian, E. (2000) Climate change and U.S. agriculture: the impacts of warming and extreme weather events on productivity, plant diseases, and pests. *Global Change and Human Health*, 2, 90–105.

Royle, D.J., Shaw, M.W., and Cook, R.J. (1986) Patterns of development of *Septoria nodorume* and *S. Tritici* in some winter wheat crops in Western Europe, 1981–1983. *Plant Pathology*, 35, 466–476.

Rudolph, K. (1993) Infection of the plant by *Xanthomonas*. In *Xanthomonas*, Swings, J.G. and Civerolo, E.L., Eds., Chapman and Hall, London, 193–264.

Runion, G.B., Curl, E.A., Rogers, H.H., Backman, P.A., Rodriguezkabana, P., and Helms, B.E. (1994) Effects of free-air CO_2 enrichment on microbial populations in the rhizosphere and phyllosphere of cotton. *Agriculture and Forest Meteorology*, 70, 117–130.

Runion, G.B., Entry, J.A., Prior, S.A., Mitchell, R.J., and Rogers, H.H. (1999) Tissue chemistry and carbon allocation in seedlings of *Pinus palustris* subjected to elevated atmospheric CO_2 and water stress. *Tree Physiology*, 19, 329–335.

Runion, G.B., Mitchell, R.J., Green, T.H., Prior, S.A., Rogers, H.H., and Gjerstad, D.H. (1999) Longleaf pine photosynthetic response to soil resource availability and elevated atmospheric carbon dioxide. *Journal of Environmental Quality*, 28, 880–887.

Sage, R.F. (1995) Was low atmospheric CO_2 during the Pleistocene a limiting factor for the origin of agriculture? *Global Change Biology*, 1, 93–106.

Sasek, T.W. and Strain, B.R. (1990) Implications of atmospheric CO_2 enrichment and climatic change for the geographical distribution of two introduced vines in the USA. *Climate Change*, 16, 31–51.

Schimel, D. et al. (1996) Radiative forcing of climate change. In *Climate Change 1995: The Science of Climate Change*, Houghton, J.T., Meira-Filho, L.G., Callander, B.A., Harris, N., Kattenberg, A., and Maskell, K., Eds., Cambridge University Press, Cambridge, U.K., 23–46.

Sicher, R.C. (1998) Yellowing and photosynthetic decline of barley primary leaves in response to atmospheric CO_2 enrichment, *Physiologia Plantarum*, 103, 193–200.

Stiling, P., Cattell, M., Moon, D.C., Rossi, A., Hungate, B., Hymus, G., and Drake, B.G. (2002) Elevated atmospheric CO_2 lowers herbivore abundance but increases leaf abscission rates. *Global Change Biology*, 8, 658–667.

Thomas, J.F. and Harvey, C.N. (1983) Leaf anatomy of four species grown under continuous CO_2 enrichment. *Botanical Gazette*, 144, 303–309.

Thompson, G.B., Brown, J.K.M., and Woodward, F.I. (1993) The effects of host carbon dioxide, nitrogen and water supply on the infection of wheat by powdery mildew and aphids. *Plant, Cell and Environment*, 16, 687–694.

Thompson, G.B. and Drake, B.G. (1994) Insects and fungi on a C_3 sedge and a C_4 grass exposed to elevated atmospheric CO_2 concentrations in open-top chambers in the field. *Plant, Cell and Environment*, 17, 1161–1167.

Tiedemann, A.V. and Firsching, K.H. (1998) Interactive effects of elevated ozone and carbon dioxide on growth and yield of leaf rust-infected versus non-infected wheat. *Environmental Pollution*, 108, 357–363.

Treharne, K. (1989) The implications of the "greenhouse effect" for fertilizers and agrochemicals. In *The Greenhouse Effect and UK Agriculture*, Bennet, R.C., Ed., Ministry of Agriculture, Fisheries and Food, U.K., 67–78.

Tremmel, D.C. and Patterson, D.T. (1993) Responses of soybean and five weeds to CO_2 enrichment under two temperature regimes, *Canadian Journal of Plant Science.* 73, 1249–1260.

Vengris, J., Colby, W.G., and Drake, M. (1955) Plant nutrient competition between weeds and corn. *Agronomy Journal*, 47, 213–216.

Ward, J.K. and Strain, B.R. (1999) Elevated CO2 studies: past, present and future. *Tree Physiology* 19, 211–220.

Watt, A.D. and Leather, S.R. (1986) The pine beauty in Scottish lodgepole pine plantations. In *Dynamics of Forest Insect Populations: Patterns, Causes, Implications*, Berryman, A.A., Ed., Plenum Press, New York, 243–266.

Watt, A.D. et al. (1996) The effects of climate change on the winter moth, *Operophtera brumata,* and its status as a pest of broadleaved trees, Sitka spruce and heather. *Aspects of Applied Biology*, 45, 307–316.

Woodward, F.I. (1988) Temperature and the distribution of plant species. In *Plants and Temperature*, Long. S.P. and Woodward, F.I., Eds., University of Cambridge Press, Cambridge, U.K., 59–75.

Woodward, F.I. and Williams, B.G. (1987) Climate and plant distribution at global and local scales. *Vegetatio*, 69, 189–197.

Woolacott, B. and Ayres, P.G. (1984) Effects of plant age and water stress on production of conidia by *Erysiphe graminis* f. sp. hordei examined by non-destructive sampling. *Transactions of the British Mycological Society*, 82, 449–454.

Zangerl, A.R. and Bazzaz, F.A. (1992) Theory and pattern in plant defense allocation. In *Plant Resistance to Herbivores and Pathogens*, Fritz, R.S. and Simms, E.L., Eds., University of Chicago Press, Chicago, 363–391.

Ziska, L.H. (2000) The impact of elevated CO_2 on yield loss from a C_3 and C_4 weed in field-grown soybean. *Global Change Biology*, 6, 899–905.

Ziska, L.H. (2001a) Growth temperature can alter the temperature dependent stimulation of photosynthesis by elevated carbon dioxide in *Abutilon theophrasti. Physiologia Plantarum*, 111, 322–328.

Ziska, L.H. (2001b) Changes in competitive ability between a C_4 crop and a C_3 weed with elevated carbon dioxide. *Weed Science*, 49, 622–627.

Ziska, L.H. (2003a) Evaluation of the growth response of six invasive species to past, present and future carbon dioxide concentrations. *Journal of Experimental Botany* 54, 395–404.

Ziska, L.H. (2003b) Evaluation of yield loss in field sorghum from a C_3 and C_4 weed with increasing CO_2. *Weed Science,* 51, 914–918.

Ziska, L.H. and Bunce, J.A. (1993) The influence of elevated CO_2 and temperature on seed germination and soil emergence. *Field Crops Research* 34,147–157.

Ziska, L.H. and Caulfield, F.A. (2000) Rising carbon dioxide and pollen production of common ragweed, a known allergy-inducing species: implications for public health, *Australian Journal of Plant Physiology*, 27, 893–898.

Ziska, L.H., Gebhard, D.E., Frenz, D.A., Faulkner, S.S., Singer, B.D., and Straka, J.G. (2003) Cities as harbingers of climate change: common ragweed, urbanization and public health, *Journal of Allergy and Clinical Immunology*, 111, 290–295.

Ziska, L.H. and Teasdale, J.R. (2000) Sustained growth and increased tolerance to glyphosate observed in a C_3 perennial weed, quackgrass (*Elytrigia repens*), grown at elevated carbon dioxide. *Australian Journal of Plant Physiology* 27, 159–164.

Ziska, L.H., Teasdale, J.R., and Bunce, J.A. (1999) Future atmospheric carbon dioxide may increase tolerance to glyphosate. *Weed Science*, 47, 608–615.

Ziska, L.H. et al. (2000) A global perspective of ground level "ambient" carbon dioxide for assessing the response of plants to atmospheric CO_2, *Global Change Biology*, 7, 789–796.

Part III

Capacity to Adapt

12 Distinguishing between Acclimation and Adaptation

Mark J. Hovenden

CONTENTS

12.1 INTRODUCTION

> Acclimation *n.*: The habituation of an individual organism to a different or
> changing environment
> Adaptation *n*: The evolutionary process by which a species becomes fitted to
> its environment

One of the major problems with predicting the responses of species to future conditions is that experiments in which plants are grown in manipulated environments are really studies of acclimatory (i.e., the plastic response of individuals) rather than of adaptive (i.e., evolutionary) responses. While acclimatory responses are of interest for long-lived plants, which are likely to survive until the future conditions arrive, and for crop plants, which are artificially selected, the evolutionary ramifications of the response to climate change are far reaching and poorly understood. Climate change scientists often confuse the processes of acclimation and adaptation (Backhausen and Scheibe 1999), which hinders our understanding of the impacts of climate change. This chapter will deal with the two processes and explain how carefully distinguishing between acclimation and adaptation can assist us in understanding and predicting the impacts of the changing climate.

The word adaptation, as defined above, has a specific, biological meaning. Organisms adapt to the environment via the process of natural selection, and adaptation is the end result of differential reproductive success over successive generations. However, the word adaptation also has several nonbiological meanings and is widely used in modern English to mean the process by which something is altered, modified, adjusted, fit or made suitable for a purpose (Moore 1997). In this context, the word features prominently in the widely publicised general literature on global climate change (Hulme 2003). Indeed, one of the sections of the Intergovernmental Panel on Climate Change (IPCC) Third Assessment Report is titled "Impacts, Adaptation and Vulnerability" (IPCC 2001). In this sense, the word describes the measures humans must take "to diminish the risk of damage from future climate change and from present climate variability" (IPCC 2001, p. ix).

Individual organisms are very responsive to their environment, as indeed they must be if they are to survive daily, seasonal, and long-term changes in local conditions. These responses to the environment are sometimes behavioural (e.g., the diurnal rhythms of animals and plants), but they can also be biochemical, physiological, morphological, anatomical, and developmental. These changes of organism structure and function in response to particular environmental conditions are termed acclimation.

The widespread use of the word adaptation in the world at large and its adoption as a catchword to describe methods of achieving sustainable use of the environment (Easterling et al. 2003) have resulted in its abuse in agricultural and biological literature. Adaptation has been widely used in biological literature to mean any organismal change in response to the environment. Acclimatory responses are distinct from adaptational responses in important and sometimes nonintuitive ways. This distinction is not purely a semantic matter because the words describe fundamental biological principles that must be properly understood in order to predict, with any accuracy, the biological and ecological consequences of the changing climate.

The aim of this chapter is, therefore, to define what is meant by adaptation, as opposed to acclimation, to describe why this distinction is important, and to demonstrate how careful attention to the two processes will aid our understanding of the impacts of climate change. The examples given will mostly be botanical because of familiarity rather than importance.

12.2 TERMINOLOGY

The quotations used in the following section were discovered through the Oxford English Dictionary (Simpson and Weiner 1989), but were checked in the primary source where possible. The oldest known published use of any form of the word *acclimate* is from the English agriculturist Arthur Young's 1792 book, *Travels in*

France (Simpson and Weiner 1989), in which he describes the appearance of individuals of English *Arbutus* sp. as being "so ac-climated as to appear native" (p. 296). Another early use that clearly demonstrates the meaning is that of Lieutenant Thomas Bacon: "English residents in India imbibe peculiarities in the process of acclimation" (Bacon 1837, p. 7). In these contexts, therefore, to acclimate is to change in response to an environment. The word acclimate is distinct from the word acclimatise, which generally implies a forced acclimation. This is exemplified in the following passage from Bartley's English translation of Topinard's *Anthropology* in which the two words are carefully defined: "The words acclimation and acclimatization are not synonymous. The former is understood of the spontaneous and natural accommodation to new climatic conditions, the latter of the intervention of man in this accommodation" (Topinard 1890, p. 393).

The term *adapt* first appears in Florio's 1611 Italian and English dictionary, in which the English translation of *addattare* is given as "to fit, to adapt, to apropriate" (Florio 1611, p. 9). Similarly, Ben Jonson states in his *Workes* "as he is adapted to it by Nature, he shall grow the perfecter Writer" (Jonson 1640). Both of these examples indicate that adapt originally meant to be fitted. Thus, Darwin's use: "The most vigorous and healthy males, implying perfect adaptation, must generally gain the victory in their contests" (Darwin and Wallace 1859, p. 50) and "[w]e see beautiful adaptations everywhere and in every part of the organic world" (Darwin 1859, p. 48). To adapt, therefore, is to become or to be fitted to an environment.

12.3 THE PROCESSES

12.3.1 ACCLIMATION

The process of acclimation is simple to understand. Acclimation is the process in which the morphology, anatomy, physiology, or biochemistry of an individual organism varies in response to environmental changes. Thus, acclimation is the expression of phenotypic plasticity. This is easily shown by growing a set of genetically identical organisms under a range of environmental conditions and observing the variations. Because the organisms are genetically identical, variation among them must be due to the influence of the environment in which they were grown. This can be seen in Figure 12.1, which shows the biomass allocation of cuttings from a single genotype of the southern beech tree, *Nothofagus cunninghamii*, at two different growth temperatures. In this simple example, cuttings were taken from a single tree and were, therefore, genetically identical. At the warmer temperature, plants had a much larger proportion of their total biomass as leaves and a substantially smaller proportion as roots, in comparison to the lower growth temperature. Thus, this genotype of *N. cunninghamii* acclimates to the increase in growth temperature by increasing its biomass allocation to the leaves at the expense of the roots. This is a morphological change in response to environmental conditions. The process is the same when a changing climate is considered in that acclimation involves the plastic response of individuals to environmental conditions.

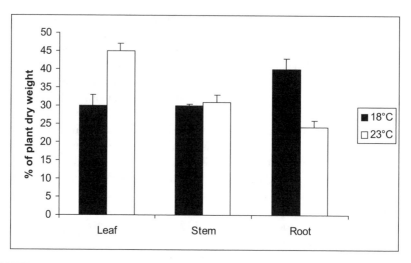

FIGURE 12.1 The percentage of whole plant dry weight that is leaf, stem, or root tissue as a function of growth temperature in cuttings from a single genotype of *Nothofagus cunninghamii*.

12.3.2 ADAPTATION

One of the fundamental tenets of modern biology is evolution through the process of natural selection, as described by Darwin (Darwin 1859). Natural selection occurs because most species have an exponential population growth rate, due to the fact that individuals tend to produce more offspring than are required to replace themselves. This must ultimately result in mortality because the Earth cannot sustain an infinite number of individuals. If within a given species, some individuals are more suited to the prevailing environmental conditions than others, they are more likely to survive and reproduce. Thus, the genes of those individuals will be present in the following generation. Over successive generations, those individuals most suited to their environment will prevail while other, less fit individuals, will perish. This is natural selection.

Individuals within a species will tend to vary in morphology, anatomy, and physiology, and this is termed *intraspecific variation*. Intraspecific variation occurs naturally as a result of sexual reproduction and chance mutations. If it so happens that individuals with a particular heritable character (or set of characters) tend to be more suited to the current environment than individuals without that character, then the individuals with this character will be more likely to survive and reproduce than others. This differential success will result in an increase in the proportion of individuals with this character in the next generation because it is inherited. So long as the environment does not change markedly, this character will become more and more common over successive generations due to the process of natural selection. This character is described as being *selected for*, and consequently the absence of this character is *selected against*. In this way, organisms become honed to specific environments. This is adaptation through natural selection. Characters that contribute to an organism's evolutionary success in a particular environment are also called adaptations. Examples of specific adaptations abound and range from the hollow bones of birds, through the long necks of giraffes to the ornate flowers of orchids.

The process of natural selection can be illustrated by a simple example (Figure 12.2). Let us assume we have a population (population A) of summer-flowering, annual plants that has a mean rooting depth, D. A representative sample of seeds from this population is sown into a competitive environment with summer drought. If we assume that those plants with a greater rooting depth are able to access more water during dry conditions and that survival to reproductive maturity is dependent on access to adequate water, then over a period of several generations, natural selection will result in an increase in mean rooting depth of the new population (population B). If a representative sample of seeds from this population (population B) are grown in the same conditions as a representative sample of seeds from the original population (population A), then the mean rooting depth of population A plants will be less than the mean rooting depth of population B plants (Figure 12.2). Population B has undergone natural selection for increased rooting depth while population A has not. Because the difference in rooting depth has arisen through differential survival and reproduction, the difference in mean rooting depth between the two populations will be genetically based and, therefore, will be maintained when plants from the two populations are grown in a similar environment.

The end result of an adaptational response to the environment is a permanent change in the range of values for a particular character within a population, as seen in Figure 12.2. This is distinct from an acclimatory response to the environment in which the changes are not permanent because they are not the result of differential survival and reproduction. If the change in mean rooting depth in Figure 12.2 were an acclimatory response rather than an adaptational response, then the differences between the two populations would disappear when they were grown in uniform conditions.

12.3.3 TYPES OF SELECTION

The hypothetical situation discussed above is an example of *directional selection*, in which selective pressure through the process of natural selection drives a particular character in one direction. Thus, following the example above, the mean rooting depth increases in the dry environment because individuals with the shortest roots tend to be selected against. Directional selection, therefore, alters the population by favouring variants of one extreme. Directional selection is not the only form of natural selection, however. Natural selection can also be *stabilising*, in which extreme variants are selected against, or *disruptive*, in which selection favors variants of the opposite extremes.

12.4 CLIMATE CHANGE EXPERIMENTS

Most climate change experiments, in which organisms are exposed to simulated future conditions (e.g., elevated atmospheric $[CO_2]$, increased temperature), are short-term. Few experiments run for longer than the life span of the organisms involved, although this is changing with an increasing number of long-term

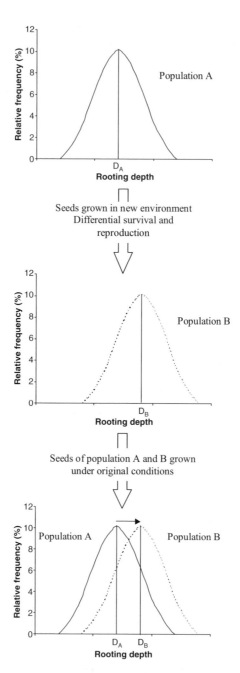

FIGURE 12.2 The impact of directional natural selection on the mean rooting depth (D) of a hypothetical plant species. Seeds from population A are grown in an environment that selects for increased rooting depth. When seeds from this selected population (B) are grown with seeds from the original population, the mean rooting depth of population B will remain larger than that of the original population, $D_B > D_A$.

experiments such as several free air CO_2 enrichment (FACE) experiments. Indeed some organisms, such as many tree species, have such long generation times that it is just not possible to follow them for even a fraction of their lifetime.

These experiments, therefore, are studies of *acclimatory* responses to climate change. These experiments provide much needed information on the responses of individuals and the relative short-term responses of different species to the changing climate. In the case of plant responses to elevated [CO_2], it is common to relate the response as *growth enhancement*. The growth enhancement for a species from a particular experiment is expressed as a ratio of the mean biomass or growth rate (or a similar variable) of individuals grown in future conditions to that of individuals grown in current or control conditions (Roumet and Roy 1996). This provides an index of response, with those species that show the greater increase in, say, growth rates showing a higher ratio of growth in future conditions to growth in current conditions. This is particularly popular in meta-analyses, which are appearing with increasing frequency (Norby et al. 1999). These measurements and calculations are fair enough in themselves, as they do provide some indication of variation among species in responses to changes in climatic variables. It is reasonably common, however, to extrapolate from these figures of mean growth enhancement to argue that a species with a larger response ratio will outcompete one with a smaller response ratio. Such comparisons of species rely on the mean response and ignore the variance of response within a species that is known to exist (Radoglou and Jarvis 1990; Ziska et al. 1998; Roumet et al. 1999; Rehfelt et al. 2002). For comparisons to be meaningful, it is important to gain some measure of intraspecific as well as interspecific variation in the response to climate change.

The major problem with extrapolating mean acclimatory results into the future is that this ignores the potential for adaptation. Perhaps the most widely measured organismal response to climate change is the plant photosynthetic response to elevated CO_2 levels (Sage 1994). There have been literally hundreds of investigations into the phenomenon of photosynthetic downregulation in response to high [CO_2], some of which even claim to be studies of adaptation to elevated [CO_2] (Backhausen and Scheibe 1999). Short-term exposure to CO_2 levels higher than normal results in an increase in the photosynthetic carbon assimilation rate in virtually all plants (Roumet et al. 2000). After an extended period of growth at elevated [CO_2], however, many plants show a reduction of this CO_2-related photosynthetic stimulation (Roumet et al. 2000). This is termed *photosynthetic downregulation* and is an acclimatory response. The reasons underlying this reduction in photosynthetic efficiency may well be related to the inability of the plant to export and utilise the extra carbohydrate produced (Stitt 1991). This leads to an increase in the concentration of nonstructural carbohydrates in the leaf, which slows photosynthesis through product inhibition. Photosynthetic downregulation is also associated with an actual reduction in the concentration and activity of photosynthetic enzymes, which affects photosynthetic performance over an extended period (Wong 1979; Stitt 1991). Thus, some meta-analyses propose an average downregulation of photosynthesis of approximately 10% (Medlyn et al. 1999). There is increasing evidence, however, that species with strong and responsive sinks for carbohydrates, or with very high growth rates, are less likely to downregulate photosynthesis at elevated $[CO_2]$ (Bryant et al. 1998; Gielen and Ceulemans 2001; Hovenden 2003).

If species with high growth rates or other strong carbohydrate sinks can grow in elevated $[CO_2]$ without photosynthetic downregulation, then it would be expected that the same comparison can be made within species. Logically, an individual that maintains high photosynthetic rates at elevated $[CO_2]$ will have more carbohydrates to invest in above- and belowground growth and in reproductive organs than one that downregulates photosynthesis. Such a plant would also have higher photosynthetic water use efficiency, higher nitrogen use efficiency, and higher growth rate in future CO_2 concentrations. It is also logical to assume that such individuals would be selected for in a competitive environment. Therefore, natural selection, with elevated $[CO_2]$ as a selective pressure, may act in such a way as to limit photosynthetic downregulation by favouring those individuals that do not acclimate.

Intraspecific variation in degree of photosynthetic acclimation has been demonstrated in species such as *Plantago lanceolata* (Curtis et al. 1996). Some family lines of *P. lanceolata* show significant photosynthetic acclimation to elevated $[CO_2]$ while others do not. Curtis and coworkers (1996) linked this variation in photosynthetic responses to intraspecific differences in the growth response to elevated $[CO_2]$. It is unlikely that this is an isolated example as many other species show genotype-dependent responses to elevated $[CO_2]$ (Radoglou and Jarvis 1990; Ziska et al. 1998; Roumet et al. 1999; Rehfelt et al. 2002).

12.4.1 Do We Have Any Evidence for Adaptational Responses?

Natural selection is occurring constantly. While there is little doubt that the global climate is changing (see Chapter 1 of this volume), it is difficult to assess the evolutionary impacts of climate change on natural populations because many selective forces are acting at the same time. Therefore, evidence for adaptational responses must come from experiments, or be indirect, such as from comparisons of genotype-specific responses. Unfortunately, there have been far fewer experiments on adaptational responses than investigating acclimatory responses.

There are two types of selection experiments that provide evidence for adaptational responses to climate change: Artificial selection experiments and controlled natural selection experiments (Conner 2003). *Artificial selection* is when the investigator chooses a particular character and the individuals with the most extreme values of this character are bred to produce the next generation. A *controlled natural selection* experiment differs in that the investigator does not directly choose the individuals that contribute to the next generation, but does manipulate some environmental factor that may result in selection. Artificial selection experiments have provided the best evidence to date of the selective pressure associated with climate change. Tousignant and Potvin (1996) grew *Brassica juncea* in control and predicted climatic conditions of increased temperature, $[CO_2]$, and climatic variability. The individuals that produced the greatest total fruit mass in each generation were selected, and their seeds formed the basis of the next generation. After seven generations, the plants were grown in a reciprocal transplant experiment. Their results showed a clear and significant effect of simulated climate change on growth of this species, such that offspring of plants grown in the predicted environment grew better in the predicted environment than did offspring of plants grown in the control environment (Tousignant and Potvin 1996). This clearly demonstrates an adaptive response of growth to the predicted environment. It is interesting, however, that fecundity of *B. juncea* was not similarly affected. Additionally, genetic correlations among traits may act in such a way as to limit adaptive evolution in some cases (Etterson and Shaw 2001).

An adaptational response was also observed in *Arabidopsis thaliana* in response to [CO_2] (Ward et al. 2000). Individuals were selected on the basis of seed production for five generations at high and low [CO_2]. Offspring were then compared in a reciprocal transplant experiment, after maternal effects were removed through the use of an additional generation at intermediate [CO_2]. In the reciprocal transplant experiment, plants produced most seed at the [CO_2] at which they had been selected. Additionally, plants selected at elevated [CO_2] flowered and senesced earlier than the low-[CO_2] plants, indicating that selection at elevated [CO_2] altered their phenology. It seems from these results that selecting for increased fecundity resulted in accelerated development and senescence. Because of this, plants selected at elevated [CO_2] did not grow any larger than plants selected at low [CO_2], they merely reached a set size more rapidly. Ward and coworkers (2000) argued that their results questioned the widely held assumption that C_3 plants will increase their carbon accumulation at elevated [CO_2]. This would have significant ramifications for the development of models of future carbon sequestration.

Further evidence of adaptation to the changing climate comes from an investigation of responses to past changes. Bunce examined the growth response of four weedy annual species to a range of [CO_2] from 90 µmol mol^{-1} below current levels to 90 µmol mol^{-1} above current levels (Bunce 2001). Using a range of measures, Bunce found that the efficiency at which the plants used CO_2 declined abruptly at CO_2 concentrations above the current level. These responses indicate that these annual weedy species are adapted to the current atmospheric [CO_2], but not to higher concentrations, which is clear evidence that adaptation has been occurring in response to the current increases in [CO_2].

12.4.2 Indirect Evidence for Adaptational Responses

Looking for actual evolutionary changes in response to climate change is very difficult in practice. Natural or artificial selection experiments such as those described above take several generations, which makes them lengthy with nearly all organisms and impossible with many. True evolutionary fitness is measured as the success with which an organism's genes are passed on to ensuing generations. The raw material of natural selection is biological variation within the population, and selection occurs when there is differential survival of individuals in that population. Therefore, one alternative method of looking for adaptational patterns, and the potential for adaptational responses, is to investigate genotype-specific responses to climate change and invent a proxy measure for true evolutionary fitness. By doing so, it should be possible to determine the responses to climate change that are likely to be selected for and those that are likely to be selected against. This provides insight into the possible mechanisms of natural selection in response to climate change as well as providing information concerning the level of variation in the response within species.

Direct measurement of fitness is difficult, particularly for long-lived species such as trees. Therefore, it becomes essential to use some proxy measure of fitness, which includes assumptions concerning the relationship between evolutionary fitness and some more easily measured character. For instance, total seed production may be an appropriate measure of fitness for annual plants (Conner 2003). If genotype-specific responses are known, then it becomes possible to compare the mean responses of a species to the response of the genotype predicted to be most successful in evolutionary terms.

While there is clear evidence of genotype-specific responses to simulated climate change (Curtis et al. 1996; Ziska et al. 1998; Rehfelt et al. 2002), few studies have used this evidence to assess the potential for adaptational responses. The southeastern Australian rainforest tree, *Nothofagus cunninghamii*, has genotype-specific responses to [CO_2] and the responses of various genotypes have been used to develop hypotheses concerning the differences between acclimational and adaptational responses to [CO_2] (Hovenden and Schimanski 2000). *N. cunninghamii* grows in a competitive rainforest environment, and individuals become established either after fire or in rainforest gaps (Read 1999). Further, it has been shown that growth rate is an important determinant of canopy composition in southeastern Australian cool temperate rainforests (Read 1995). Hence, it is likely that growth rate may be the best proxy for evolutionary fitness in this species because those individuals with high growth rates will be likely to shade out competitors and form part of the canopy. Short-term growth cabinet experiments indicated that the response of stomatal density (i.e., the number of stomata per unit area of leaf surface) to [CO_2] in experiments was the opposite of that obtained from the fossil record (Hovenden and Schimanski 2000). Stomatal density is usually seen to decrease with increasing [CO_2] (Woodward 1987; Woodward and Kelly 1995), but the average stomatal density of *N. cunninghamii* leaves increased with increasing [CO_2] in growth experiments (Hovenden and Schimanski 2000). Additionally, the stomatal density of fossil *N. cunninghamii* leaves from a glacial climate is higher than in extant plants, despite the fact the atmospheric [CO_2] was likely to have been much lower than at present.

These apparently contradictory results can be explained by considering adaptational responses. There were clear differences among genotypes in the level of growth retardation by low [CO_2] (Table 12.1). Total plant weight was reduced by 76% in genotype number 1, but only by 29% in genotype number 4, with the other genotypes being intermediate (Table 12.1). There was a similar degree of variability in the response of many other variables to a reduction in [CO_2] (Table 12.1). The overall mean stomatal density for the species decreased slightly (9%), but significantly ($P < 0.001$) in response to low [CO_2]. However, stomatal density decreased most in the genotype with the largest growth reduction (#1 Table 12.1) and actually increased in the genotype with the slightest reduction of growth (#4 Table 12.1). Thus, it seems highly likely that genotypes in which the stomatal density increased with decreasing

TABLE 12.1
**The Relative Response of Biomass and Morphology of *Nothofagus*
cunninghamii to Exposure to Half-Current [CO$_2$]**

	Mean Species Response	Genotype-Specific Responses			
		Most Growth Reduction #1	#2	#3	Least Growth Reduction #4
Total plant wt	0.44	0.24	0.37	0.54	0.71
Leaf wt	0.49	0.29	0.42	0.53	0.81
Stem wt	0.42	0.23	0.38	0.55	0.57
Root wt	0.41	0.21	0.32	0.52	0.82
Stem length	0.54	0.30	0.49	0.89	0.64
No. of leaves	0.70	0.62	0.63	0.55	1.07
Total leaf area	0.60	0.36	0.58	0.64	0.90
SLA	1.35	1.23	1.81	1.19	1.11
LAR	1.45	1.49	1.79	1.19	1.26
Weight / leaf	0.68	0.47	0.57	0.98	0.76
Area / leaf	0.88	0.58	0.93	1.15	0.86
Stomatal density	0.91	0.80	0.86	0.92	1.06

Notes: Values are the ratio of the mean value for plants grown at 170 mmol CO$_2$ mol^{-1} with respect to controls the mean species response is the average value for all plants, ignoring genotype. Genotypes are ranked from left to right, with that genotype in which depleted [CO$_2$] caused the greatest growth retardation on the left.

[CO$_2$] would grow better and, therefore, be selected for, in conditions of low [CO$_2$]. Hence, changes in stomatal density over evolutionary periods may well be opposite to those observed in short-term acclimatory experiments.

Following the trend for stomatal density, it can be seen that average acclimatory responses are different to predicted adaptational responses for many variables (Table 12.1). For instance, the mean acclimatory response of *N. cunninghamii* to depleted [CO$_2$] is a marked increase in specific leaf area (SLA) and leaf area ratio (LAR). It would also appear likely that there would be a large reduction in the number of leaves per plant and the total leaf area per plant, as well as in the mean weight and area of individual leaves (Table 12.1). However, if genotype number 4 is selected for, then we might expect that the number of leaves per plant and total leaf area per plant would be largely unaffected by a halving of [CO$_2$] (Table 12.1). The increase in SLA and LAR would also be less than is expected from estimation of the mean response of the species because those genotypes that experienced a significant growth reduction (genotype numbers 1 and 2) had large increases in both SLA and LAR.

12.4.2.1 Taking Intraspecific Variation into Account When Comparing Species

Intraspecific variation is important when considering the potential for evolutionary change in response to climate change (Kingsolver 1996; Thomas and Jasienski 1996). Intraspecific variation within one species can, however, also affect a competing species directly without the requirement for microevolution to occur (Schmid et al. 1996). This is important when predicting the impact of climate change on the survival of species in communities because ignoring potential variability within species makes comparisons among species flawed. This could prove important, for instance, in selecting pasture species.

Take for instance, two currently cooccurring species that exist in a state of competitive balance, have similar growth rates, and have a distribution of responsiveness to elevated $[CO_2]$ as given in Figure 12.3. Let the growth response be the ratio of growth at elevated $[CO_2]$ to that of controls and, therefore, a measure of the level of growth increase expected under the altered environment. Species A is on average more responsive to the increase in $[CO_2]$ than is species B, so species A might be predicted to grow faster than B under conditions of global climate change and, therefore, outcompete species B (Navas 1999; Greer 2000). However, if the environment is competitive, then selection is bound to play a role immediately. When establishment becomes possible (such as after a fire or upon formation of a gap) and resources are limited, those individual plants that grow most quickly and are able to establish first are likely to be those that survive to reproductive maturity. In the hypothetical example illustrated in Figure 12.3, the highly responsive individuals of species B would grow quickly enough to maintain the species within the community. The degree of competition and the carrying capacity of the site are important factors in this scenario. Let us assume that when a particular site becomes vacant and conditions are suitable for establishment, there are 100 propagules of each of species A and B present. The two species have the distribution of growth response shown in Figure 12.3. If the vacant site can support 20 individuals to reproductive maturity, then on average 12.5 (62.5%) of these will be of the more responsive species A and 7.5 individuals of species B. However, if the vacant site can only support 10 individuals to reproductive maturity, then on average, three-quarters of the successful individuals will be species B. This demonstrates that in environments in which survival to reproductive maturity is low, the mean responsiveness of the component species becomes unimportant and it is the actual growth rates of individual plants that determines the outcome. It is also possible that because the most responsive individuals of species B are more responsive than any individuals of species A, if the competition were intense enough, species A may become locally extinct. In our hypothetical example, this occurs if only five plants or fewer can be supported in the vacant site. While this hypothetical example is highly simplistic, it does illustrate some of the shortcomings of research in which species' average responsiveness is investigated with no consideration of intraspecific variation. It also provides a possible reason that smaller-scale experiments fail to predict outcomes from larger competition experiments.

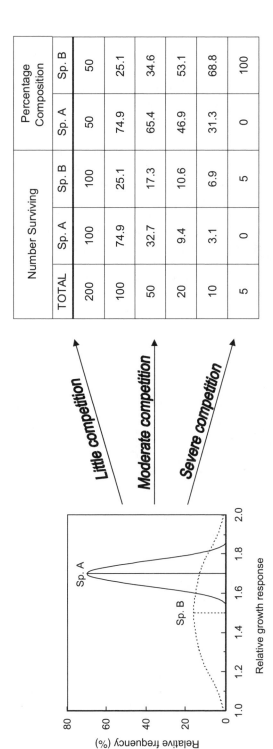

Number Surviving			Percentage Composition	
TOTAL	Sp. A	Sp. B	Sp. A	Sp. B
200	100	100	50	50
100	74.9	25.1	74.9	25.1
50	32.7	17.3	65.4	34.6
20	9.4	10.6	46.9	53.1
10	3.1	6.9	31.3	68.8
5	0	5	0	100

FIGURE 12.3 The effect of intensity of competition and variation in the growth stimulation of two hypothetical species on community composition. Species A has a relative stimulation of 1.7 and low variance, whereas species B has a relative stimulation of 1.5 and large variance. The composition of the community depends on both the level of competition and the intraspecific variation.

12.5 CONCLUSIONS

One of the main interests in studying responses to global climate change is the variation among species' responses. This is important from an agricultural perspective because some species may become more or less profitable than others, but these changes may also produce changes in community and ecosystem dynamics. Analyses in which the mean response of a species is extrapolated to long-term changes is fairly routine when comparisons of the responsiveness to climate change are made among species (Greer et al. 1995). One of the major problems with predicting the responses of species to future conditions is that experiments in which plants are grown in manipulated environments are really studies of acclimatory (i.e., the plastic response of individuals) rather than of adaptive (i.e., evolutionary) responses. While acclimatory responses are of interest for long-lived plants (which are likely to survive until the future conditions arrive) and for crop plants (which are artificially selected), the evolutionary ramifications of the response to climate change are far reaching and poorly understood.

Artificial selection and controlled natural selection experiments have shown that adaptation does occur in response to simulated climate change. Other work has shown that short-lived plants have been adapting to the increase in atmospheric $[CO_2]$ since the industrial revolution. Thus, there is a high likelihood of adaptational responses in any system where there is differential survival and reproduction. This is important because it means that the average response, as estimated in acclimatory experiments, is not relevant because selection may skew the mean by selecting against some individuals and selecting for others.

Adaptation may alter the way in which organisms respond to the changing climate, and such changes can have large implications for models of future impacts. Models of carbon sequestration depend upon current information regarding photosynthetic carbon assimilation and allocation. If these patterns change because of the selective pressure of climate change, then such models may be inaccurate.

Intraspecific variation in the responses to climate change is known to exist in many species. The impacts of natural selection, and thus the adaptational responses to climate change, are known for only a few, short-lived plant species. Our ability to predict and understand the long-term responses to climate change depends upon the improvement of our knowledge of species' adaptational responses. Such knowledge must come from both direct means (selection experiments) and indirect means (genotypic variation). Without this information, our ability to predict climate change impacts accurately will be severely limited.

REFERENCES

Backhausen, J.E. and Scheibe, R. (1999) Adaptation of tobacco plants to elevated CO_2: influence of leaf age on changes in physiology, redox states and NADP-malate dehydrogenase activity. *Journal of Experimental Botany,* 50, 665–675.

Bacon, T. (1837) *First Impressions of Hindostan, Vol. I.* W.H. Allen & Co., London.

Bryant, J., Taylor, G., and Frehner, M. (1998) Photosynthetic acclimation to elevated CO_2 is modified by source: sink balance in three component species of chalk grassland swards grown in a Free Air Carbon Dioxide Enrichment (FACE) experiment. *Plant, Cell and Environment,* 21, 159–168.

Bunce, J.A. (2001) Are annual plants adapted to the current atmospheric concentration of carbon dioxide? *International Journal of Plant Science,* 162, 1261–1266.

Conner, J.K. (2003) Artificial selection: a powerful tool for ecologists. *Ecology,* 84, 1650–1660.

Curtis, P.S., Klus, D.J., Kalisz, S., and Tonsor, S.J. (1996) Intraspecific variation in CO_2 responses in *Raphanus rahpanistrum* and *Plantago laceolata*: assessing the potential for evolutionary change with rising atmospheric CO_2. In *Carbon Dioxide, Populations and Communities*, Körner, C. and Bazzazz, F.A., Eds., Academic Press, San Diego, CA, 13–22.

Darwin, C.R. (1859) *On the Origin of Species by Means of Natural Selection or the Preservation of Favoured Races in the Struggle for Life,* John Murray, London.

Darwin, C.R. and Wallace, A.R. (1859) On the tendency of species to form varieties; and on the perpetuation of varieties and species by natural means of selection. *Journal of the Proceedings of the Linnean Society (Zoology),* 3, 45–62.

Easterling, W.E., Chhetri, N., and Niu, X.Z. (2003) Improving the realism of modeling agronomic adaptation to climate change: simulating technological substitution. *Climatic Change,* 60, 149–173.

Etterson, J.R. and Shaw, R.G. (2001) Constraint to adaptive evolution in response to global warming. *Science,* 294, 151–154.

Florio, J. (1611) *Queen Anna's New World of Words, or Dictionarie of Italian and English Tongues*, Melch and Bradwood, London.

Gielen, B. and Ceulemans, R. (2001) The likely impact of rising atmospheric CO_2 on natural and managed *Populus*: a literature review. *Environmental Pollution,* 115, 335–358.

Greer, D. (2000) The effect of perturbations in temperature and photon flux density on the growth and photosynthetic responses of five pasture species to elevated CO_2. *Australian Journal of Plant Physiology,* 27, 301–310.

Greer, D.H., Laing, W.A., and Campbell, B.D. (1995) Photosynthetic responses of thirteen pasture species to elevated CO_2 and temperature. *Australian Journal of Plant Physiology,* 22, 713–722.

Hovenden, M.J. (2003) Photosynthesis of coppicing poplar clones in a Free Air CO_2 Enrichment (FACE) experiment in a short rotation forest. *Functional Plant Biology,* 30, 391–400.

Hovenden, M.J. and Schimanski, L.J. (2000) Genotypic differences in growth and stomatal morphology of Southern Beech, *Nothofagus cunninghamii*, exposed to depleted CO_2 concentrations. *Australian Journal of Plant Physiology,* 27, 281–287.

Hulme, M. (2003) Abrupt climate change: can society cope? *Philosophical Transactions of the Royal Society of London Series A Mathematical, Physical, and Engineering Sciences,* 361, 2001–2019.

IPCC (Intergevernmental Panel on Climate Change) (2001) *Climate Change 2001: Intergovenmental Panel on Climate Change Third Assessment Report*, Cambridge University Press, Cambridge, U.K.

Jonson, B. (1640) *Workes,* Vol. 2. R. Meighen, London.

Kingsolver, J.G. (1996) Physiological sensitivity and evolutionary responses to climate change. In *Carbon Dioxide, Populations and Communities*, Körner, C. and Bazzazz, F.A., Eds., Academic Press, San Diego, CA, 3–12.

Medlyn, B.E., Badeck, F.W., De Pury, D.G.G., et al. (1999) Effects of elevated CO_2 on photosynthesis in European forest species: a meta-analysis of model parameters. *Plant, Cell and Environment*, 22, 1475–1495.

Moore, B. (1997) *The Australian Concise Oxford Dictionary of Current English*. Oxford University Press, Melbourne, Australia.

Navas, M-L. (1999) Effect of competition on the responses of grasses and legumes to elevated atmospheric CO_2 along a nitrogen gradient: differences between isolated plants, monocultures and multi-species mixtures. *New Phytologist*, 143, 323–331.

Norby, R.J., Wullschleger, S.D., Gunderson, C.A., Johnson, D.W., and Ceulemans, R. (1999) Tree responses to rising CO_2 in field experiments: implications for the future forest. *Plant, Cell and Environment*, 22, 683–714.

Radoglou, K. and Jarvis, P. (1990) Effects of CO_2 enrichment on four poplar clones I. Growth and leaf anatomy. *Annals of Botany*, 65, 617–626.

Read, J. (1995) The importance of comparative growth rates in determining the canopy composition of Tasmanian rainforest. *Australian Journal of Botany*, 43, 243–271.

Read, J. (1999) Rainforest ecology. In *Vegetation of Tasmania*, Reid, J.B., Hill, R.S., Brown, M.J., and Hovenden, M.J., Eds., Australian Biological Resources Study, Canberra, Australia, 160–197.

Rehfelt, G.E., Tchebakova, N.M., Parfenova, Y.I., Wykoff, W.R., Kuzmina, N.A., and Milyutin, L.I. (2002) Intraspecific responses to climate in *Pinus sylvestris*. *Global Change Biology*, 8, 912–919.

Roumet, C., Garnier, E., Suzor, H., Salager, J., and Roy, J. (2000) Short and long-term responses of whole-plant gas exchange to elevated CO_2 in four herbaceous species. *Environmental and Experimental Botany*, 43, 155–169.

Roumet, C., Laurent, G., and Roy, J. (1999) Leaf structure and chemical composition as affected by elevated CO_2: genotypic responses of two perennial grasses. *New Phytologist*, 143, 73–81.

Roumet, C. and Roy, J. (1996) Prediction of the growth response to elevated CO_2: a search for physiological criteria in closely related grass species. *New Phytologist*, 134, 615–621.

Sage, R. (1994) Acclimation of photosynthesis to increasing atmospheric CO_2: the gas exchange perspective. *Photosynthesis Research*, 39, 351–368.

Schmid, B., Birrer, A., and Lavigne, C. (1996) Genetic variation in the response of plant populations to elevated CO_2 in a nutrient poor, calcareous grassland. In *Carbon Dioxide, Populations and Communities*, Körner, C. and Bazzazz, F.A., Eds., Academic Press, San Diego, CA, 31–50.

Simpson, J.A. and Weiner, E.S.C. (1989) *The Oxford English Dictionary*, Clarendon Press, Oxford, U.K..

Stitt, M. (1991) Rising CO_2 levels and their potential significance for carbon flow in photosynthetic cells. *Plant, Cell and Environment*, 17, 465–487.

Thomas, S.C. and Jasienski, M. (1996) Genetic variability and the nature of microevolutionary responses to elevated CO_2. In *Carbon Dioxide, Populations and Communities*, Körner, C. and Bazzazz, F.A., Eds., Academic Press, San Diego, CA, 51–81.

Topinard, P. (1890) *Anthropology*, Chapman and Hall, London.

Tousignant, D. and Potvin, C. (1996) Selective responses to global change: experimental results on *Brassica juncea* (L.) Czern. In *Carbon Dioxide, Populations and Communities*, Körner, C. and Bazzazz, F.A., Eds., Academic Press, San Diego, CA, 23–30.

Ward, J.K., Antonovics, J., Thomas, R.B., and Strain, B.R. (2000) Is atmospheric CO_2 a selective agent on model C_3 annuals? *Oecologia*, 123, 330–341.

Wong, S-C. (1979) Elevated atmospheric partial pressure of CO_2 and plant growth I. Interactions of nitrogen nutrition and photosynthetic capacity in C_3 and C_4 plants. *Oecologia*, 44, 68–74.

Woodward, F.I. (1987) Stomatal numbers are sensitive to increases in CO_2 from pre-industrial level. *Nature*, 327, 617–618.

Woodward, F.I. and Kelly, C.K. (1995) The influence of CO_2 concentration on stomatal density. *New Phytologist*, 131, 311–327.

Ziska, L.H., Bunce, J.A., and Caulfield, F. (1998) Intraspecific variation in seed yield of soybean (*Glycine max*) in response to increased atmospheric carbon dioxide. *Australian Journal of Plant Physiology*, 25, 801–807.

13 Plant Breeding for a Changing Environment

Paul C.D. Newton and Grant R. Edwards

CONTENTS

13.1 INTRODUCTION

As the direction and pace of environmental change becomes clearer, the possibility of taking prophylactic steps to either mitigate or adapt to these changes becomes more realistic. In this respect, there is clearly a much greater opportunity to intervene in managed agroecosystems than in natural ecosystems, assuming that relevant interventions are available and correct choices are made. In this chapter we consider one of these options — the development of new cultivars — and consider whether it makes sense to consider breeding to take advantage of the increasing concentration of CO_2 in the atmosphere (the most predictable of the changes [Newton et al., Chapter 1, this volume] and, therefore, arguably potentially the most attractive to plant breeders) and whether we have the understanding necessary to make this breeding a success. The new technologies are widely touted (White et al. 2004) as offering new, previously unobtainable, possibilities to plant breeders, and we will conclude by considering their contribution to mitigating or adapting to climate change.

13.2 IS THERE AN ADVANTAGE IN BREEDING FOR A HIGHER CO₂ WORLD?

The possibility of breeding for global change has been considered before by Hall and Allen (1993) and more recently by Hall and Ziska (2000). Reading these two papers, it is evident that rather little has been attempted along these lines; so are breeders taking this possibility seriously? Perhaps the first question we need to ask is whether it is possible to show that a significant advantage can be gained from breeding specifically for change, and in this case, for an atmosphere with a higher concentration of CO_2 The first question we consider is: Will breeding for a higher-CO_2 world result in production gains of any consequence? Note that this is not the same as asking whether elevated CO_2 increases crop yields, because here we are interested in potential gains above any conferred by a CO_2 growth response in plant material adapted to current or precurrent CO_2 concentration.

Let us consider the information we have available from elevated CO_2 studies to address this question. First there are numerous studies of CO_2 effects on plant performance conducted in a range of experimental situations and summarized in reviews such as that of Poorter and Navas (2003). These data are not directly relevant to our question as the plant material used is not adapted to the higher CO_2 environment, which is the case we wish to explore. However, the data are informative about interspecific differences in response to CO_2 and, thus, offer some guidance as to how a change in crop type might be used to adapt to a change in CO_2. Some relevant generalities that emerge are that C_3 crops are generally more responsive than C_4 crops, and those with large sinks, such as tubers, do relatively better in an increasingly CO_2-rich environment (Figure 13.1).

The second type of data we have are comparisons of CO_2 response among genotypes of the same species. These can involve studies within populations or the more agronomic approach using different cultivars. The standard format for these experiments is to grow the material at current and elevated CO_2, observe reaction norms, and analyse for genotype (cultivar) $\times CO_2$ interactions. If the interaction is nonsignificant then the reaction norms are parallel (Figure 13.2a) and the ranking of genotypes does not change from current to elevated CO_2. In this case we assume there is no "inherited differences in response to elevated CO_2," (Klus et al. 2001 p. 1081). Where the interaction is significant, this implies that some of the genotypes do relatively better (or worse) at elevated CO_2 and that there is genetic variation in response (Figure 13.2b). From such studies it has become evident that heritable

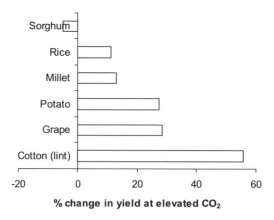

FIGURE 13.1 Responsiveness of various crop plants to elevated CO_2. (Adapted from Kimball, B.A., Kobayashi, K., and Bindi, M. (2002) *Advances in Agronomy*, 77, 293–368.)

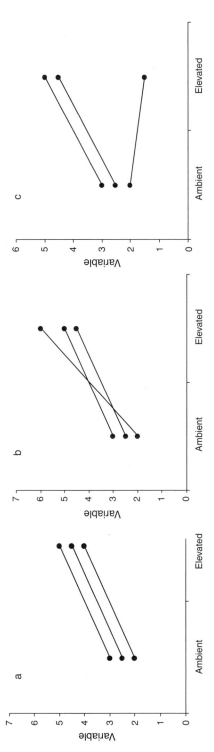

FIGURE 13.2 Reaction norms for hypothetical responses of plants to elevated CO_2. (a) The case where there is no CO_2–genotype interaction, (b) the case when a CO_2–genotype interaction exists indicating genetic variation in response to CO_2, and (c) the case where there is increased variation in response at elevated compared to ambient CO_2.

variation in response to elevated CO_2 is present in many species for a range of traits including photosynthesis (Curtis et al. 1996), stomatal characteristics (Case et al. 1998), germination and growth (Wulff and Alexander 1985; Schmid et al. 1996; Mohan et al. 2004) and fecundity (Curtis et al. 1994). A recent review of this literature can be found in Ward and Kelly (2004). These data suggest that plant breeders may have material with which to work. Tests using existing cultivars have also demonstrated significant cultivar $\times CO_2$ interactions indicating a change in rank order of cultivars between ambient and elevated CO_2 and potentially useful information about cultivars that may perform well in the future (e.g., Moya et al. 1998; Ziska et al. 2001; Baker 2003).

Additional information can be extracted from these genotype and cultivar response experiments relating to the variation available for selection in current and elevated-CO_2 environments. The interpretation of Figure 13.2c is that variation for selection is greater under a future CO_2 environment; this is what one would expect if plants were adapted to the current environment but not to the future environment (Schmid et al. 1996). If the slopes of the lines in Figure 13.2c were reversed, this would imply that plants were better adapted to the future environment; an unlikely situation as selection would not yet have been occurring (Schmid et al. 1996). Figure 13.3a shows data on the seed yield of 17 rice (*Oryza sativa* L.) cultivars (from Ziska et al. 1996) grown at ambient and elevated CO_2. The reaction norms show the pattern from Figure 13.2b demonstrating that there is genetic variation in response to elevated CO_2. We can also use these data to consider the range of variation for selection across the two environments in Figure 13.3b — the case considered in Figure 13.2c. There is little evidence for reduced variation at ambient CO_2 implying that these cultivars are not well adapted to current CO_2. We will return to this subject in the section on plant traits.

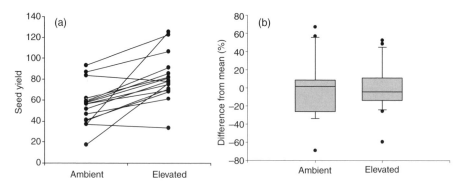

FIGURE 13.3 (a) Reaction norms of 17 soybean cultivars to ambient and elevated CO_2 concentration (Adapted from Table 5 of Ziska, L.H., Manalo, P.A., and Ordonez, R.A. (1996) *Journal of Experimental Botany*, 47, 1353–1359). (b) The same data recalculated to show the proportional difference in yield between each cultivar and the grand mean within each CO_2 environment displayed as a percentile box plot (median, 10th, 25th, 75th, and 90th percentiles) of seed yield data of rice (g plant^{-1}).

The information above still does not place us in a position to determine whether a significant advantage could be gained from breeding specifically for an elevated-CO_2 environment because the plant germplasm used is not necessarily adapted to the higher CO_2 concentration. The true test of benefits would only come from testing a high CO_2-adapted genotype or population at high CO_2, mimicking the outcome of plant breeding. One line of research that offers some insights is the study of plant populations growing around natural CO_2 springs (e.g., Raschi and Miglietta 1997). These populations have been exposed to elevated CO_2 over evolutionarily relevant time spans, and in the absence of confounding biotic or edaphic factors, provide a source of CO_2-adapted material. We have sampled such populations of the grass *Dactylis glomerata* L. from a natural CO_2 spring in the Northland region of New Zealand (Newton et al. 1996). Seed was collected from high and low CO_2 sites and plants grown in a common garden at ambient CO_2. Plants from high and low sites were then grown independently in pollen exclusion houses to create two synthetic populations — one adapted to ambient or preambient CO_2 and the other to elevated CO_2; this polycross process was repeated for 2 years. While the procedure is regarded as generating genotypically different populations (Ward et al. 2000), maternal effects can carry over through several generations (Durrant 1962), and it cannot be stated unequivocally that these were not present. We determined that the main selection pressure for these plant populations was CO_2, as changes in soil and plant properties around the spring were strongly correlated with CO_2 concentration and not with other factors such as soil pH or moisture content (Ross et al. 2000). Seed of the high and ambient populations from the second polycross was then grown in current and future environments. In an important extension of this work from previous studies (e.g., Fordham et al. 1997), the future environment included not only an elevated CO_2 atmosphere (475 ppm), but also soil collected from the natural CO_2 springs that had developed under plants in a high CO_2 atmosphere (Ross et al. 2000; Newton et al. 2001; Edwards et al. 2003). The outcome, shown in Figure 13.4, was that (1) home populations did better in their home environment than the away population, indicating that adaptation had occurred (Ward et al. 2000); (2) both populations grew better in elevated CO_2; and most significantly for our question on plant breeding for the future; and (3) in the high CO_2 environment of the future, the high CO_2-adapted population produced 30% more biomass than the population adapted to current or precurrent CO_2. The percentage change in biomass per ppm of CO_2 was 0.28%; if we assume the CO_2 to be increasing at a rate of 1.5 ppm per year, then the gain in biomass per decade is 4.2%. To put this in perspective, the calculated increase in forage yield from U.S. breeding programmes for *Dactylis glomerata* between 1955 and 1997 was 4.5% per decade (Casler et al. 2000).

We do not want to lean too heavily on one set of data and make general statements about the potential gains to be made from breeding for high CO_2 across all crops. However, we do suggest that there is sufficient evidence to seriously consider breeding objectives around elevated CO_2 or at least encouragement for further experiments of the kind we have described, which remain very scarce (Kohut 2003; Ward and Kelly 2004). If we are to proceed on the assumption that there are benefits to be gained from breeding for elevated CO_2, can we identify

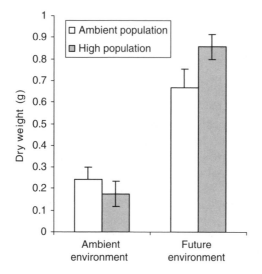

FIGURE 13.4 Total dry matter of plants (mean, sem) of *Dactylis glomerata* from populations adapted to ambient or elevated CO_2 and grown either in ambient (current atmospheric CO_2 and soil developed under ambient and preambient CO_2) or future (atmospheric CO_2 concentration of 475 ppm and soil developed under elevated CO_2 conditions. (From Newton, P.C.D., Clark, H., Edwards, G.R., and Ross, D.J. (2001) *Ecology Letters*, 4, 344–347.) There was a significant population by environment interaction (p = 0.04).

traits that we should be selecting for, and is this an opportunity to apply the new technologies?

13.3 TRAITS FOR ELEVATED CO_2

After 10 years of intensive research on CO_2 effects on plants, Thomas and Jasienski (1996, p. 73) concluded that "no strong *a priori* theory exists for allowing for predictions of what traits will be favoured under rising CO_2." After another 10 years of research, the position is still not markedly improved, and we find Ward and Kelly (2004, p. 433) commenting that "[v]ery little is known about the actual changes that may occur as plants respond to selective pressure from changes in CO_2." Not only does this lack of understanding prevent us from picking "winners and losers" (Poorter and Navas 2003, p. 175), but it leaves plant breeding with rather little specific to work with. This task is made even more difficult when we appreciate that the selection pressures arising from changes in CO_2 could be wide ranging, including indirect effects of elevated CO_2 on soil properties, the level of herbivore feeding, and community composition.

Hall and Allen (1993) and Hall and Ziska (2000) concluded from CO_2 response studies that plants that will be favoured at elevated CO_2 will avoid downregulation of photosynthesis and have a high harvest index (a high sink/source ratio). There is some evidence to support this view. For example, in two studies with wheat (Manderscheid and Weigel 1997; Ziska et al. 2004), it has been found that older cultivars, circa early 20th century, respond significantly more strongly to elevated CO_2 than modern varieties; the reason being that the older cultivars increase both their dry matter and seed yield by having a greater capacity to tiller. Similar results have been reported for soybean, in which increased seed yield at elevated CO_2 could be ascribed to seed production on lateral branches (Ziska et al. 2001) and rice, where the newer plant material with a fixed tiller number was unable to respond to elevated CO_2 (Moya et al. 1998). It should be noted that this work has been largely confined to pot trials in controlled environments, and it remains unknown whether a CO_2 response expressed through increased lateral branching would be expressed in the field at high planting densities (Hall and Ziska 2000). The notion of testing the CO_2 responsiveness of cultivars or genotypes that have distinct trait differences is attractive, but has rarely been executed. The Ziska et al. (2001) tests of soybean cultivars was a reasonably direct test of the sink–source theory as it involved cultivars that were either determinate or indeterminate. In this case there was no clear difference in responsiveness among these two forms. An even stronger test of trait effects is possible if we compare isolines that vary only in the trait of interest. The first paper describing an experiment of this type in a field setting looked at source–sink relationships in soybean (Ainsworth et al. 2004) and provided qualified support for the suggestions of Hall and Allen (1993) and Hall and Ziska (2000). Downregulation of photosynthesis was observed in a determinate genotype of the indeterminate cultivar Williams; however, increased responsiveness to elevated CO_2 was not observed in an indeterminate form of the determinate cultivar Elf, probably because Elf had been bred to produce large numbers of pods on branches as well as the main stem, and likely had strong sink activity (Ainsworth et al. 2004). This point is also probably relevant to the results of the Ziska et al. (2001) comparisons.

An interesting insight into growth at elevated CO_2 can be gained from a study of adaptation to low CO_2 as there is strong evidence to suggest that plants are currently adapted to subambient CO_2 (Sage and Coleman 2001). Bunce (2001) grew four C_3 weedy species at 280, 370 (the current concentration), and 460 ppm CO_2; in every case the increase in yield due to elevated CO_2 was more strongly expressed between 280 and 370 ppm than between 370 and 460 ppm. In two cases the yield actually went down in the 460 compared to 370 ppm environment. The average increase in dry matter over the period 280 to 370 ppm was in the order of 60%; assuming an increasing CO_2 of 1.5 ppm per year, this equates to an increase in plant growth of about 9% per decade — an increase that stands comparison with those achieved through plant breeding (Goodman 2004). Bunce (2001) concluded from his experiment that plants are adapted to the current level of CO_2 and, therefore, will not necessarily respond to further increases. Sage and Coleman (2001, p. 22) also identify this response and conclude that advances might be made in crop breeding through overcoming "constraints linked to adaptations to low levels of $[CO_2]$." Traits they identify are: A high allocation to leaf rather than root biomass, expression of high levels of Rubisco, high carbonic anhydrase activity, high expression of Rubisco activase, and low investment in sucrose synthase. To this list we can add high stomatal density (Ward and Kelly 2004).

13.4 TOOLS AND APPROACHES

Effective progress has been made in plant breeding by simple recurrent selection. This has some advantage in relation to global change as selections are made in the total environment, so that selection for resistance to, say root-feeding pathogen may also include selection for growth at the current CO_2, although this is not likely to be as effective as direct selection for this environmental condition (Hall and Ziska 2000). Even if plants were selected for optimal performance at their ambient CO_2, the plant breeding process for crop plants is likely to extend over 10 to 15 years; the cultivar may then spend some time on the recommended lists extending the period from first selection over perhaps 20 years. Of course, long-lived species such as trees may have a much longer development stage and necessarily a longer period of growth and use on the farm. Reselection may offer some possibility of adjusting the cultivar to a new higher level of CO_2, but note here that it is a requirement of plant variety rights that cultivars are maintained in their original genotypic condition (i.e., the genotype selected in 1970 remains the same genotype sold in 1990), unless deliberate reselection has occurred.

The new technologies, genomics, and molecular transformation have been seen as offering tremendous scope for widening the pool of variation that breeders can exploit, as well as increasing the precision with which changes can be made and reducing the time taken to develop new cultivars. It is clearly advantageous for breeders to have more weapons in their armoury, but are these new weapons going to make a substantive contribution to mitigating or taking advantage of climate change? Sinclair et al. (2004) have described in detail how difficult it is to translate an understanding at the molecular level into effective cultivar development. A continuing difficulty is translating gains at the bottom of the yield hierarchy into gains of similar magnitude at the top of the yield hierarchy (yield) (Sinclair et al. 2004). For example, a major effort to identify lines of grass plants (*Lolium perenne*) with reduced rates of mature leaf respiration has not resulted in the development of new cultivars, as respiration rate was not consistently related to differences in yield (Kraus and Lambers 2001). Molecular techniques, of course, go even further back down the hierarchy extending the possibilities for tradeoffs and diluting potential gains.

The possibility of unlocking new sources of variation — perhaps through transgenic technologies that may offer previously unobtainable traits — is an attractive prospect. Clearly, the desirable traits need to be well characterized; unfortunately, in regard to elevated CO_2 we have seen that this is not the case. So our lack of mechanistic understanding of plant responses to elevated CO_2 remains a difficulty for this line of breeding progress. A further issue is the multifaceted nature of global change; we are not only considering changes to CO_2 but also potential changes in rainfall and temperature, the responses to which are likely to be controlled by multiple genes. Whether genetic engineering can meet this challenge is still uncertain (Knight 2003).

13.5 CONCLUSIONS

There is evidence to suggest that substantial gains might be made in crop yields by taking into account the future CO_2 concentration in the atmosphere, which has the advantage of being reasonably predictable and global in extent. At the simplest level, this might involve identifying types of crops more likely to respond to elevated CO_2 or selecting from existing cultivars those that show a more positive response. In addition, we have argued that the evidence suggests that at current rates of change in CO_2 concentration, and with the time taken to develop new cultivars, it appears likely that substantive gains (as judged relative to current gains from breeding) might be achieved by specifically developing plant material for a higher CO_2 environment. This process would be greatly assisted by better information on the traits that confer responsiveness to CO_2 either through comparative experiments of the Ainsworth et al. (2004) kind or from identification of differences in the biology of plants growing at natural CO_2 springs compared to plants from ambient CO_2. Finally, it will be necessary to test the CO_2 selections for their performance in environments that have altered temperature and moisture conditions in order to ensure that there are no negative interactions with potential climatic changes.

REFERENCES

Ainsworth, E.A., Rogers, A., Nelson, R., and Long, S.P. (2004) Testing the "source-sink" hypothesis of down-regulation of photosynthesis in elevated [CO_2] in the field with single gene substitutions in *Glycine max*. *Agricultural and Forest Meteorology*, 122, 85–94.

Baker, J.T. (2003) Yield responses of southern U.S. rice cultivars to CO_2 and temperature. *Agricultural and Forest Meteorology*, 122, 129–137.

Bunce, J.A. (2001) Are annual plants adapted to the current atmospheric concentration of carbon dioxide? *International Journal of Plant Sciences* 162, 1261–1266.

Case, A.L., Curtis, P.S., and Snow, A.A. (1998) Heritable variation in stomatal responses to elevated CO_2 in wild radish, *Raphanus raphanistrum* (Brassicaceae). *American Journal of Botany* 85, 253–258.

Casler, M.D., Fales, S.L., McElroy, A.R., Hall, M.H., Hoffman, L.D., and Leath, K.T. (2000) Genetic progress from 40 years of orchard grass breeding in North America measured under hay management. *Crop Science* 40, 1019–1025.

Curtis, P.S., Klus, D.J., Kalisz, S., and Tonsor, S.J. (1996) Intraspecific variation in CO_2 responses in *Raphanus raphanistrum* and *Plantago lanceolata*: assessing the potential for evolutionary change with rising atmospheric CO_2. In *Carbon Dioxide, Populations and Communities*, Körner, C. and Bazzaz, F.A., Eds., Academic Press, San Diego, CA, 13–22.

Curtis, P.S., Snow, A.A., and Miller, A.S. (1994) Genotype-specific effects of elevated CO_2 on fecundity in wild radish (*Raphanus raphanistrum*). *Oecologia* 97, 100–105.

Durrant, A. (1962) The environmental induction of heritable change in *Linum*. *Heredity*, 17, 27–61.

Edwards, G.R., Clark, H., and Newton, P.C.D. (2003) Soil development under elevated CO_2 affects plant growth responses to CO_2 enrichment. *Basic and Applied Ecology*, 4, 185–195.

Fordham, M.C., Barnes, J.D., Bettarini, I., Miglietta, F., and Raschi, A. (1997) The impact of elevated CO_2 on the growth of *Agrostis canina* adapted to contrasting CO_2 concentrations. In *Carbon Dioxide Springs and Their Use in Biological Research*, Raschi, A. and Miglietta, F., Eds., Cambridge University Press, Cambridge, U.K., 174–196.

Goodman, M.M. (2004) Plant breeding requirements for applied molecular biology. *Crop Science* 44, 1913–1914.

Hall, A.E. and Allen, L.H., Jr. (1993) Designing cultivars for the climatic conditions of the next century. *International Crop Science 1*. Crop Science Society of America, Madison, WI.

Hall, A.E. and Ziska, L.H. (2000) Crop breeding strategies for the 21st century. In *Climate Change and Global Crop Productivity*, Reddy, K.R. and Hodges, H.F., Eds., CAB International, Wallingford, U.K., 407–423.

Kimball, B.A., Kobayashi, K., and Bindi, M. (2002) Responses of agricultural crops to free-air CO_2 enrichment. *Advances in Agronomy*, 77, 293–368.

Klus, D.J., Kalisz, S., Curtis, P.S., Teeri, J.A., and Tonsor, S.T. (2001) Family- and population-level responses to atmospheric CO_2 concentration: gas exchange and the allocation of C, N and biomass in *Plantago lanceolata* (Plantaginaceae). *Americal Journal of Botany*, 88, 1080–1087.

Knight, J. (2003) Crop improvement: a dying breed. *Nature*, 421, 568–70.

Kohut, R. (2003) The long-term effects of carbon dioxide on natural systems: issues and research needs. *Environment International*, 29, 171–180.

Kraus, E. and Lambers, H. (2001) Leaf and root respiration of *Lolium perenne* populations selected for contrasting leaf respiration rates are affected by intra- and interpopulation interactions. *Plant and Soil*, 231, 267–274.

Manderscheid, R. and Weigel, H.J. (1997) Photosynthetic and growth responses of old and modern spring wheat cultivars to atmospheric CO_2 enrichment. *Agriculture, Ecosystems and Environment*, 64, 65–73.

Mohan, J.E., Clark, J.S., and Schlesinger, W.H. (2004) Genetic variation in germination, growth and survivorship of red maple in response to subambient through to elevated atmospheric CO_2. *Global Change Biology* 10, 233–247.

Moya, T.B., Ziska, L.H., Namuco, O.S., and Olszyk, D. (1998) Growth dynamics and genotypic variation in tropical, field-grown paddy rice (*Oryza sativa* L.) in response to increasing carbon dioxide and temperature. *Global Change Biology*, 4, 645–656.

Newton, P.C.D., Bell, C.C., and Clark, H. (1996) Carbon dioxide emissions from mineral springs in Northland and the potential of these sites for studying the effects of elevated carbon dioxide on pastures. *New Zealand Journal of Agricultural Research*, 39, 33–40.

Newton, P.C.D., Clark, H., Edwards, G.R., and Ross, D.J. (2001) Experimental confirmation of ecosystem model predictions comparing transient and equilibrium plant responses to elevated atmospheric CO_2. *Ecology Letters*, 4, 344–347.

Poorter, H. and Navas, M-L. (2003) Plant growth and competition at elevated CO_2: on winners, losers and functional groups. *New Phytologist* 157, 175–198.

Raschi, A. and Miglietta, F. (1997) *Carbon Dioxide Springs and Their Use in Biological Research*, Cambridge University Press, Cambridge, U.K.

Ross, D.J., Tate, K.R., Newton, P.C.D., Wilde, R.H., and Clark, H. (2000) Carbon and nitrogen pools and mineralization in a grassland gley soil under elevated carbon dioxide at a natural CO_2 spring. *Global Change Biology*, 6, 779–790.

Sage, R.F. and Coleman, J.R. (2001) Effects of low atmospheric CO_2 on plants: more than a thing of the past. *Trends in Plant Science* 6, 18–24.

Schmid, B., Birrer, A., and Lavigne, C. (1996) Genetic variation in the response of plant populations to elevated CO_2 in a nutrient-poor, calcareous grassland. In *Carbon Dioxide, Populations and Communities*, Körner, C. and Bazzaz, F.A., Eds., Academic Press, San Diego, CA, 31–50.

Sinclair, T.R., Purcell, L.C., and Sneller, C.H. (2004) Crop transformation and the challenge to increase yield potential. *Trends in Plant Science* 9, 70–75.

Thomas, S.C. and Jasieski, M. (1996) Genetic variability and the nature of microevolutionary responses to elevated CO_2. In *Carbon Dioxide, Populations and Communities*, Körner, C. and Bazzaz, F.A., Eds., Academic Press, San Diego, CA, 51–81.

Ward, J.K., Antonovics, J., Thomas, R.B., and Strain, B.R. (2000) Is atmospheric CO_2 a selective agent on model C_3 annuals? *Oecologia*, 123, 330–341.

Ward, J.K. and Kelly, J.K. (2004) Scaling up evolutionary responses to elevated CO_2: lessons from *Arabidopsis*. *Ecology Letters* 7, 427–440.

White, J.W., McMaster, G.S., and Edmeades, G.O. (2004) Genomics and crop responses to global change: what have we learned? *Field Crops Research* 90, 165–169.

Wulff, R.D. and Alexander, H.M. (1985) Intraspecific variation in the response to CO_2 enrichment in seeds and seedlings of *Plantago lanceolata* L. *Oecologia* 66, 458–460.

Ziska, L.H., Bunce, J.A., and Caulfield, F.A. (2001) Rising atmospheric carbon dioxide and seed yield of soybean genotypes. *Crop Science* 41, 385–391.

Ziska, L.H., Manalo, P.A., and Ordonez, R.A. (1996) Intraspecific variation in the response of rice (*Oryza sativa* L.) to increased CO_2 and temperature: growth and yield response of 17 cultivars. *Journal of Experimental Botany*, 47, 1353–1359.

Ziska, L.H., Morris, C.F., and Goins, E.W. (2004) Quantitative and qualitative evaluation of selected wheat varieties released since 1903 to increasing atmospheric carbon dioxide: can yield sensitivity to carbon dioxide be a factor in wheat performance. *Global Change Biology*, 10, 1810–1819.

Part IV

Special Examples

Special Example 1

Impacts of Climate Change on Marginal Tropical Animal Production Systems

Chris Stokes and Andrew Ash

CONTENTS

SE 1.1 CLIMATE CHANGE IN TROPICAL RANGELANDS VS. TEMPERATE ANIMAL PRODUCTION SYSTEMS

This example is aimed at complementing the chapter of Carran and Allard (2006, Chapter 6, this volume) (which focused on the impacts of climate change on productive temperate animal production systems) with a perspective from more marginal environments, particularly in tropical rangelands. There are a number of important distinctions in climatically variable tropical rangeland systems that would be expected to differentiate their responses to climate change from those in more mesic temperate systems:

1. Forage quality is generally poor and forage supply is highly variable in tropical rangeland systems. Any reduction in forage quality (associated with increased carbon assimilation and nutrient dilution) under elevated atmospheric CO_2 ($e[CO_2]$) is likely to be detrimental to pastoral enterprises. In these water-limiting environments, interactions among CO_2,

323

rainfall, and temperature under climate change will greatly influence forage supply.
2. Tropical pastures, dominated by C4 grasses, would be expected to respond
    differently to C3-dominated temperate grasslands.
3. In tropical regions, current temperatures are already close to the optimum
    for physiological processes, so rising temperatures will largely have a
    negative effect by increasing heat stress for plants and animals, and will
    produce climatic conditions without recent precedent anywhere else on
    earth, limiting the opportunities for agricultural and ecological adaptation.
4. The capacity of human communities in tropical areas is further constrained
    by their relative socioeconomic disadvantage and the paucity of informa-
    tion about the likely impacts of climate change and adaptation options in
    the tropics (Rosenzweig and Liverman 1992).

We briefly consider the effects of rising $[CO_2]$ on forage production and forage
quality in tropical rangeland systems, with specific reference to responses of C4
grasses in a coastal tropical savanna in the Australian savanna free air CO_2 enrichment (OzFACE) experiment (Stokes et al., forthcoming).

SE 1.2   FORAGE PRODUCTION:
MOISTURE-MEDIATED RESPONSES

Early expectations had been that C_3 plants would be substantially advantaged by the
stimulation of carbon fixation under $e[CO_2]$ relative to C_4 species, in which photosynthetic rates are close to saturation at the current atmospheric $[CO_2]$ (Gifford
1989). However, evidence from temperate grasslands suggests that much of the
response to $e[CO_2]$ is mediated by the effects of increased water-use efficiency (rather
than directly stimulated carbon fixation) (Morgan et al. 2004) and, correspondingly,
this has lead to a growing recognition of the role C_4 grasses could play in vegetation
responses (Owensby et al. 1993; Anderson et al. 2001). This was highlighted in a
recent review that found biomass increases of 44% for C_3 grasses (at 550 to 750
μmol mol[1] vs. ambient $[CO_2]$) and of 33% for C_4 grasses (Wand et al. 1999). Tropical
pastures, dominated by C_4 grasses, could also respond substantially to $e[CO_2]$. There
are a number of mechanisms by which altered water fluxes under $e[CO_2]$ could
benefit plants: Reduced water stress associated with increased plant water potentials;
enhanced growth associated with increased water-use efficiency; increases in soil
moisture associated with reduced plant water demand; an increase in the duration
of the growing season, if water savings through the wet season accumulate in the
soil delaying the onset of the dry season; and improved root systems (Field et al.
1997; Owensby et al. 1997; Jackson et al. 1998; Niklaus et al. 1998; Wullschleger
et al. 2002; Nelson et al. 2004). In temperate grasslands, relative growth responses
to $e[CO_2]$ are typically stronger in drier years than in wetter ones (Nowak et al.
2004; Morgan et al. 2004), which would tend to partially buffer forage production
from the effects of low rainfall years (Howden et al. 1999). These interactions are
of practical significance for management of marginal animal production systems

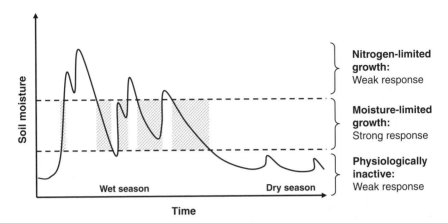

FIGURE SE 1.1 Proposed interactive effects of elevated CO_2 and seasonal variation in soil moisture on grass production in a tropical savanna. Moisture-mediated CO_2 responses would be expected to be strongest in years where soil moisture is maintained in the band of moisture-limited growth for extended periods.

such as tropical rangelands, in which rainfall variability is a major constraint (Ash and McIvor 2005). Results from arid and semiarid environments, however, serve as a caution against extrapolating this benefit of $e[CO_2]$ to extremely water-stressed conditions: In both the Nevada Desert FACE (Smith et al. 2000) and OzFACE (Stokes et al. 2005) experiments there were no measurable growth responses to $e[CO_2]$ in very dry years. In rangelands, it may well be that under strongly water-stressed conditions, plants are so physiologically inactive that there is little opportunity for CO_2 responses to occur. We suggest a working hypothesis for moisture-mediated responses in rangelands whereby grass growth is most strongly affected by $e[CO_2]$ during periods of intermediate water availability, when plants are growing under moisture-limited conditions (Figure SE 1.1) (Howden et al. 1999). At low levels of soil moisture, plants would be too physiologically inactive to respond to $e[CO_2]$, and under very moist conditions, other factors (particularly nitrogen availability) would limit growth rates. Responses of rangeland pastures to $e[CO_2]$ would therefore be expected to vary strongly between years depending on the distribution of rainfall and fluctuations in soil moisture through the year. In general it might be expected that enhanced growth of pastures in tropical rangelands would be strongest in moderately dry years, with little response in either severe droughts or wetter-than-normal years. Measured responses of pasture production to $e[CO_2]$ over the first four years of the OzFACE experiment (Stokes and Ash unpublished data) and cross-site comparisons between FACE environments (Nowak et al. 2004) are consistent with this proposed trend.

SE 1.3   FORAGE QUALITY

The enhanced carbon assimilation of plant shoots under $e[CO_2]$ generally exceeds the capacity of root systems to provide nutrients. This usually leads to a decline in

leaf nitrogen concentration, which may involve physiological acclimation in allocation of limited nitrogen resources (Long et al. 2004). Changes in assimilation and
metabolism of carbon and nitrogen under $e[CO_2]$ can affect various aspects of forage
quality, for example, protein levels, nonstructural carbohydrates, and digestibility
(Poorter et al. 1997; Scholes and Howden 2003). In the OzFACE experiment, forage
quality of two dominant C_4 grasses, *Themeda triandra* Forrsk. and *Chrysopogon
fallax* S.T. Blake, have been monitored by measuring leaf nitrogen levels and *in
vitro* digestibility at 4-week intervals. Results from the first 3 years of the experiment
show an approximate 4% decline in nitrogen levels and 2% decline in *in vitro*
digestibility for both species (550 µmol mol[1] vs. ambient) (Stokes and Ash unpublished data). This will increase the duration of the period each year where dietary
protein falls below maintenance levels for livestock, increasing dry season weight
losses or requiring greater expenditure on feed supplements.

SE 1.4   CONCLUSIONS

Climate change will likely bring both benefits (increased forage production in some
years) and new challenges (reduced forage quality) for tropical rangeland systems.
Further changes in the quantity and quality of forage produced are likely to occur
from changes in species composition, both within the grass community and through
any shift in balance between woody plants and grasses (Campbell and Smith 2000;
Howden et al. 2001). These changes are likely to occur through the interactive effects
of management and $e[CO_2]$ on ecological processes (Le Houérou 1996). The limited
opportunities for adaptation will leave enterprises in rangelands vulnerable if the
net effects of climate change on animal production are negative (Scholes and Howden
2003).

ACKNOWLEDGMENTS

OzFACE is a collaborative experiment involving James Cook University, Commonwealth Scientific and Industrial Research Organisation (CSIRO), and Queensland
Nickel Pty. Ltd., with additional support from the Australian Research Council, the
Australian Greenhouse Office, and the Cooperative Research Centre for Management
of Tropical Savannas.

REFERENCES

Anderson, L.J., Maherali, H., Johnson, H.B., Polley, H.W., and Jackson, R.B. (2001) Gas
        exchange and photosynthetic acclimation over subambient to elevated CO_2 in a C-3-
        C-4 grassland. *Global Change Biology,* 7, 693–707.
Ash, A.J. and McIvor, J.G. (2005) Constraints to pastoral systems in marginal environments.
        In *Pastoral Systems in Marginal Environments*, Milne, J.A., Ed., Wageningen Aca-
        demic Publishers, 17–28.

Campbell, B.D. and Smith, D.M.S. (2000) A synthesis of recent global change research on pasture and rangeland production: reduced uncertainties and their management implications. *Agriculture Ecosystems & Environment,* 82, 39–55.

Carran, R.A. and Allard, V. (2006) Grazing animals in agroecosystems. I: *Agroecosystems in a Changing Climate*, Newton, P.C.D., Carran, R.A., Edwards, G.R., and Niklaus, P.A., Eds.), CRC Press, Boca Raton, FL, 145–160.

Field, C.B., Lund, C.P., Chiariello, N.R., and Mortimer, B.E. (1997) CO_2 effects on the water budget of grassland microcosm communities. *Global Change Biology,* 3, 197–206.

Gifford, R.M. (1989) Direct effects of higher carbon dioxide concentration on vegetation. In *Greenhouse: Planning for Climate Change*, Pearman, G.I., Ed., Commonwealth Scientific and Industrial Research Organisation, Melbourne, Australia, 506–519.

Howden, S.M., Mckeon, G.M., Walker, L., et al. (1999) Global change impacts on native pastures in south-east Queensland, Australia. *Environmental Modelling & Software,* 14, 307–316.

Howden, S.M., Moore, J.L., Mckeon, G.M., and Carter, J.O. (2001) Global change and the mulga woodlands of southwest Queensland: greenhouse gas emissions, impacts, and adaptation. *Environment International,* 27, 161–166.

Jackson, R.B., Sala, O.E., Paruelo, J.M., and Mooney, H.A. (1998) Ecosystem water fluxes for two grasslands in elevated CO_2: a modeling analysis. *Oecologia,* 113, 537–546.

Le Houérou, H.N. (1996) Climate change, drought and desertification. *Journal of Arid Environments,* 34, 133–185.

Long, S.P., Ainsworth, E.A., Rogers, A., and Ort, D.R. (2004) Rising atmospheric carbon dioxide: plants face the future. *Annual Review of Plant Biology,* 55, 591–628.

Morgan, J.A., Pataki, D.E., Korner, C., et al. (2004) Water relations in grassland and desert ecosystems exposed to elevated atmospheric CO_2. *Oecologia,* 140, 11–25.

Nelson, J.A., Morgan, J.A., Lecain, D.R., Mosier, A., Milchunas, D.G., and Parton, B.A. (2004) Elevated CO_2 increases soil moisture and enhances plant water relations in a long-term field study in semi-arid shortgrass steppe of Colorado. *Plant and Soil*, 259, 169–179.

Niklaus, P.A., Spinnler, D., and Korner, C. (1998) Soil moisture dynamics of calcareous grassland under elevated CO_2. *Oecologia,* 117, 201–208.

Nowak, R.S., Ellsworth, D.S., and Smith, S.D. (2004) Functional responses of plants to elevated atmospheric CO_2 — do photosynthetic and productivity data from FACE experiments support early predictions? *New Phytologist*, 162, 253–280.

Owensby, C.E., Coyne, P.I., Ham, J.M., Auen, L.M., and Knapp, A.K. (1993) Biomass production in a tallgrass prairie ecosystem exposed to ambient and elevated CO_2. *Ecological Applications,* 3, 644–653.

Owensby, C.E., Ham, J.M., Knapp, A.K., Bremer, D., and Auen, L.M. (1997) Water vapour fluxes and their impact under elevated CO_2 in a C4-tallgrass prairie. *Global Change Biology,* 3, 189–195.

Poorter, H., VanBerkel, Y., Baxter, R., et al. (1997) The effect of elevated CO_2 on the chemical composition and construction costs of leaves of 27 C-3 species. *Plant Cell and Environment*, 20, 472–482.

Rosenzweig, C. and Liverman, D. (1992) Predicted effects of climate change on agriculture: a comparison of temperate and tropical regions. In *Global Change: Implications, Challenges and Mitigation Measures*, Majundar, S.K., Ed., Pennsylvania Academy of Sciences, Pennsylvania.

Scholes, R. and Howden, S.M. (2003) Rangeland vulnerability and adaptation in a changing world: a review. *Proceedings of the VII International Rangeland Congress*, Allsopp, N., Palmer, A.R., Milton, S.J., Kirkman, K.P., Kerley, G.I.H., Hurt, C.R., and Brown, C.J., Eds., Document Transformation Technologies, Irene, South Africa, 1021–1029.

Smith, S.D., Huxman, T.E., Zitzer, S.F., et al. (2000) Elevated CO_2 increases productivity and invasive species success in an arid ecosystem. *Nature,* 408, 79–82.

Stokes, C., Ash, A., Tibbett, M., and Holtum, J. (2005) OzFACE: the Australian savanna free air CO_2 enrichment facility and its relevance to carbon cycling issues in a tropical savanna. *Australian Journal of Botany,* 53, 677–687.

Stokes, C.J., Ash, A.J., and Holtum, J.A.M. (Forthcoming) Short-term responses of an Australian tropical savanna to elevated CO_2. Ascona, Switzerland.

Wand, S.J.E., Midgley, G.F., Jones, M.H., and Curtis, P.S. (1999) Responses of wild C_4 and C_3 grass (Poaceae) species to elevated atmospheric CO_2 concentration: a meta-analytic test of current theories and perceptions. *Global Change Biology,* 5, 723–741.

Wullschleger, S.D., Tschaplinski, T.J., and Norby, R.J. (2002) Plant water relations at elevated CO_2 — implications for water-limited environments. *Plant Cell and Environment,* 25, 319–331.

Special Example 2

Climate Change and Biological Control

Stephen L. Goldson

It is probable that climate change has the potential to impact on biological control stability. Indeed, concern about the potential effect of such change is not new; Prestidge and Pottinger (1990) compiled a series of analyses 15 years ago for New Zealand.

A useful way to assess the potential impacts of climate change on biological control is to consider the dynamics of biological pest suppression comprising genetically identical (or at least very similar) pest- and biocontrol agents in different climate zones. This is particularly so where ecosystem function can be relatively easily interpreted as a result of the species paucity such as that which occurs in many Australian and New Zealand pastoral systems (e.g., Goldson et al. 1997). To this effect, the contrast between Australia and New Zealand offers significant opportunity for such an analysis. A particularly good example is the interrelationship between *Sitona discoideus* Gyllenhal (Coleoptera: Curculionidae), the serious pest of lucerne (*Medicago sativa* L.), and its natural enemy the endoparasitoid *Microctonus aethiopoides* Loan (Hymenoptera: Braconidae, Euphorinae). The weevil was first identified in New Zealand in 1974 (Esson 1975), after which it rapidly spread throughout the country (Kain and Trought 1982). There is every probability that the founding population arrived from Australia. The New Zealand parasitoid population was deliberately imported into New Zealand from Australia (Stufkens et al. 1987) where it had earlier been introduced from Morocco (Aeschlimann 1983). Thus, there is good reason to believe that the populations of both the parasitoid and weevil are the same in both countries. The parasitoid kills the weevil by laying its eggs in the weevil's body. The larvae go through five stages sequestering the weevil's metabolic energy, after which the prepupa emerges and pupates in the soil.

In spite of the presence of effectively identical populations of weevils and parasitoids, the behaviours in both locations have been found to be substantially different. Significantly, useful biological control has occurred in New Zealand where sustained levels of parasitism have been higher than in Australia (e.g., Goldson et al. 1990) where little, if any, useful suppression has been observed (e.g., Hopkins

1989). The suppression of the weevil in New Zealand has been attributed to a substantial shift in the behaviour of *M. aethiopoides* (Goldson et al. 1990).

In general, the host *S. discoideus*, shows strong migratory behaviour. In New Zealand much of the emergent spring and summer population migrates out of the lucerne crops and spends the summer months at the base of hedgerows, etc., in a quiescent (aestivatory) state (Goldson et al. 1984). Irrespective of whether individual weevils migrate out or not, it is generally understood that they only become reproductive after this obligatory aestivation, and it is only then that the parasitoid can develop through its larval stages to produce adult wasps. In view of these circumstances, there is an absence of adult parasitoid wasps in the field during the (aestivatory) summer months, and the parasitoid's reproductive activity is thus more or less restricted to the cooler parts of the year. Based on day-degree calculations above a development temperature threshold of 8.2°C, it has been calculated that about two parasitoid generations should occur per season in New Zealand, which is not notably different from Australia (Goldson et al. 1990).

As it happened however, the observed impact of the parasitoid on the weevil in New Zealand was far greater than would have been predicted by this modest generational activity. Postaestivatory autumnal ground densities of the weevil in the lucerne were found to be low at 10 and 25 m². Indeed, of these at least 50% were parasitised and therefore effectively sterilized (Loan and Holdaway 1961), thus bringing the reproductive populations down to only around 5 to 13 m⁻². This was in marked contrast to the earlier autumnal densities of reproductive weevils (measured prior to the parasitoid's release), which were frequently of the order of 70 m² (Goldson et al. 1988). Further, earlier work had shown that initial postaestivatory populations of less than 20 to 25 weevils m² usually resulted in no lucerne yield loss (Goldson et al. 1985).

Investigation revealed that this positive biological control result in New Zealand was because of a very subtle shift in the parasitoid–host behaviour. This involved the uncoupling of the rule that the parasitoid could only develop in reproductive weevils after postaestivatory. Dissection of weevils collected from the lucerne in the late spring and early summer (i.e., that portion of the population that had not migrated out to aestivate after emerging) showed that about half of them unexpectedly contained developing larvae and about a quarter contained eggs. These findings indicated the presence of ongoing ovipositional activity by adult parasitoid wasps, which theoretically should not have been present. Overall, it was estimated that about 3% of the total weevil population supported parasitoid development at a time that should have shown no parasitoid developmental activity whatsoever (Goldson et al. 1990). Further, because of the warm summer temperatures this ongoing parasitoid activity resulted in an extra three parasitoid generations, and this contributed to a parasitoid buildup sufficient to cause the very high levels of impact among the weevils that returned from the aestivation sites in the autumn (Goldson et al. 1990). This was the basis for the biological control success.

It may be argued that there is a quirk in the New Zealand climate that permits this success to occur. It has been further speculated that the mechanism for the atypical development is related to very high titres of juvenile hormone that prevail ephemerally in newly enclosed weevils. When attacked during this very brief interval, the metabolism of these weevils permits full parasitoid larval development. However, once the titres of hormone have rapidly reduced to their normal preaestivatory levels, the larval development is thereafter arrested at the first instar, as usual during the aestivatory period (Goldson et al. 1990).

Whatever the mechanism, the case in this example demonstrates how subtle environmental differences can impact very markedly on biological control performance. In this case, the New Zealand system would seem to be fragile and based on some kind of fluke and indicates that climate change would very probably disrupt the processes that lead to successful and valuable pest suppression.

Conversely and broadly, it could probably be argued that failed control systems could suddenly be enabled by climate change through similar serendipity. The issue here, however, is that pests that evaded earlier, failed biological control attempts have sometimes been managed using other means such as plant breeding. Under these circumstances, the unpredicted appearance of effective biological control opportunity may not be apparent or be worth as much as the cost of the relatively rapid collapse of existing systems (resulting from climate change), for which no other management systems have been developed or sought.

REFERENCES

Aeschlimann, J-P. (1983) Sources of importation. establishment and spread in Australia, of *Microctonus aethiopoides* Loan (Hymenoptera: Braconidae) a parasitoid of *Sitona discoideus* Gyllenhal (Coleoptera: Curculionidae). *Journal of the Australian Entomological Society*, 22, 325–331.

Esson, M.J. (1975) Notes on the biology and distribution of three recently discovered exotic weevil pests in Hawkes Bay. In *Proceedings of the 28th New Zealand Weed and Pest Control Conference*, Hartley, M.J., Ed., Hamilton, NZ, 208–212.

Goldson, S.L., Dyson, C.B., Proffitt, J.R., Frampton, E.R., and Logan, J.A. (1985) The effect of *Sitona discoideus* Gyllenhal (Coleoptera: Curculionidae) on lucerne yields in New Zealand. *Bulletin of Entomological Research*, 75, 429–442.

Goldson, S.L., Frampton, E.R., Barratt, B.I.P., and Ferguson, C.M. (1984). The seasonal biology of *Sitona discoideus* Gyllenhal (Coleoptera: Curculionidae), an introduced pest of New Zealand lucerne. *Bulletin of Entomological Research*, 74, 249–259.

Goldson, S.L., Frampton, E.R., and Proffitt, J.R. (1988) Population dynamics and larval establishment of *Sitona discoideus* (Coleoptera: Curculionidae) in New Zealand lucerne. *Journal of Applied Ecology*, 25, 177–195.

Goldson, S.L., Phillips, C.B., McNeill, M.R., and Barlow, N.D. (1997) The potential of parasitoid strains in biological control: observations to date on *Microctonus* spp. intraspecific variation in New Zealand. *Agriculture Ecosystems and Environment*, 64, 115–124.

Goldson, S.L., Proffitt, J.R., and McNeill, M.R. (1990) Seasonal biology and ecology in New Zealand of *Microctonus aethiopoides* (Hymenoptera: Braconidae), a parasitoid of *Sitona* spp. (Coleoptera: Curculionidae), with special emphasis on atypical behaviour. *Journal of Applied Ecology*, 27, 703–722.

Hopkins, D.C. (1989) Widespread establishment of the Sitona weevil parasite Microctonus aehtiopoides and its effectiveness as a control agent in South Australia. In *Proceedings of the 5th Australasian Conference on Grassland Invertebrate Ecology*, Stahle, P.P., Ed., Melbourne, Australia, 49–54.

Kain, W.M. and Trought, T.E.T. (1982) Insect pests of lucerne. In *Lucerne for the 80s*, Wynn-Williams, R.B., Ed., Agronomy Society of New Zealand Special Publication No. 1, 49–57.

Loan, C.C. and Holdaway, F. (1961) *Microctonus aethiops* (Nees) auctt. and *Perilitus rutilus* (Nees) (Hymenoptera: Braconidae), European parasites of *Sitona* weevils (Coleoptera: Curculionidae). *Canadian Entomologist*, 93, 1057–1079.

Prestidge, R.A. and Pottinger, R.P. (1990) *The Impact of Climate Change on Pests, Diseases, Weeds and Beneficial Organisms*. Plant protection group, Ruakura Agriculture Centre, Hamilton, New Zealand.

Stufkens, M.W., Farrell, J.A., and Goldson, S.L. (1987) Establishment of *Microctonus aethiopoides* a parasitoid of the Sitona weevil in New Zealand, *In Proceedings of the 40th New Zealand Weed and Pest Control Conference*, Popay, A., Ed., Nelson, pp. 31–32.

Special Example 3

Efficacy of Herbicides under Elevated Temperature and CO_2

Daniel J. Archambault

The dynamics of competition between weed and crop plants are affected by environmental conditions and have been shown to change with CO_2 enrichment (Patterson and Flint 1980). Differential responses of C_3 plants (carbon from CO_2 initially fixed into 3-carbon compounds; e.g., *Hordeum vulgare*, *Avena fatua*, *Amaranthus retroflexus*) and C_4 plants (carbon from CO_2 initially fixed into 4-carbon compound; e.g., *Setaria viridis*) to elevated CO_2 and temperature may cause shifts in their competitive interactions. These changes have particular significance given that most of the world's crop species are C_3 plants, many of the major weed species are C_4 plants, and that C_3 plants are expected to benefit more from elevated CO_2 than C_4 plants. While this might suggest that crops will gain a competitive advantage over most weeds, other factors, such as changes in herbicide efficacy, may come into play and limit this advantage and decrease potential yield increases in crops. While many studies have examined the effects of environmental change on crop and weed interactions, relatively few have included the effects of environmental change on herbicide efficacy.

Environmental factors such as temperature, precipitation, wind, and relative humidity influence the efficacy of herbicides (Hatzios and Penner 1982; Muzik 1976). Elevated temperatures and metabolic activity tend to increase uptake, translocation, and efficacy of many herbicides (Patterson et al. 1999), while moisture deficit, especially when severely depressing growth, tends to decrease efficacy of postemergence herbicides, which generally perform best when plants are actively growing. High concentrations of starch in leaves, which commonly occurs in C_3 plants grown under CO_2 enrichment (Wong 1990), might interfere with herbicide activity (Patterson et al. 1999). Although the effects of elevated atmospheric CO_2 concentrations on weed–crop–herbicide interactions are not well known, Ziska et al. (1999) showed that elevated CO_2 levels diminished the efficacy of the widely used herbicide glyphosate. If these effects are common and widespread, they will have a significant impact on agriculture.

There is a need to evaluate the eff   of elevated CO_2 and temperature (including their interactive effects) on herbicide efficacy to develop strategies for the agricultural industry in the face of global climate change. This information would allow the identification and prediction of effects of elevated CO_2 and temperature on herbicide efficacy.

In controlled-environment studies, herbicide efficacy was frequently negatively affected by elevated CO_2, and effects were dependent on the mode of action of herbicides, on weed species, and on competition (Archambault et al. 2001). While double-ambient CO_2 caused a decrease of 57% in efficacy of the herbicide fluazifop-butyl + fenoxyprop (blocks the activity of ACCase, an enzyme involved in fatty acid biosynthesis) applied to *A. fatua* (C_3), no effects of elevated CO_2 were found when the herbicide was applied to *S. viridis* (C_4). CO_2-related reduction in efficacy of glyphosate (inhibits the activity of EPSP, an enzyme involved in the production of several amino acids) applied to *Cirsium arvense* was reversed when weeds were grown in competition with *Brassica napus*. Dose response experiments showed that the efficacy of certain herbicides could be adversely affected at CO_2 levels approximately 160 ppm above ambient. Based on these findings, we designed an experiment to study CO_2–temperature interactions on growth of *A. fatua* and herbicide efficacy using either ambient levels of CO_2 or ambient + 160 ppm and a daytime temperature of 23, 26, or 29°C. Daytime temperatures above 23°C decreased growth in both control and herbicide-treated plants. Increasing daytime temperature from 23 to 29°C caused decreased efficacy in the herbicides fluazifop-butyl + fenoxyprop and glufosinate-ammonium (inhibits the activity of glutamine synthase, a key enzyme in the incorporation of ammonium into amino acids), but not in imazemethabenz (inhibits the activity of ALS synthase, an enzyme involved in the production of essential branch-chain amino acids) (Figure SE 3.1). Decreases in efficacy were greatest at ambient CO_2 for fluazifop-butyl + fenoxyprop and greatest at ambient + 160 ppm CO_2 in glufosinate-ammonium. While analysis of variance did not detect a significant interaction between CO_2 and temperature, both elevated CO_2 and temperature caused decreased efficacy of the herbicide glufosinate-ammonium on *A. fatua*.

An economic analysis performed using plant growth and herbicide efficacy changes suggested that potential monetary losses due to CO_2-induced decreases in herbicide efficacy could be partially or totally overcome by increases in crop yields caused by elevated CO_2. Nonetheless, the results also suggest that weed control will be crucial in realizing potential increases in economic yield of crops as atmospheric CO_2 concentrations increase.

Herbicide efficacy was found to increase, decrease, or remain the same when plants were grown at elevated CO_2 and efficacy changes were species specific. Differences in growth response and effects of CO_2 on herbicide efficacy between *A. fatua* (C_3) and *S. viridis* (C_4) serve to illustrate the complexity of the issue. The results suggest that assumptions as to CO_2-induced effects on herbicide efficacy cannot be made. Effects of competition on CO_2-induced changes in herbicide efficacy further complicate the issue.

Increased efficacy of herbicides on weeds grown in competition may have resulted from the length of the experiment, where the competitive advantage of herbicide resistant crops increased over time. The herbicide–competition combination appeared to abolish the detrimental effects of elevated CO_2 on herbicide efficacy.

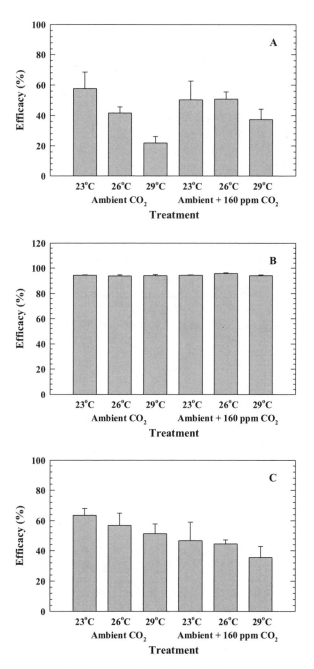

FIGURE SE 3.1 Herbicide efficacy of fluazifop-butyl + fenoxyprop (A), imazemethabenz (B), and glufosinate-ammonium (C) applied to *A. fatua* grown at different temperatures at either ambient or ambient + 160 ppm CO_2. Bars represent means +/– standard errors. N = 5.

Studies on effects of environmental conditions on herbicide efficacy on *Agropyron repens* have shown that increased light, temperature, and humidity immediately following application increased the efficacy of fluazifop-butyl, but that prolonged exposure of plants to increased temperature and light decreased efficacy (Coupland 1986). In this study, increased daytime temperature also caused decreased herbicide efficacy, but no interactions were found between CO_2 level and temperature. The results of our study (Archambault et al. 2001) indicate that both CO_2 level and temperature can affect herbicide efficacy, and that in most cases elevated CO_2 and temperature, when they did have an effect, caused decreased efficacy. It seems likely that interactive effects of these two parameters would occur under certain conditions, perhaps in other weed species or on other herbicides.

From this it may be possible to prepare for environmental changes through development of adaptive agricultural practices for existing herbicides, especially those involving rate and timing of application to control weeds in a high-CO_2, high-temperature environment. If existing products cannot provide adequate control of weeds in a high-CO_2, high-temperature environment, this work will also provide useful information for the development of new products by identifying the herbicides with modes of action that are least affected. Decreases in herbicide efficacy began to appear when plants were grown at 160 ppm above current levels of atmospheric CO_2, which represents predicted levels for the middle of this century.

REFERENCES

Archambault, D.J., Li, X., Robinson, D., O'Donovan, J.T., and Klein, K.K. (2001) The effects of elevated CO_2 and temperature on herbicide efficacy and weed/crop competition. Report to the Prairie Adaptation Research Collaborative, 29.

Coupland, D. (1986) The effects of environmental factors on the performance of fluazifop-butyl against *Elymus repens*. *Annals of Applied Biology*, 108, 353–363.

Hatzios, K.K. and Penner, D. (1982) Metabolism of herbicides in higher plants, CEPCO iv., Burgess Publications, Edina, MN.

Muzik, T.J. (1976) Influence of environmental factors on toxicity to plants. In *Herbicides: Physiology, Biochemistry, Ecology*, 2nd ed., Audus, L.J., Ed., Academic Press, New York, 204–247.

Patterson, D.T. and Flint, E.P. (1980) Potential effects of global atmospheric CO_2 enrichment on the growth and competitiveness of C_3 and C_4 weed and crop plants. *Weed Science*, 28(1), 71–75.

Patterson, D.T., Westbrook, J.K., Joyce, R.J.V., Lingren, P.D., and Rogasik, J. (1999) Weeds, insects, and diseases. *Climate Change*, 43, 711–727.

Wong, S.C. (1990) Elevated atmospheric partial pressure of CO_2 and plant growth. II. Non-structural carbohydrate content in cotton plants and its effect on growth parameters. *Photosynthesis Research*, 23, 171–180.

Ziska, L.H., Teasdale, J.R., and Bunce, J.A. (1999) Future atmospheric carbon dioxide may increase tolerance to glyphosate. *Weed Science*, 47, 608–615.

Special Example 4

Evolution of Pathogens under Elevated CO$_2$

Sukumar Chakraborty

While there is extensive literature on plant responses to elevated atmospheric CO$_2$, most do not consider the interacting effect of plant pathogens (Coakley et al. 1999; Chakraborty 2001; Chakraborty and Pangga 2004), and until recently, the limited data on pathogens had come mostly from *in vitro* studies (Manning and Tiedemann 1995). Of the 26 diseases where CO$_2$ effects have been considered so far, severity has increased in 13, decreased in 9, and remained unchanged in 4 (Chakraborty and Pangga 2004). Unlike factors such as temperature, where the impact of climate change can be modelled using well-established quantitative relationships (Teng et al. 1996), we are far from discovering key rules that govern the influence of CO$_2$ on host–pathogen interactions. Elevated CO$_2$ can influence host resistance, pathogen life cycle, host–pathogen interaction, and disease epidemiology. High CO$_2$ changes anatomy, morphology, and phenology to alter host resistance, but some changes, including phenolic production (Hartley et al. 2000) and disease resistance (Pangga et al. 2004), can be transient and may not persist when plants are grown at elevated CO$_2$ for a long period of time or over a number of generations. Among pathogen attributes, conidia germination, germtube growth, and the production of infection structure are all influenced at high CO$_2$ as fewer conidia germinate and produce appressoria and the growth of germtube and appressoria production are delayed by several hours (Hibberd et al. 1996; Chakraborty et al. 2000). These contribute to a longer incubation period (time between inoculation and symptom expression) and a reduced disease severity at high CO$_2$, but the latent period (time for the host tissue to become infectious) remains unchanged, as growth within host tissue is unaffected.

Two important trends have emerged from the scant literature: (1) The enlarged canopy of plants at high CO$_2$, with an average 30% increase in biomass and yield (Idso and Idso 1994), offers a microclimate that is often highly conducive to disease development, and contains many more potential infection sites; and (2) some fungal pathogens produce more spores at high CO$_2$ (Hibberd et al. 1996; Chakraborty et al. 2000) but increased resistance in some plants (Hibberd et al. 1996; Pangga et al. 2004) slows host invasion. While infection efficiency is reduced in high CO$_2$ as a consequence of enhanced resistance and a slowing down of the prepenetration phase of the pathogen, up to twice as many lesions can be produced in the high-CO_2 plants, as the enlarged canopy traps many more pathogen spores (Pangga et al. 2004). Thus, the overall impact of high CO_2 is determined by a balance between the effect of enhanced host resistance and that of increased canopy size and fecundity in a favourable microclimate. This combination of enlarged canopy (with its associated microclimate and infection sites) and increased pathogen fecundity raises a particular concern as to whether pathogens could rapidly evolve to erode the usefulness of disease resistance in plants. Increased fecundity produces a population size that is several orders of magnitude larger than at ambient CO_2 on which mutation and selection can act. In agricultural systems where plant breeders largely select and deploy host resistance genes, it is the fitness of a pathogen population that determines how quickly it adapts to a new source of resistance. There are many examples where host resistance is overcome by matching pathogen virulence, and the adaptation to resistant cultivars can occur after only a few asexual cycles (Newton and McGurk 1991). Although so far only a single study has examined pathogen microevolution under elevated CO_2, examples abound in the literature of rapid genetic adaptation to various biotic and abiotic influences (Travis and Futuyama 1993).

Plant pathogen evolution under elevated CO_2 has been studied in *Colletotrichum gloeosporioides* using two strains that differ in aggressiveness on two *Stylosanthes scabra* varieties over 25 sequential infection cycles (Chakraborty and Datta 2003). The two varieties had different levels of resistance to this pathogen. Conidial suspension of the fungus grown on artificial media was used as inoculum for the first cycle, and thereafter conidia from lesions were multiplied separately to reinoculate fresh batches of plants at subsequent infection cycles. While aggressiveness at ambient CO_2 increased steadily on both varieties for most cycles, this increased only after a few initial cycles at 700 ppm CO_2, as the pathogen strains adapted to combat induced host resistance (Figure SE 4.1). The genetic fingerprint and molecular karyotype of isolates changed in some CO_2 variety combinations, but these changes were not linked to increased aggressiveness. Pathogen fecundity of both strains steadily increased at elevated CO_2 with each cycle. Of the two strains, fecundity of the more aggressive strain increased more rapidly, indicating that reproductive fitness is a component of its overall fitness (Chakraborty and Datta 2003).

The scant literature based on findings from growth-room studies make it impossible to generalise on the evolutionary consequences of elevated CO_2 on plant pathogens. These studies ignore changes in the host plant and its interaction with pathogen virulence. Most research has compared two or more set levels of CO_2 on host–pathogen interaction, although in reality, change in atmospheric CO_2 concentration is very gradual where the rate depends on many physical and anthropogenic factors. Experimental and modelling studies of interacting host–pathogen populations in polycyclic epidemics under gradual increases in CO_2 concentration are necessary to enlarge our understanding. Studies of plant pathogens in natural ecosystems using free air CO_2 enrichment (FACE) environments are only just starting to emerge (Mitchell et al. 2003).

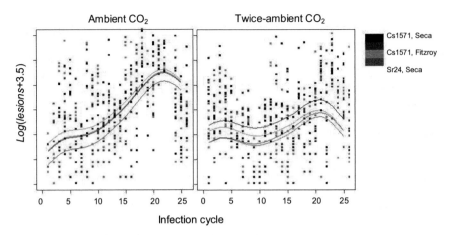

FIGURE SE 4.1 Changing aggressiveness of *C. gloeosporioides* isolates CS1571 and SR24 on *S. scabra* varieties Fitzroy (susceptible) and Seca (partially resistant) at ambient and 700 ppm CO$_2$ over 25 sequential infection cycles. The solid squares are original data to illustrate the generally variable aggressiveness data, and lines represent fitted regression models with smooth nonparametric terms for the cycle effect estimated separately for each CO$_2$ level. (Reproduced with permission from Chakraborty, S. and Datta, S. [2003] *New Phytologist*, 159, 733–742.)

REFERENCES

Chakraborty, S. (2001) Effects of climate change. In *Plant Pathologists' Pocketbook*, 3rd. ed., Waller, J.M., Lenne, J., and Waller, S.J., Eds., CABI Publishing, Wallingford, U.K., 203–207.

Chakraborty, S. and Datta, S. (2003) How will plant pathogens adapt to host plant resistance at elevated CO$_2$ under a changing climate? *New Phytologist*, 159, 733–742.

Chakraborty, S. and Pangga, I.B. (2004) Plant disease and climate change. In *Plant Microbiology*, Holmes, A. and Gillings, M., Eds., Bios Scientific Publishers, Abingdon, U.K., 163–180.

Chakraborty, S., Pangga, I.B., Lupton, J., Hart, L., Room, P.M., and Yates, D. (2000) Production and dispersal of *Colletotrichum gloeosporioides* spores on *Stylosanthes scabra* under elevated CO$_2$. *Environmental Pollution*, 108, 381–387.

Coakley, S.M., Scherm, H., and Chakraborty, S. (1999) Climate change and plant disease management. *Annual Review of Phytopathology*, 37, 399–426.

Hartley, S.E., Jones, C.J., Couper, G.C., and Jones, T.H. (2000) Biosynthesis of plant phenolic compounds in elevated atmospheric CO$_2$. *Global Change Biology*, 6, 497–506.

Hibberd, J.M., Whitbread, R., and Farrar, J.F. (1996) Effect of elevated concentrations of CO$_2$ on infection of barley by *Erysiphe graminis*. *Physiological and Molecular Plant Pathology*, 48, 37–53.

Idso, K.E. and Idso, S.B. (1994) Plant response to atmospheric CO$_2$ enrichment in the face of environmental constraints: a review of the past 10 years' research. *Agricultural and Forest Meteorology*, 69, 153–203.

Manning, W.J. and Tiedemann, A.V. (1995) Climate change: potential effects of increased atmospheric Carbon Dioxide (CO_2), Ozone (O_3), and Ultraviolet-B (UVB) radiation on plant diseases. *Environmental Pollution*, 88, 219–245.

Mitchell, C.E., Reich, P.B., Tilman, D., and Groth, J.V. (2003) Effects of elevated CO_2, nitrogen deposition, and decreased species diversity on foliar fungal plant disease. *Global Change Biology*, 9, 438–451.

Newton, A.C. and McGurk, L. (1991) Recurrent selection for adaptation of *Erysiphe graminis* fsp *hordei* to partial resistance of barley. *Journal of Phytopathology*, 132, 328–338.

Pangga, I.B., Chakraborty, S., Room, P.M., and Yates, D. (2004) Resistance and canopy size in *Stylosanthes scabra* determine anthracnose severity at high CO_2. *Phytopathology*, 94, 221–227.

Teng, P.S.K.L.K., Heong, M.J., Kropff, K.J., Nutter, F.W., and Sutherst, R.W. (1996) Linked pest-crop models under global change. In *Global change and terrestrial ecosystems*, Walker, B. and Steffen, W., Eds., Cambridge University Press, Cambridge, U.K., 291–316.

Travis, J. and Futuyama, D.J. (1993) Global change: lessons from and for evolutionary biology. In *Biotic interactions and global change*, Kareiva, P.M., Kingslover, J.G., and Huey, R.B., Eds., Sinauer, Sunderland, U.K., 251–265.

Special Example 5

Adapting United Kingdom Agriculture to Climate Change

Jo E. Hossell

This contribution examines the issue of adaptation to climate change within United Kingdom agriculture. It begins by summarising possible impacts that may occur in the United Kingdom based on modelling of future conditions and experience of recent extreme events. It examines the sort of adaptation that may be necessary, and then moves on to consider the sort of information that farmers and the agricultural industry may need in order to respond to such changes and the way that this should be presented.

Table SE 5.1 shows the effects of different elements of climate change on aspects of U.K. agriculture and relates them to the uncertainty associated with projecting future patterns of climate change in the United Kingdom. The main impacts of climate change will be through warmer temperatures, changing rainfall totals, and seasonality (in particular drier summer and autumn conditions but wetter winters), elevated atmospheric CO_2 levels, and sea level rise. Such changes will affect both average and extreme weather events. The impacts on the agricultural industry could be direct and immediate, such as crop yields affected by increased temperatures or field operations affected by heavy autumn rainfall. Indirect impacts could include, for example, changes in market demand and United Kingdom and international prices through weather-related impacts on crop yields.

There is general agreement in studies in the United Kingdom that many crop yields may increase under climate change (Downing et al. 2000; Ministry of Agriculture Fisheries and Food [MAFF] 2000; Holman and Loveland 2002). At this latitude, the negative effects of temperature increases on determinate crops such as wheat are offset by the CO_2 fertilisation effect. But this assessment assumes that farmers will make autonomous responses to climate change, such as changing sowing and harvesting dates and cultivars. New or novel crops are also likely to be introduced (Holloway and Ibery 1997; MAFF 2000), but the net effect of all these changes on farming systems is less clear.

The issue is further complicated by the combined impacts of both average and extreme events. Changes in average climate conditions are likely to be gradual enough within the time horizon of the industry to allow successful adaptation to occur. However, some aspects of agriculture are vulnerable to extreme events. In

TABLE SE 5.1
Impacts of the Different Elements of Climate Change on Agriculture and the Confidence Level Associated with Predicting Each Variable

Climate	Expected Change by 2050s[a] Based on 1961–90 Baseline	Confidence Level	Effects on U.K. Agriculture
CO_2	489–593ppm[b]	Very high	Beneficial effects for most U.K. crops: Increased rate of photosynthesis, reduced water use.
Sea level rise	14–18cm[c]	Very high	Loss of land, salinisation of groundwater.
Annual average temperature	0.5–3°C	High	Accelerated growth, shorter and earlier growing season, expand suitability northward and to higher elevations, risk of adverse temperatures, reduced grain yield through more rapid development season, higher potential evapotranspiration.
Precipitation total	Summer: 0 to 40% Winter: 0 to +25%	Medium–High	Effects depend on extent of precipitation change, but could increase drought risk, water logging, and transpiration, and reduce supply of irrigation water and soil workability.
Storminess	40% by 2080s (Medium high emissions scenario)	Low	Risk of lodging, soil erosion, increased leaching and pesticide/nutrient runoff, reduced infiltration of rainfall.
Climatic extremes	Varies by event	Very low	Potential changes in the risk of damaging events (heat waves, frost, drought, floods, intense rainfall events) affecting crop yields and quality, timing of farming operations.

[a] Figures based on UKCIP02 Low – High emission scenarios unless otherwise stated.

[b] Absolute rather than change value of atmospheric CO_2 concentration.

[c] Global sea level rise – U.K. values will vary depending on isostatic responses locally.

Source: Adapted from Ministry of Agriculture Fisheries and Food (MAFF) (2000). MAFF, London, 65, but updated to UKCIP02 scenarios; Hulme, M., Jenkins, G.J., Lu, X. et al. (2002) *The UKCIP02 Scientific Report,* School of Environmental Science, University of East Anglia, Norwich, U.K., 120.

recent years, farmers in the United Kingdom have been affected by the hot summer conditions of 1995 (ADAS 1999) and 2003, and particularly wet conditions such as the autumn and winter of 2000 and 2001 (Shepherd 2001). In 1995 hot, dry conditions at the end of July occurred late enough not to affect wheat yields, but potato quality (and, therefore, saleable yields) was badly affected by scab. This differential impact is an indication of how sensitive the production levels could be to relatively small changes in the timing of extreme events such as the start of a high temperature/drought period (Hossell 2003).

A number of the impacts identified from the hot U.K. summer of 1995 can be responded to rapidly, such as improving crop monitoring to detect changes in pest levels (aphids, cutworms, and red spider mite infestations increased) or a wider spread of crop cultivars, but longer-term planning is required to implement better (and more) irrigation systems or altered designs of livestock housing to deal with warmer and drier summer conditions and wetter winter conditions (ADAS 1999). Yet it is important to consider even short-term adaptations at an early stage; for example, a reduction in heating usage within glasshouse production systems as winter temperatures warm, is an adaptation that may be rapidly adopted because it requires no additional costs and may be achieved as required. However, the reduced need for winter heating may affect the cost-to-benefit ratio of investing in combined heat and power (CHP) or other waste heat supplies and so a longer-term implication could be expected for the economy of the farm. It should also be recognised that adaptations to rarer events will also help in responding to less dramatic extremes such as the warm conditions of 1999 (1.2°C warmer than average), which could occur 1 in 4 years by the 2020s.

Results from integrated crop and farm modelling work suggest that radical adaptations (e.g., large changes between arable and livestock production) may not be necessary by the 2020s, although partial adaptations may be needed in the short term (e.g., the mix of break crops may need to change on arable farms due to a decline in the yield of oilseed rape and an increase in sunflower yields) (Hossell et al. 2001). Many of the changes modelled were relatively small, needing to be adopted only by the 2050s, but the more sensitive farm types may need to make adaptations by the 2020s. The early adaptations include changes in both working capital (e.g., variable costs) and investment capital (e.g., crop storage and irrigation) (Table SE 5.2).

The degree to which the projected changes will be taken up will depend on the value farmers see in them in the context of their own situation. The study also suggested that climate change might increase farm profitability on most farms. Moreover, the incentive to adapt was always positive (in relation to not adapting), though relatively small for many farm types. However, experience of the agricultural industry suggests that responses to losses from climate change will be more rapidly adopted than responses to gains (Hossell et al. 2001).

Yet while awareness of climate change is high, acceptance of it as an issue for the industry is not. U.K. agricultural policy does not include adaptation measures, and individual farmers do not seem to be widely considering how to adapt to climate change. Yet increasingly, decisions are being made and policies implemented that will be affected by climate change. It is important to ensure that knowledge of

TABLE SE 5.2
Key Adaptations and the Time Scale for Their Adoption

Time Scale	Adaptation
2020s	Increase irrigation capacity
	Substitute sunflowers for oilseed rape
	Introduction or increase of forage maize
	Increased need for storage
	Increase in arable production on livestock farms
	Change in spring work days
2050s	Substitute sunflowers for oilseed rape
	Change in spring work days
	Change in autumn work days

Source: From Hossell, J.E., Ramsden, S.J., Gibbons, J., Harris, D., Pooley, J., and Clarke, J. (2001) Wolverhampton, U.K. *ADAS Final Report to MAFF for Project cc0333.*

climate change impacts is widespread within the industry, and that this information is regularly updated in order to guarantee that any adaptations are considered within future business strategies and farm plans. The industry needs to understand both the nature of climate change (i.e., the sort of changes that may be expected, both to average as well as extreme conditions), and more importantly to realise that these changes are already happening and cannot now be avoided.

So what do farmers need to know about the impacts of climate change? Much of the information provided by scenarios is of little value unless it is translated into terms of significance within the industry. The most effective adaptations involve actions that can be taken by the farmer without assistance (e.g., changes in timing of farm operations in response to the changing growing season length, and changes in cropping patterns and cultivars). But where no new investment or know-how is needed, growers will take advantage of an adaptation only if they are aware of the likely persistence of the altered conditions. Hence, information on the level of climate change impacts, their persistence (for mean changes) or return frequency (for extreme events) is needed. Reference to past weather events makes it easier to understand the implications of possible changes, but it is also important to illustrate the changes in risk as well as absolute levels of change. For example, the hot summer of 1995 was 3.4°C warmer than usual in the United Kingdom. Currently this weather extreme has an expected frequency of 1 in 100 years and is not something for which the industry should plan. However, the UKCIP02 (United Kingdom Climate Impacts Programme) scenarios (Hulme et al. 2002) indicate that it may be a 1 in 5 year event by the 2050s. This is well within the planning horizon of farmers and the industry in general.

It is not enough just to tell the agricultural industry that the growing season will increase by *x* many days over the next 20 years. The interactions of climate change with the farming calendar need to be explored in more detail. Some aspects, such as fertiliser application, which are linked to crop growth and canopy closure, may need to be brought forward under warmer winter conditions, but the windows of opportunity for spraying may be reduced due to wetter winter conditions. The effective use of limited amounts of labour and machinery for field operations will become increasingly important; the alternative of maintaining high levels of labour and machinery would be a costly insurance policy. Similarly, with the introduction of cross-compliance measures as a result of European Union Common Agricultural Policy (CAP) reform, fixed windows of time for farm operations need to be set with climate change in mind.

But in responding to climate change, the assessment of the importance of climate change impacts needs to be weighed against other driving forces. Maladaptation (actions taken that can restrict the ability to adapt to climate change in the future) can be as bad as underadaptation (insufficient weight given to the need to adapt or insufficient adaptation action) and overadaptation (overemphasis given to the need to adapt) (United Kingdom Climate Impacts Programme [UKCIP] 2003). Moreover, adaptation is not the sole responsibility of farmers. There are some risks that are beyond their influence to affect, and there are considerations that need to be made by other parts of the industry. For example, notwithstanding the effects of CO_2 on yields, large temperature increases or high temperatures at key times in the crop growth cycle may reduce wheat yields. Investment in extra wheat breeding to overcome yield losses would provide a cost-effective means of adapting to this issue (Hossell et al. 2002). While farmers can switch cultivars, they will rely ultimately on seed producers to produce varieties better adapted to the changing conditions in the United Kingdom.

Similarly, some adaptations are currently unviable economically (e.g., changes to storage and handling of pig slurry under wetter winter conditions). However, they may be necessary to reduce environmental damage such as N pollution. Hence government assistance may be needed to encourage their uptake. Shifts in cropping areas may require early consideration by the industry because they may only be viable if the savings made through not renewing existing equipment in current production areas are balanced against potential moving costs.

Understanding the need to adapt to climate change is part of the process of change that needs to be adopted within the industry. But adaptation need is often overshadowed by discussion of the uncertainty of climate change predictions. Agricultural systems are constantly evolving and adapting to external drivers, such as policy changes, technological improvement, and market demands. Climate change in that sense is just another driver to be considered with the others. Accepting climate change as inevitable should allow for the consideration of its interactions with policies, particularly long-term policy frameworks such as the EU Water Framework Directive (due to be implemented by 2029). As yet, little work has examined how climate change may conflict with or support the aims of such policies. Yet there is a need to recognise that in setting baselines against which policy success may be measured, climate change impacts may create a moving target.

The degree to which adaptations will be taken up will depend on the value farmers and the industry see in them in the context of their own situation. In particular, responses to losses from climate change will be more rapidly adopted than responses to gains. It is key that all aspects of agriculture have real-world knowledge of the impacts of climate change on which to base their decision making. There may be considerable scope for influencing the perception of these changes and communicating the information should be undertaken with care to ensure that a message is delivered that encourages both appropriate and timely adaptation.

REFERENCES

ADAS (1999) The review of the direct effects of the dry and hot summer of 1995 on decision making of the individual farmer. ADAS, Cambridge, U.K.

Downing, T.E., Harrison, A.P., Butterfield, R.E., and Lonsdale, K.G. (2000) Climate change, climatic variability and agriculture in Europe: an integrated assessment. *Research Report no. 21. 441.* Oxford, Environmental Change Institute, Oxford, U.K.

Holloway, L.E. and Ilbery, B.W. (1997) Global warming and Navy beans: decision making by farmers and food companies in the U.K. *Journal of Rural Studies*, 13, 343–355.

Holman, I. and Loveland, P. (2002) Regional climate change impact and response studies in East Anglia and North West England (RegIS). *Final Report to MAFF on Project cc0337,* Soil Survey and Land Research Centre, Cranfield, U.K.

Hossell, J.E. (2003) Validating the impacts of climate extremes. *Final Report to Defra on Project cc0360,* ADAS, Cambridge, U.K.

Hossell, J.E., Ramsden, S.J., Gibbons, J., Harris, D., Pooley, J., and Clarke, J. (2001) Timescale of farm level adaptations and responses to climate change. *ADAS Final Report to MAFF for Project cc0333*, Wolverhampton, U.K.

Hossell, J.E., Temple, M.L., Finlay, I., Gay, A., Oakley, J., Symmonds, W., and Moorhouse, D. (2002) Identifying and costing agricultural responses under climate change scenarios (ICARUS). *Final Report to DEFRA on Project cc0357*, ADAS, Wolverhampton, U.K..

Hulme, M., Jenkins, G.J., Lu, X. et al. (2002) Climate change scenarios for the United Kingdom. *The UKCIP02 Scientific Report,* School of Environmental Science, University of East Anglia, Norwich, U.K., 120.

Ministry of Agriculture Fisheries and Food (MAFF) (2000). Climate change and agriculture in the United Kingdom. MAFF, London, 65.

Shepherd, M.A. (2001). A review of the impact of the wet autumn of 2000 on the main agricultural and horticultural enterprises in England and Wales. *Final Report to Department of Environment, Food and Rural Affairs for Project CC0372*, ADAS. 151, Wolverhampton, U.K.

United Kingdom Climate Impacts Programme (UKCIP) (2003) Climate adaptation: risk, uncertainty and decision-making. *Technical Report,* UKCIP, Oxford, U.K., 154.

Index

A